T0366502

Essentials of Electromyography

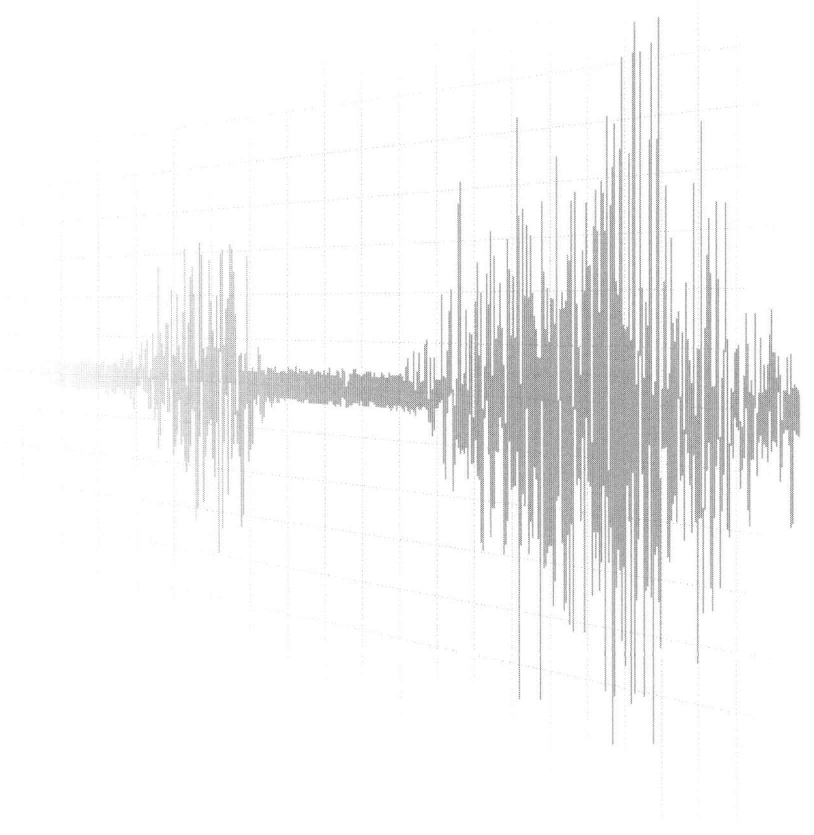

Gary Kamen, PhD
University of Massachusetts, Amherst

David A. Gabriel, PhD
Brock University, St. Catharines, Ontario

Human Kinetics

Library of Congress Cataloging-in-Publication Data

Kamen, Gary.
Essentials of electromyography / Gary Kamen, David A. Gabriel.
p. ; cm.
Includes bibliographical references and index.
ISBN-13: 978-0-7360-6712-6 (hard cover)
ISBN-10: 0-7360-6712-4 (hard cover)
1. Electromyography. I. Gabriel, David A., 1961- II. Title.
[DNLM: 1. Electromyography. WE 500 K15e 2010]
RC77.5.K36 2010
616.7'407547--dc22
2009015382

ISBN-10: 0-7360-6712-4 (print)
ISBN-13: 978-0-7360-6712-6 (print)
ISBN-10: 0-7360-8550-5 (Adobe PDF)
ISBN-13: 978-0-7360-8550-2 (Adobe PDF)

The Web addresses cited in this text were current as of March, 2009, unless otherwise noted.

Acquisitions Editor: Loarn D. Robertson, PhD; **Developmental Editor:** Elaine H. Mustain; **Managing Editor:** Katherine Maurer; **Assistant Editors:** Elizabeth Evans and Steven Calderwood; **Copyeditor:** Joyce Sexton; **Proofreader:** Pamela Johnson; **Indexer:** Michael Ferreira; **Permission Manager:** Martha Gullo; **Graphic Designer:** Bob Reuther; **Graphic Artists:** Yvonne Griffith and Angela K. Snyder; **Cover Designer:** Bob Reuther; **Photo Production Manager:** Jason Allen; **Art Manager:** Kelly Hendren; **Associate Art Manager**: Alan L. Wilborn; **Illustrator:** Tim Brummett; **Printer:** Sheridan Books.

Unless otherwise noted, all art in chapters 2-4 has been redrawn from images conceptualized and executed by David A. Gabriel.

Printed in the United States of America 10 9 8 7 6 5 4 3 2 1

The paper in this book is certified under a sustainable forestry program.

Human Kinetics
Web site: www.HumanKinetics.com

United States: Human Kinetics
P.O. Box 5076
Champaign, IL 61825-5076
800-747-4457
e-mail: humank@hkusa.com

Canada: Human Kinetics
475 Devonshire Road Unit 100
Windsor, ON N8Y 2L5
800-465-7301 (in Canada only)
e-mail: info@hkcanada.com

Europe: Human Kinetics
107 Bradford Road
Stanningley
Leeds LS28 6AT, United Kingdom
+44 (0) 113 255 5665
e-mail: hk@hkeurope.com

Australia: Human Kinetics
57A Price Avenue
Lower Mitcham, South Australia 5062
08 8372 0999
e-mail: info@hkaustralia.com

New Zealand: Human Kinetics
Division of Sports Distributors NZ Ltd.
P.O. Box 300 226 Albany
North Shore City
Auckland
0064 9 448 1207
e-mail: info@humankinetics.co.nz

E3896

To Bobbie and Suzanne

Contents

Preface

Imagine that you are living in an apartment with rather thin walls and your neighbor is throwing a party. From your apartment it seems like there are groups of conversations next door, and you're wondering who's at the party, how many people there are, whether they are men or women, and so on. The conversations closer to the wall are easier to hear, and the voices sound a bit different from those deeper in the room. A radio is playing so it is somewhat difficult to hear the conversations, and as more people enter the party, everything gets louder.

The challenge of recording and interpreting electromyographic activity (EMG) is analogous to the task you face in this thin-walled apartment. If you record from the skin surface (the wall), the superficial muscle fibers nearest the skin (voices closer to the wall) contribute greater activity than those farther from the surface electrodes. Groups of motor units (analogous to groups of human conversations) make unique contributions to the EMG signal. As more motor units participate in the muscle contraction (more people enter the room), the EMG signal increases in amplitude. Numerous sources of noise (like background music) can make the interpretation of the EMG signal difficult.

Ever since Luigi Galvani discovered "animal electricity" in frog muscle, researchers and practitioners have found numerous clinical and research uses for the EMG signal. Applications for the use of EMG include biofeedback, gait analysis, and clinical diagnosis for neuromuscular disorders. Moreover, numerous kinesiological researchers have reported their results of EMG studies involving many issues such as spinal reflexes, the action of specific muscles in various movements, muscle fatigue, the use of EMG for rehabilitation, and ergonomic design.

Only a handful of research articles using EMG techniques were published in the early 1950s. Today, over 2,500 research publications appear each year (figure 1). The growth of the EMG literature and the availability of appropriate instrumentation and techniques might suggest that our understanding of the procedures used to record the EMG signal and the relevant analysis methods must be complete. Yet the

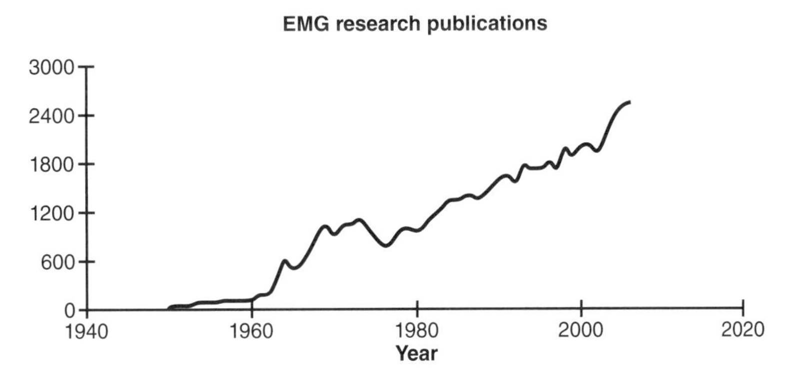

FIGURE 1 The growth in the number of EMG-related publications since the mid-1940s is a clear indication of the significant growth in the interest in and use of EMG in the past six decades.

interpretation of the signal remains controversial; and there are few sources available to help the novice electromyographer understand the physiological and biophysical basis of EMG, characteristics of the instrumentation, signal analysis techniques, and appropriate EMG applications.

This book is written for the novice who is just beginning to discover EMG and is considering its use for clinical or research purposes. Our intent is *not* to review cutting-edge research in the field. Rather, we hope to provide a starting point from which individuals who plan to use EMG can understand the underlying physiological basis of the signal and basic principles of the technology, and be able to apply appropriate analysis techniques to avoid pitfalls in interpretation.

What You Will Find in This Text

We start with a review of the physiological basis of the EMG signal (chapter 1). Since the EMG signal is ultimately a signal based on physiology, it is important to understand the origin and generation of the muscle fiber action potential, the numerous factors that determine muscle fiber conduction velocity, various physiological mechanisms responsible for the gradation of muscular force, and the many physiological factors that affect the electromyogram.

The bioelectricity chapter (chapter 2) then proceeds with a review of fundamental biophysical principles. Here we lead the reader from the very elementary ideas of electric charge to the recording of muscle action potentials. The basics of electric charge are linked to the electric potential recorded at the electrode and an explanation of the use of two electrodes to detect a potential difference at the muscle. Electric fields are an important part of the explanation that connects these concepts. We review volume-conducted potentials in the same manner. This is an important topic because action potential shape determines amplitude and frequency content of the signal. The geometry of action potential "appearance," based on its position relative to the electrode, is described qualitatively using explanatory figures. Our discussion of bioelectricity then closes with an introductory treatment of alternating current (AC) since EMG is treated as an AC signal and is subject to many of the same measurement conventions.

The chapter on EMG instrumentation (chapter 3) is unique in the inclusion of topics frequently omitted in introductory treatments. For example, most surveys of electrode types omit the mechanistic events underlying signal transduction from the muscle action potential to a voltage recorded at the amplifier. Electrode configuration is also a common topic; however, the effect of interelectrode distance on the amplitude and frequency content of the EMG is discussed only in more advanced texts. We have developed figures and qualitative explanations to convey the same information as might be presented in a more mathematical treatment. The formulas that are used to present these concepts are simplified and explained in detail. For example, Kirchhoff's loop law is mentioned in more advanced treatments in discussions of amplifier input impedance. In the present text, we first describe input impedance and why it is important. The electrical circuit and associated formulas are then described in detail, assuming only the background presented in the previous chapter. New concepts in electrode placement relative to the motor point, innervation zone, and tendon are presented with recommendations that depend on the goal(s) of the study. We also introduce bode plots, demonstrate how they are generated, and use them to describe both analog and digital filtering.

The EMG signal-processing presentation (chapter 4) is a unique blend unavailable anywhere else. It is based on traditional signal-processing theory, experience, and recommendations for practical applications. The material combines concepts from more advanced texts in signal-processing theory, communications theory, and papers published on EMG methodological issues. There is a heavy reliance on figures to illustrate physical concepts. Elaborate qualitative explanations are then reinforced with the basic formula associated with the concept or methodology. For example, the origins of linear envelope detection in communications theory are reviewed so that the reader understands its predominant use as a signal-processing method. Of course, the traditional amplitude and frequency measures are described. However, since the frequency measures are often more difficult to understand, calculate, and apply properly without violating the basic assumptions of their use, a significant portion of the chapter is dedicated to reviewing the principles of frequency analysis.

The appropriate location to extract the EMG signal in a discrete trial has not been presented in previous texts but is discussed here. Similarly, the inclusion of specific recommendations and procedures for handling noise contamination of the EMG signal is unique to this text. This chapter integrates both theory and practice to describe the extraction of useful measures from the EMG signal. Area, slope, and variability of the EMG signal are discussed. The interaction between linear envelope detection and low-pass cutoff frequency is described with respect to the detection of EMG onset. Power spectral analysis and the calculation of frequency measures from the EMG signal based on Fourier analysis are discussed. The chapter concludes with a basic explanation of digital filtering.

The last two chapters of the text (chapters 5 and 6) provide examples of the use of EMG techniques with numerous references to the existing literature. The relationship between EMG activity and muscular force has considerable relevance for the development of prosthetic devices and other applications. Much of the research on this issue is discussed, as well as the important work related to the characteristics of the EMG signal that accompany muscular fatigue. We also discuss the use of EMG techniques for recording evoked potentials such as the M-wave, H-reflex, and motor-evoked potentials using transcranial magnetic stimulation (TMS). Chapter 6 also includes an overview of the use of EMG techniques for gait analysis with examples from the extant research literature.

Although simplified explanations of concepts in communication theory, signal processing, electronics, and other issues are presented in the chapters, we include extensive appendixes for those readers interested in more advanced topics and derivations. For example, basic electrode geometry is described in the text, but we include two associated appendixes that provide a computational understanding of the topic through detailed examples. Modeling and simulation of the EMG signal has moved to prominence in the physiological literature, and we include an appendix providing information relevant to understanding that material. Fundamental concepts of EMG frequency measures are provided in the text, but an appendix is then provided with a worked example to help readers internalize this important methodology. A list of acronyms precedes the first chapter for those unfamiliar with the typical acronyms used in the field of electromyography. The text also includes a glossary of new terms introduced throughout the book, as well as brief lists of suggested readings for each chapter, including classic readings in the field. Terms defined in the glossary appear in bold type in the text.

The reference list is far from complete, but it does provide a starting point for the individual intent on gaining additional knowledge. EMG is a rapidly changing technology. Advances in instrumentation, such as array electrodes, and in analysis techniques, such as nonlinear analysis and pattern classification, mean that the future for this field is bright, particularly for those with a thorough understanding of the underlying concepts.

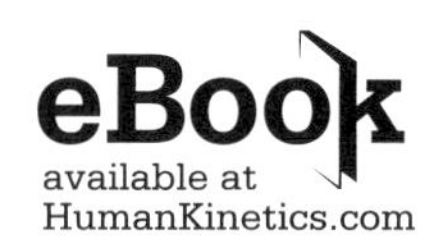

Additional Unique Contributions of This Text

Relatively few previous texts have been made available, and these have provided valuable information about electrode placement, the relationship between applied anatomy and EMG, and EMG applied to clinical areas such as biofeedback and neuromuscular diagnosis. In this text, we provide updated information from the latest available sources. Much of the information in this text is available only in scattered books and papers among the numerous disciplines that use EMG as a tool. For example, although basic electrophysiology is covered in several EMG texts and neurophysiology texts, it is difficult to find summarized materials that relate muscle architecture to EMG. The bioelectricity chapter is particularly novel in how it relates concepts of electric charge to EMG action potentials. The chapter on EMG analysis includes concepts that would otherwise be available only in journal review articles, such as techniques for defining EMG onset and issues relevant to electromechanical delay.

The text is written for individuals with a wide variety of backgrounds, including engineers, physical therapists, kinesiologists, physicians, biofeedback practitioners, and ergonomists. The level of the book is aimed at a fourth-year undergraduate or entry-level graduate student who has a modest background in science. The book relies heavily on the use of figures and qualitative explanations to convey important concepts to bridge any gap that may exist in the background preparation of the reader. However, mathematical derivations have been included in the appendixes to allow able readers to work through equations associated with detecting, filtering, and processing the EMG signal. These mathematical skills require only a first-year calculus course. The algebraic steps have been included because this is often the first skill to go "rusty." Grasping these basic equations contributes to the depth of understanding of the physical side of EMG. For example, certain changes in the EMG signal are predictable based on the physical properties of the electrode detection system and are of no physiological consequence. Once the physical effects have been identified, a clear understanding of anatomy and physiology is then necessary for valid interpretation of the EMG signal.

As a resource for instructors using this text in their courses, an image bank is provided at www.HumanKinetics.com/essentialsofelectromyography. The image bank contains most of the figures and tables from the text, sorted by chapter. These images can be used to develop a custom presentation based on specific course requirements. A blank PowerPoint template and instructions are also included.

We hope you will find this book a useful primer and frequent reference as you begin your exploration of the field of electromyography.

Gary Kamen
David A. Gabriel

Acknowledgments

No text of this scope could be complete without acknowledging the many contributions from professional and research colleagues as well as from technical personnel. We would like to thank our mentor, Walter Kroll, for providing our initial introduction to electromyography and the inspiration to seek new knowledge in this important field. In addition, our graduate students and faculty colleagues, too numerous to mention without mistakenly excluding some, raised a never-ending string of intriguing and challenging questions and encouraged us to engage in the important research needed to obtain the answers.

Several staff at Human Kinetics were helpful in ensuring completion of the book in an accurate and timely manner. These individuals included Loarn Robertson, Elaine Mustain, Kate Maurer, Martha Gullo, and Dalene Reeder. Many obstacles to the timely completion of the book were overcome with the help of Greig Inglis.

Acronyms and Symbols

A	area
A/D	analog-to-digital
AC	alternating current
ARV	average rectified value
C	coulomb
CMAP	compound muscle action potential
CMRR	common mode rejection ratio
CV	conduction velocity
DC	direct current
DFT	Discrete Fourier transform
E	electric field
ECG	electrocardiogram
EMD	electromechanical delay
EMG	electromyography
f	frequency
F	force
FFT	Fast Fourier transform
FT	Fourier transform
G	gain
i	current
ICC	intraclass correlation
IED	interelectrode distance
IEMG	integrated electromyography
IFFT	inverse Fast Fourier transform
IPA	interference pattern analysis
J	current density
m	meter
M-wave	massed action potential
MDF	median power frequency
MEP	motor evoked potential
MEPP	miniature end-plate potential
MFAP	muscle fiber action potential
MFCV	muscle fiber conduction velocity
MNF	mean power frequency
MSA	mean spike amplitude
MSF	mean spike frequency
MU	motor unit
MUAP	motor unit action potential
MVC	maximal voluntary contraction
N	newton
PDF	probability density function
PSD	power spectral density
P-P	peak-to-peak
Q	charge
Q_{30}	The area on the EMG–time curve computed between the onset of EMG activity and a point 30 ms following EMG onset
QE	quantization error
r	radial distance
R	resistance
RMS	root mean square
SD	standard deviation
sEMG	surface electromyography
SI	International System of Units
SNR	signal-to-noise ratio
TMS	transcranial magnetic stimulation
TP	total power
U	potential energy
V	volt
VR	variance ratio
W	work
X_C	reactive capacitance
Z	impedance
$\mathcal{E}$	electromotive force
λ	length constant
ρ	resistivity to the flow of charge
σ	conductivity
Ω	ohm

chapter 1

Anatomy and Physiology of Muscle Bioelectric Signals

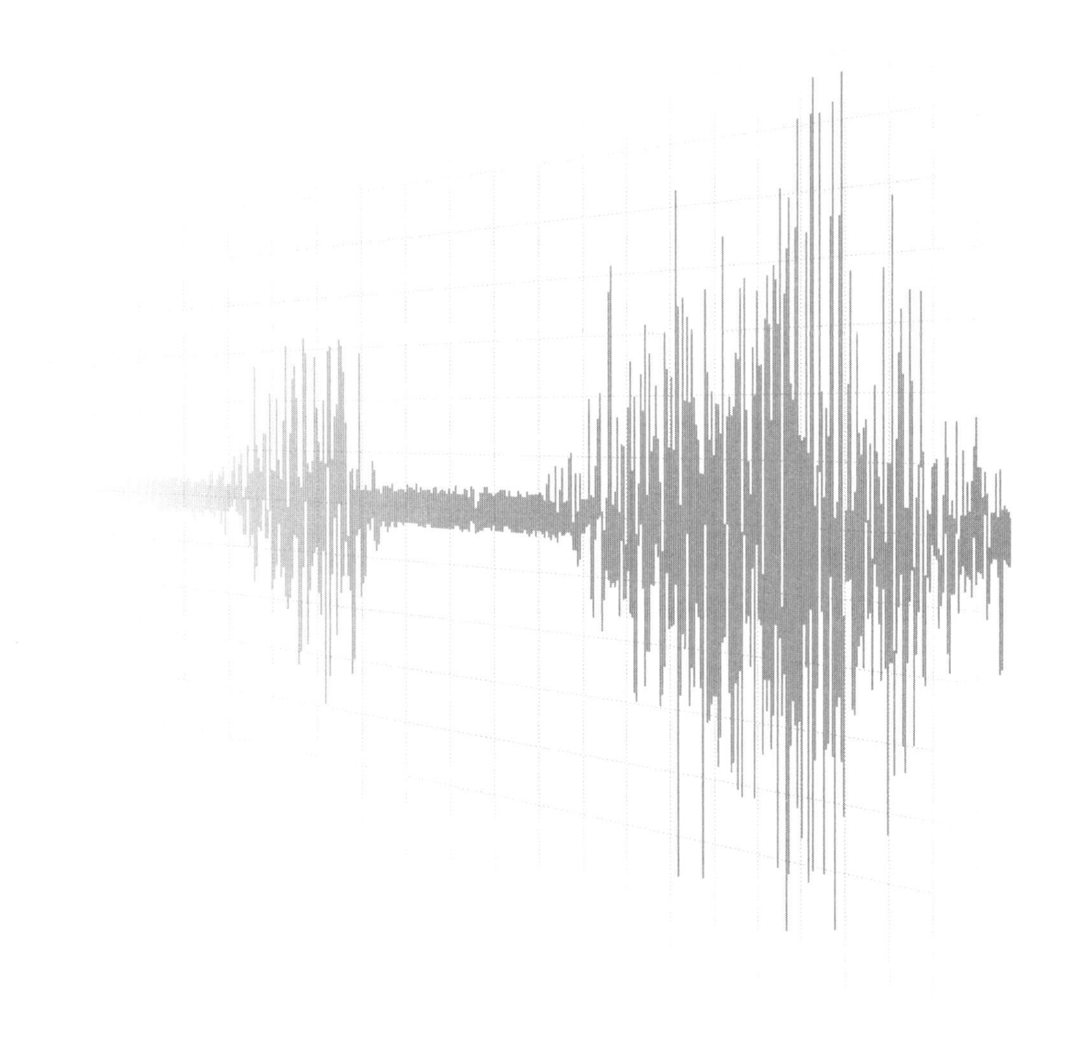

Electromyography (EMG) is a valuable technique for studying human movement, evaluating mechanisms involving neuromuscular physiology, and diagnosing neuromuscular disorders. However, there are many potential pitfalls in the use of EMG as a tool. The question that a researcher is asking may not be amenable to solution using EMG techniques. The researcher could err in the selection of recording electrodes, the recording site, or the data acquisition specifications. Furthermore, the interpretation of the EMG signal requires a thorough knowledge of the origin of the signal.

Although researchers frequently evaluate electromyographic waveforms as an electrical signal whose characteristics can be assessed using traditional signal-processing techniques, the EMG signal has physiological origins in individual fibers or groups of muscle fibers. The anatomical features of individual fibers, the architectural features of whole muscle, and the physiological origins of action potentials are key to understanding how to record, analyze, and interpret the EMG signal. In this chapter, we study the origins of the EMG signal, including relevant muscle physiological concepts.

Anatomical Features of Muscle

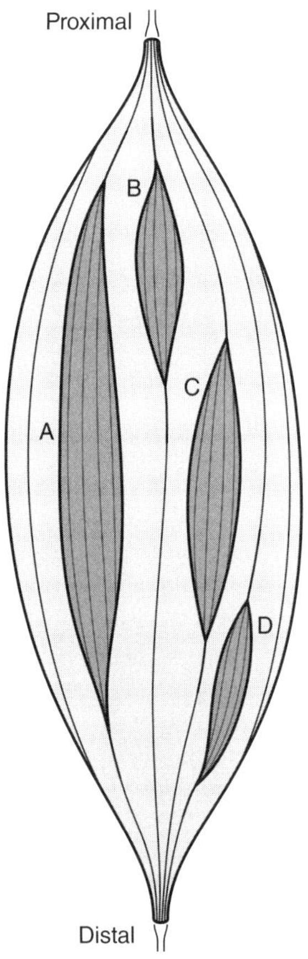

FIGURE 1.1 Muscle fibers vary in length. Some muscle fibers range from proximal tendon to distal tendon *(A)*. Other fibers lie chiefly in proximal *(B)* or distal *(D)* portions of the muscle. Still other muscle fibers may range from proximal to distal tendon but vary considerably in length *(C)*.

The salient anatomical features that affect the EMG signal include variations in muscle fiber length and fiber type composition, muscle partitioning, and variations in the distribution of sensory receptors. These anatomical and architectural muscle features differ among muscles and even within and among individual subjects. Thus, they need to be considered to ensure proper EMG recording and interpretation.

- **Muscle fiber length.** Although it is frequently assumed that muscle fibers run continuously from distal to proximal tendon, this is not always the case. Some muscle fibers are short and may lie at proximal, distal, or middle portions of the muscle (Gans and de Vree 1987; Heron and Richmond 1993; van Eijden and Raadsheer 1992). The human hamstrings, for example, is composed of fibers that range from 4 to 20 cm in length, and some muscle fibers may be tapered at one or both ends (Heron and Richmond 1993). Surface electrodes placed longitudinally on either a distal or proximal part of the muscle, then, will record only from those muscle fibers underlying the electrode. Action potentials may differ at different portions of a tapered muscle fiber (figure 1.1).
- **Muscle fiber architectural characteristics.** The characteristics of fibers may vary between deeper and more superficial portions of the muscle (Dwyer et al. 1999; Lexell et al. 1983; Pernus and Erzen 1991; Roeleveld et al. 1997). Deeper muscle fibers seem to comprise a greater proportion of slow-twitch fibers, while muscle fibers lying more superficially comprise a greater proportion of larger, fast-twitch fibers (Polgar et al. 1973). Electrophysiological evidence in human muscle using a technique termed **macro-EMG** supports this idea (Knight and Kamen 2005). The variance in fiber type composition in different areas of the muscle could be due to the greater access to the blood supply afforded to slow-twitch fibers lying deeper in the muscle, though this speculation has yet to be corroborated. Since the global EMG signal recorded from surface EMG electrodes presents a biased estimate of

activity closer to the electrodes (which we will discuss later), this anatomical feature is important.

■ **Muscle partitioning.** Another factor concerning gross muscle structure that affects the interpretation of the EMG signal relates to *muscle partitioning* (figure 1.2). Many human and animal muscles are partitioned, and each partition may have a specific role in the function of a particular muscle (Blanksma and van Eijden 1990; English et al. 1993; Segal 1992; Segal et al. 1991, 2002; van Eijden and Raadsheer 1992). For example, the flexor carpi radialis consists of three major architectural divisions based on both muscle architecture and the innervation pattern (Segal et al. 1991). A lateral partition functions during radial deviation, while both lateral and medial partitions function during pure wrist flexion. If one obtains multiple recordings from the human extensor carpi radialis longus, proximal and distal portions of the muscle are shown to be selectively active, depending upon whether the movement is pure extension or extension and radial deviation (English et al. 1993). Even small facial muscles like the orbicularis oris can have specific divisions (Abbs et al. 1984). Thus, the electromyographer needs to be aware of whether the recording is representative of the entire muscle or is characteristic of a specific muscle partition.

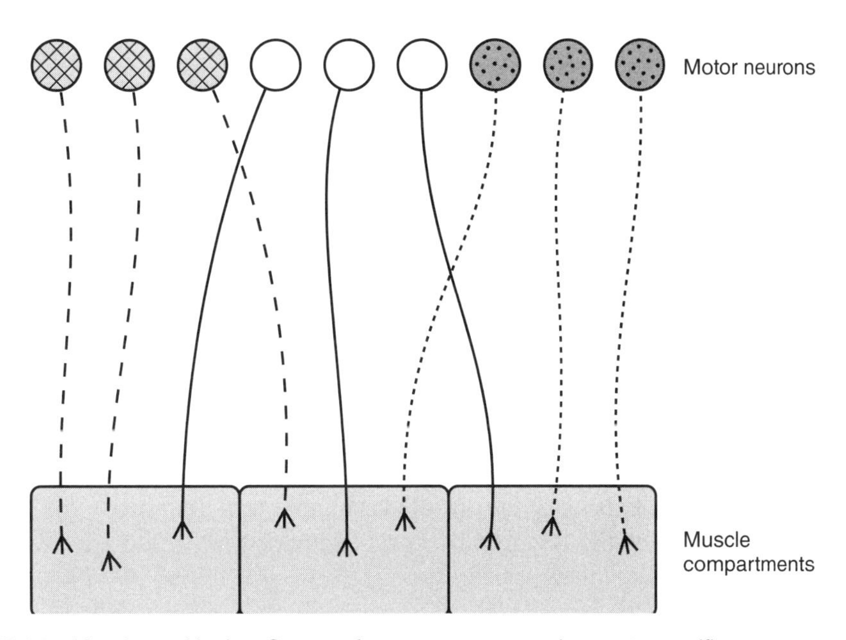

FIGURE 1.2 Muscle partitioning. Groups of motor neurons may innervate specific compartments. Note that one population of motor neurons may innervate more than one compartment.

■ **Neuromuscular compartment partitioning.** Neuromuscular compartments may also be partitioned, such that specific receptors like muscle spindles and tendon organs may be sensitive to the activity of a specific, localized group of motor units. This neuromuscular partitioning has been well demonstrated in the cat but is also present in some human muscles (Kamibayashi and Richmond 1998; Windhorst et al. 1989). One way to identify partitioning may be to observe EMG responses to stimuli presented to different parts of a muscle. The human tibialis anterior may not be compartmentalized, for example, since vibration and tendon tapping, which serve as strong inputs to Ia receptors, failed to identify localized reflex responses (McKeon et al. 1984).

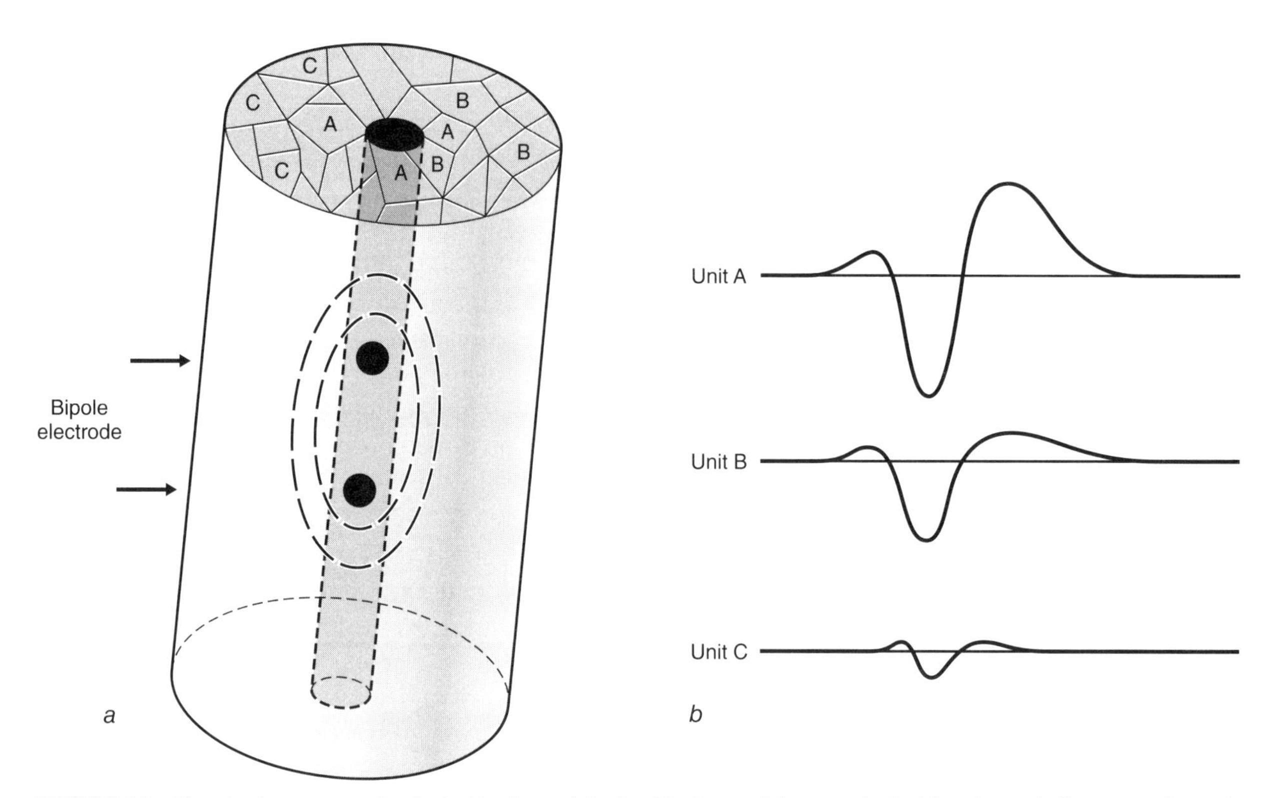

FIGURE 1.3 The electromyogram is affected by the architectural features of the muscle. In this schematic figure, an electrode placed in the center of the muscle would record the largest action potential from motor unit A, the next largest from unit B, and the smallest from unit C.

Reprinted, by permission, from G.E. Loeb and C. Gans, 1986, *Electromyography for experimentalists* (Chicago: University of Chicago Press), 51.

- **Sensory receptor distribution.** In feline muscle and perhaps in human muscle, the distribution of sensory receptors within the muscle may not be homogeneous. Thus, the muscle region with the greatest density of receptors may provide regional information about localized changes in muscle length, force, and limb displacement (Richmond and Stuart 1985). Clearly, a knowledge of the specific anatomy of the muscle(s) of interest is necessary before the EMG signal is recorded. The function revealed by EMG may be the function of a specific partition (figure 1.3).

KEY POINTS

- The length of individual muscle fibers within the whole muscle varies, and the characteristics of the muscle fiber action potential change at different fiber sites, rendering the EMG signal dependent on the specific location at which the electrodes are placed.
- Larger, type II muscle fibers tend to lie more superficially, while deeper muscle fibers tend to be smaller, type I fibers. Since the surface EMG signal is biased toward the fibers closest to the electrode, this means that the action potentials from the superficial type II fibers are disproportionally represented in the EMG signal. In other words, a greater proportion of the EMG signal is derived from superficial fibers than from deeper fibers.
- Due to compartment partitioning, the interpretation of the EMG signal may depend on the part of the muscle from which recordings are made. Hence, knowledge of neuromuscular architecture is critical.

Physiology of the Muscle Fiber

Muscle is a tissue constantly bathed in an ionic medium. Like all living cells, muscle is surrounded by a membrane—the **sarcolemma,** which is about 75 angstroms (Å) thick. At regular intervals, the **transverse tubular system** interrupts the membrane. In some places, the transverse tubules (T-tubules) run longitudinally, connecting with other T-tubules and the **sarcoplasmic reticulum** (SR) network (Hayashi et al. 1987). T-tubules serve as important structures for carrying the action potential deep transversely into the myofibrils to fully activate all portions of the muscle fiber.

Resting Membrane Potentials

Under resting conditions, a voltage gradient exists across the muscle fiber membrane such that the inside of the fiber lies about –90 mV with respect to the outside. The voltage gradient arises from the different concentrations of sodium (Na^+), potassium (K^+), and chloride (Cl^-) and other anions across the membrane. Under resting conditions, the concentration of Na^+ is relatively high outside the membrane and relatively low inside the fiber. On the other hand, the concentration of K^+ is relatively low outside the membrane and relatively high on the inside of the muscle fiber. The size of the resting membrane potential is slightly more positive in slow-twitch fibers. The greater positivity arises from the enhanced Na^+ permeability and higher intracellular Na^+ activity in slow-twitch fibers than in fast-twitch fibers (Hammelsbeck and Rathmayer 1989; Wallinga-De Jonge et al. 1985). Also, the resting membrane potential can be changed by exercise training (Moss et al. 1983).

Generation of the Muscle Fiber Action Potential

Muscle fibers are excitable tissues. When the muscle fiber is depolarized by about 10 mV or more, the **membrane potential** reacts in a stereotypical and predictable fashion, producing a response we call the **muscle fiber action potential (MFAP),** or just action potential. The action potential generated at the neuromuscular junction proceeds along the muscle fiber in both directions from the neuromuscular junction. In the first phase of the action potential, Na^+ permeability increases and Na^+ rushes into the cell, ultimately reversing the polarity of the cell so that the cell is momentarily about 10 mV positive. As the Na^+ permeability increases, so does the membrane permeability to K^+, and it is the outflow of K^+ that ultimately results in the return of the membrane potential to its resting state (figure 1.4).

The sodium permeability exerts considerable control over the time course of the action potential. A **refractory period** follows the nerve or muscle impulse, during which time there is a decrease in the excitability of the membrane. For a brief period of time, the membrane is *absolutely* refractory and all the Na^+ channels are closed, and so the membrane cannot respond with an action potential regardless of the size of the excitatory stimulus. There follows a *relative* refractory period during which some Na^+ channels are open, and an action potential can be generated so long as the excitatory stimulus is sufficiently large to overcome the increase in the threshold needed for excitation.

The main spike portion of the muscle fiber action potential (MFAP) is followed by a **terminal wave,** produced by the termination of the action potential at the muscle–tendon junction (McGill et al. 2001). Muscle fiber action potentials also have a unique feature called the slow **afterwave,** also termed the slow **afterpotential** (Lang and

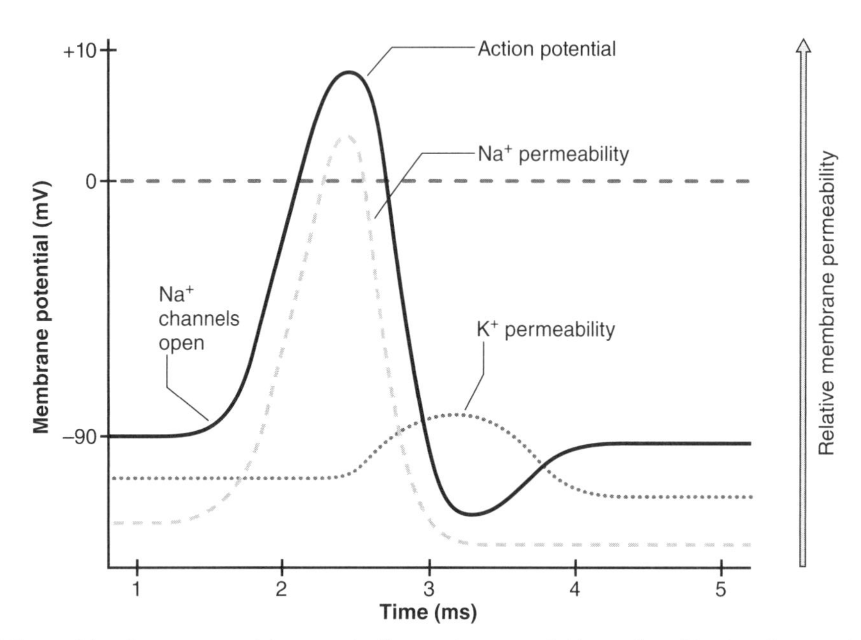

FIGURE 1.4 The time course of the muscle fiber action potential is mediated by the changes in membrane permeability to Na^+ and K^+ ions.

Vaahtoranta 1973). Following the main portion of the action potential, the return of the membrane potential to baseline follows a very slow time course. In clinical and quantitative studies, this can make it difficult to quantify the duration of the MFAP. This slow afterwave reflects the negative phase of the muscle fiber's action potential. It seems to be due to the repolarization of the T-tubule system (MacFarlane and Meares 1958). Recordings obtained close to the neuromuscular junction produce more exaggerated afterwaves (Lateva and McGill 1998). The frequency characteristics of the slow afterwave are in the 2 to 40 Hz band, so high-pass filtering the EMG signal at upper frequencies will depress the appearance of the slow afterwave.

At random intervals, small potentials are released at the neuromuscular junction (Zigmond et al. 1999). These **miniature end-plate potentials (MEPPs)** are high-frequency spikes and are sometimes recorded as spikes in the electromyogram (Simons 2001). The spikes decay considerably with length, and so these MEPP spikes are much more evident if the electrodes are placed close to an end-plate zone.

Muscle Fiber Conduction Velocity

The electromyogram is largely influenced by the features of the MFAP as it propagates along the muscle fiber. Relative to nerve conduction velocity rates, which can be as high as 100 m/s, **muscle fiber conduction velocity (MFCV)** is relatively slow, on the order of 2 to 6 m/s. There are numerous techniques available to compute MFCV (figure 1.5), and several useful reviews have been published (Arendt-Nielsen and Zwarts 1989; Farina and Merletti 2004; Zwarts and Stegeman 2003). Some of the important analytical issues are discussed in chapter 4. However, from a physiological perspective, MFCV is dependent upon a number of muscle fiber characteristics:

- **Intramuscular milieu.** Using isolated mouse soleus and extensor digitorum longus muscle, Juel (1988) showed that MFCV decreases with higher extracellular K^+ concentrations and decreases with lower intracellular pH, but is independent of Na^+ concentration and extracellular pH. The decrease of MFCV with low pH values is a major reason that MFCV decreases with fatigue.

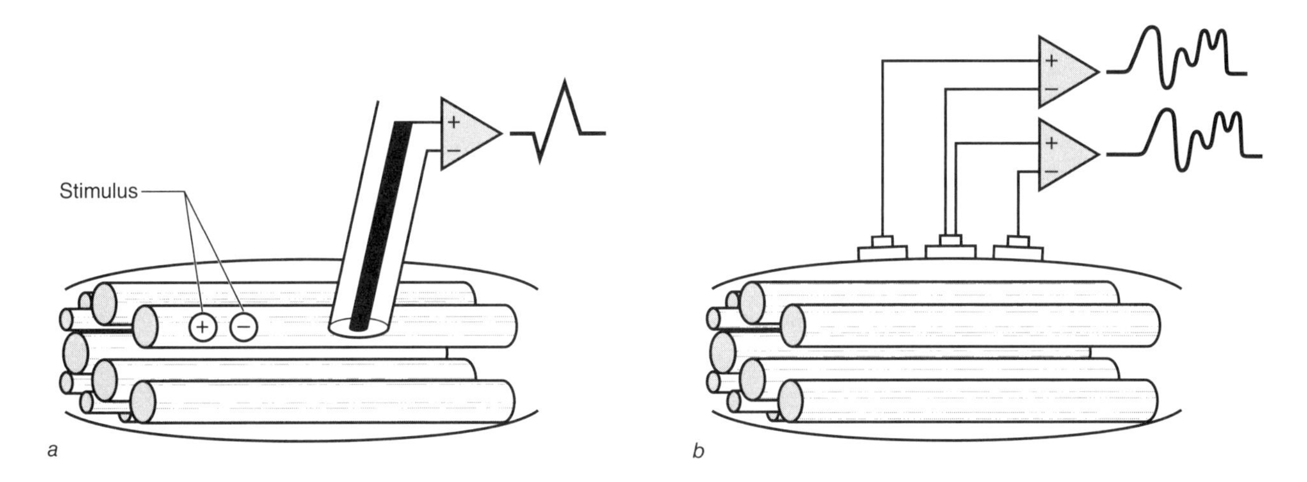

FIGURE 1.5 *(a)* One can obtain muscle fiber conduction velocity (MFCV) by inserting a pair of stimulating electrodes in the muscle and measuring the response at a known distance (Troni et al. 1983). *(b)* Alternatively, two or more channels can be recorded from the skin surface during voluntary activation, and MFCV can be determined using cross-correlation techniques. These techniques are further detailed in chapter 4.

- **Temperature.** Muscle fiber conduction velocity changes with muscle temperature, generally increasing with increases in temperature and decreasing with decreasing temperature (Stålberg 1966). Thus, it is important to maintain constant temperature in the examination room or laboratory.

- **Muscle fiber diameter and muscle morphology.** Håkansson (1956) also reported a linear relationship between diameter and conduction velocity (CV) of frog fibers. Stålberg (1966) found a positive relationship between the circumference of the upper arm and the CV. Conduction velocity also increases with the recruitment threshold of the motor unit (Gantchev et al. 1992) and is greatest near the end-plate region and lowest near the tendon (Li and Sakamoto 1996a; Sakamoto and Li 1997).

- **Muscle length.** When the muscle fiber is lengthened (stretched), the CV is expected to decrease due to a decrease in the effective diameter of the fiber. This was originally demonstrated by Håkansson (1956). In human muscle, MFCV also decreases with increased muscle length (Arendt-Nielsen et al. 1992; Morimoto 1986; Trontelj 1993). The effect is observed more dramatically in superficial muscle fibers than in deeper fibers (Kossev et al. 1992), and this could be due to the greater change of muscle length during passive stretching in superficial fibers.

- **Fiber type.** As early as 1912, the CV of fast-twitch fibers was known to be higher than that of slow-twitch muscle fibers (Kohlrausch 1912), and this has since been verified in human muscle fibers (Hopf et al. 1974). A correlation of $r = 0.84$ was reported between vastus lateralis fiber type and MFCV measured during maximal-effort contraction, suggesting that fiber type could be well predicted from a noninvasive measurement of CV (Sadoyama et al. 1988). Juel (1988) showed that the difference in CV could not be attributed to fiber diameter; he speculated that instead it may be due to differing ion channel density between fast-twitch and slow-twitch fibers.

- **Muscle fatigue.** During light exercise, MFCV can actually increase, presumably due to increases in temperature, muscle swelling, or changes in other membrane properties (Van Der Hoeven et al. 1993; Van Der Hoeven and Lange 1994). Numerous studies have documented the decline in MFCV with fatigue (Sadoyama et al. 1985; Stålberg 1966; Zwarts and Arendt-Nielsen 1988).

■ **Neuromuscular pathology.** Although nerve CV has long been used as a clinical technique for neuromuscular diagnosis, little clinical use has been made of MFCV measurements. However, several studies point to the potential diagnostic value of MFCV assessment (Blijham et al. 2004; Van Der Hoeven et al. 1993, 1994; Yaar and Niles 1992; Yamada et al. 1991). Diseases of the muscle, such as various dystrophies and polymyositis, tend to decrease CV (Hong and Liberson 1987), though it also decreases with motor neuron disease (Gruener et al. 1979).

■ **Other factors.** Muscle fiber conduction velocity increases with age, being faster in adults than in children (Cruz Martinez and López Terradas 1992). It is reduced by hypoxia (Gerilovsky et al. 1991) and increases with joint torque (Masuda et al. 2001), probably due to the recruitment of new motor units with faster CV. There has also been some suggestion that MFCV of individual muscle fibers increases with increasing force (Masuda et al. 1996; Mitrovic et al. 1999; Sadoyama and Masuda 1987). Mitrovic and colleagues (1999) cited changes in membrane channel configuration or electrical resistance as possible explanations for the increasing CV. Increasing firing rate also seems to result in faster CV (Morimoto and Masuda 1984); and this may be due to the presence of a supernormal period of the action potential, previously reported in frog nerve and in human muscle (Stålberg 1966), facilitating CV for a brief period after the action potential. Muscle fiber conduction velocity correlates well with individual motor unit twitch torque (Nishizono et al. 1990), prompting the suggestion that CV could be another component of the Henneman size principle (Andreassen and Arendt-Nielsen 1987).

Anatomically, the location of the innervation zone can be approximated using multiple electrodes to identify the site at which the polarity of the MFAP reverses (Masuda and Sadoyama 1989). In the biceps brachii, for example, the neuromuscular junctions were found to lie in a discrete zone in the middle of the muscle (Masuda et al. 1983). Some subjects, however, seemed to have multiple myoneural junctions, and this is consistent with other suggestions that some muscle fibers may be multiply innervated (Jarcho et al. 1952; Lateva et al. 2002).

Electrodes must be placed longitudinally along the muscle fibers. Otherwise, inaccurate CV estimates can be obtained. Errors in orientation of up to 10% can result in measured changes in CV of up to 10% (Sadoyama et al. 1985; Sollie et al. 1985a). Placement of the electrodes near the motor point or near the tendon can result in errors in CV measurement. In general, the **innervation zone** lies near the middle of each muscle fiber. However, there frequently are extreme cases in which the innervation zone can be found at proximal or distal muscle fiber locations (Saitou et al. 2000).

It is clear that a number of physiological and technical issues affect the assessment of MFCV. Thus, researchers attempting to record MFCV should standardize conditions, including muscle length, muscular force level, and intramuscular temperature.

KEY POINTS

- The physiology of the muscle fiber determines the amplitude, shape, and time course of each MFAP, which determine, as an ensemble, the characteristics of the EMG signal.

- Differences in ionic concentrations produce voltage gradients across the sarcolemma. These voltage gradients account for the resting membrane potential that varies in slow- and fast-twitch muscle fibers.
- The electrical message to initiate muscle contraction is transmitted through the muscle fiber T-tubules via the MFAP.
- Specific phases of the action potential such as the main spike, terminal wave, and afterpotential have been defined; and some features of the electromyogram have been interpreted using these phases.
- The afterhyperpolarization limits the frequency of MFAPs.
- As the action potential is propagated along the muscle fiber, it proceeds at a rate measured as the CV. This MFCV is affected by ionic concentrations, temperature, muscle fiber length and diameter, fiber type, fatigue, various neuromuscular pathologies, and other factors such as hypoxia and age.

Motor Unit Features

The **motor unit (MU)** concept was originally described by Sherrington as consisting of a single motoneuron and all the muscle fibers innervated by that motoneuron (see figure 1.6). Sherrington (1906) labeled the motor unit as the "final common pathway"; barring any pathology, all of the muscle fibers innervated by the motoneuron are activated when an action potential appears in the motoneuron.

There have been occasional challenges to the motor unit idea. For example, the suggestion has been made that a few fibers may be multiply innervated (Lateva and McGill 2001). From a motor control viewpoint, this would complicate the ability of the central nervous system to predict muscular force output based on motoneuron activation. Additional confirmation of a multiple-innervation suggestion seems prudent.

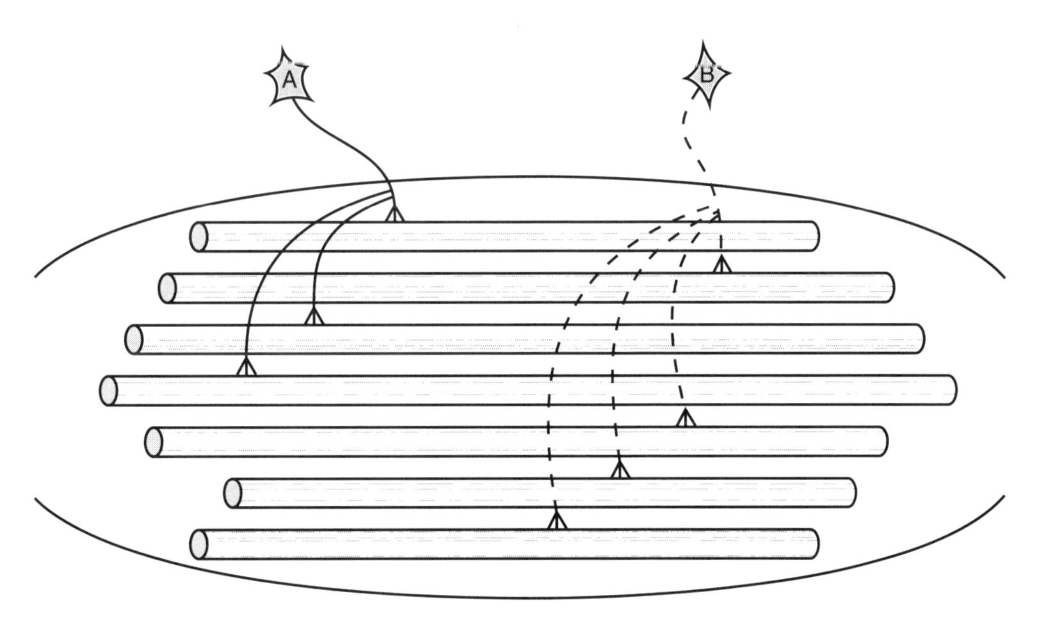

FIGURE 1.6 Motor units differ in their innervation ratio (number of muscle fibers per motor unit). Also, some motor units are located mostly in superficial regions of the muscle, while other units are located in deeper regions.

Fiber Organization

Muscle fibers can vary in their histochemical characteristics and in size. Groups of motor units may vary in size and in other organizational features. In this section, we discuss some of these anatomical and physiological factors.

- **Fiber type.** As discussed earlier, there are numerous differences between fast-twitch and slow-twitch fibers that affect the electromyogram, and the resting membrane potential is more positive in slow-twitch fibers than in fast-twitch fibers. Because of the differences in fiber diameter, fast-twitch fibers can be expected to produce larger action potentials than slow-twitch fibers, and the CV is faster in fast-twitch fibers than in slow-twitch fibers. Consequently, the electromyogram can be expected to be different in fast-twitch and slow-twitch muscle.
- **Motor unit organization.** The number of motor units and the **innervation ratio** (number of muscle fibers per motor unit) differ among muscles. Large muscles like the gastrocnemius have about 600 motor units with an innervation ratio of 2000 muscle fibers per motor unit. Small eye muscles like the external rectus may have as many as 3000 motor units but an innervation ratio of 9 muscle fibers per motor unit (Feinstein et al. 1955). Various other innervation ratios for different muscles have been reported (Gath and Stålberg 1982). Thus the innervation ratio is one morphological means of controlling the precision of muscle contraction. More subtle gradations in muscular force can be achieved through increasing or decreasing motor unit firing rate than through the activation or deactivation of an entire motor unit. Techniques are available for estimating the number of motor units in human muscle, and these will be discussed in a later section.
- **Fiber grouping.** There is some evidence that the distribution of fibers may be such that fibers within a specific type or a specific motor unit may be localized (Bodine-Fowler et al. 1990); however, most of the evidence favors a random distribution of fibers within the muscle territory (Buchthal and Rosenfalck 1973; Dubowitz and Brooke 1973; Edstrom and Kugelberg 1968; Gates and Betz 1993). If fibers are grouped, a surface or needle recording from a single area will comprise signals from a select and relatively restricted group of motor units. Motor unit architecture and fiber grouping change with advancing age. Muscle fibers belonging to some motor units lose their innervation as motoneuron "death" occurs. Some of these muscle fibers are reinnervated by neighboring motoneurons, producing large motor units. In these muscles, fiber density is not random. Rather, muscle fibers belonging to the same motor unit are frequently found in localized sections within the muscle (Andersen 2003; Lexell 1995).

Motor Unit Action Potential

Since multiple muscle fibers are innervated by a single motoneuron, the firing of a motoneuron results in the near-simultaneous discharge of many muscle fibers. The summed activity of all these muscle fibers culminates in the generation of a **motor unit action potential** or **MUAP** (figure 1.7). The amplitude of the MUAP is determined by the individual MFAPs, summed at the recording site both temporally and spatially (see figure 1.8). Some muscle fibers within a motor unit may have long

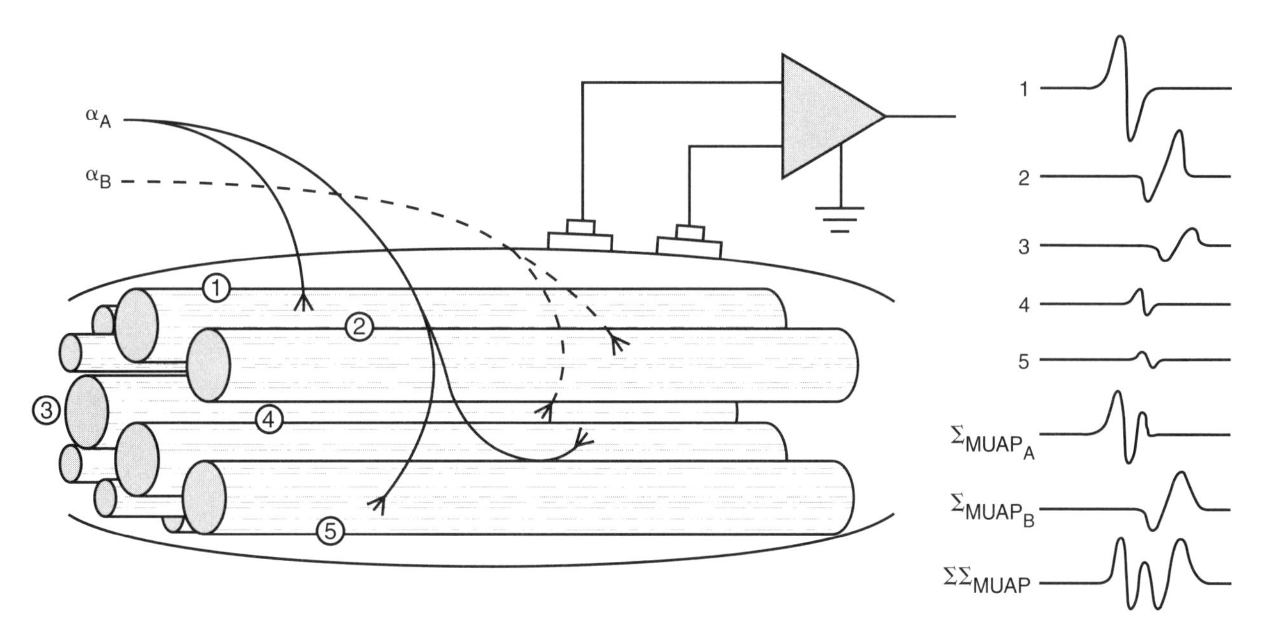

FIGURE 1.7 The surface electromyogram is composed of the algebraic sum of all motor unit action potentials.

Reprinted, by permission, from G. Kamen, 2004, Electromyographic kinesiology. In *Research methods in biomechanics,* edited by D.G.E. Robertson, G.E. Caldwell, J. Hamill, G. Kamen, and S.N. Whittlesey (Champaign, IL: Human Kinetics), 165.

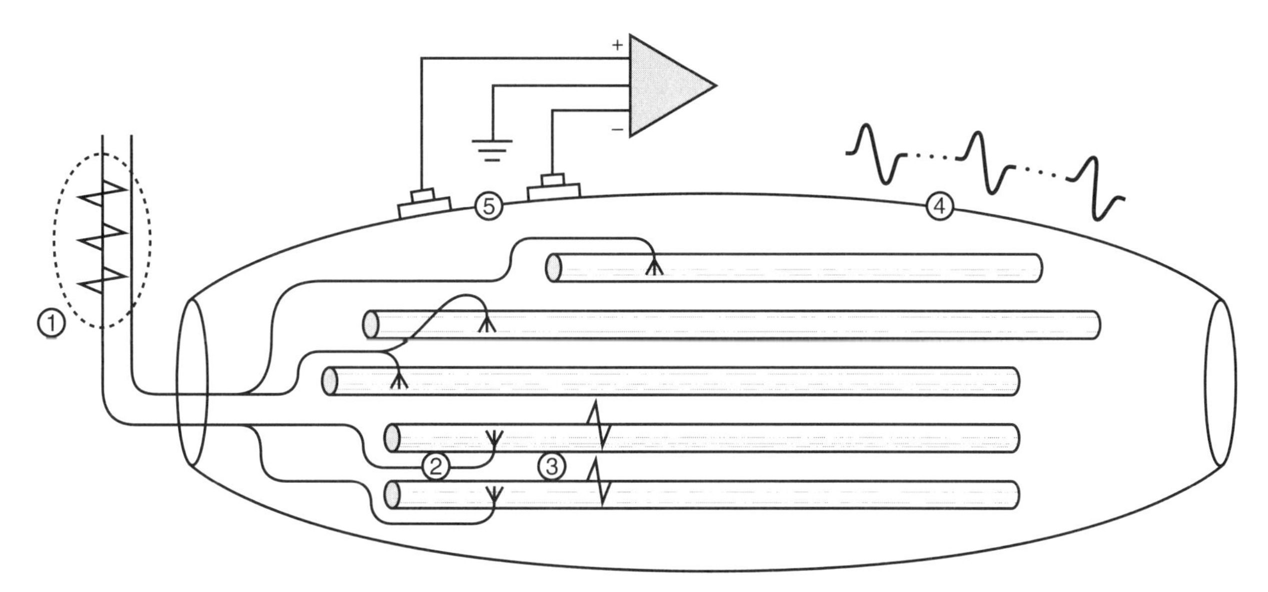

FIGURE 1.8 The process of recording motor unit action potentials (MUAPs) begins with the generation of motoneuron APs (1). The motoneuron APs arrive at each muscle fiber end plate (2) and result in the production of muscle fiber APs (3). The sum of all individual muscle fiber APs produces a motor unit AP (4) that can be recorded with appropriate electrodes and amplifiers (5).

Reprinted, by permission, from G. Kamen, 2004, Electromyographic kinesiology. In *Research methods in biomechanics,* edited by D.G.E. Robertson, G.E. Caldwell, J. Hamill, G. Kamen, and S.N. Whittlesey (Champaign, IL: Human Kinetics), 164.

axon “twigs” (terminal axon segments). These contribute to the later components of the MUAP and may result in complex MUAP shapes with numerous spikes. On the other hand, if the terminal axon segments are of equal length and all muscle fibers within the motor unit fire simultaneously, then the MUAP may be of short duration and high amplitude.

KEY POINTS

- The motor unit is the fundamental unit of control in the neuromuscular system. Since motor units are composed of a single motoneuron and a number of muscle fibers, we are unable to activate a single fiber. Instead, we activate groups of muscle fibers via the motor unit.
- Motor unit features can vary considerably. For example, the contribution to the EMG signal made by slow-twitch fibers differs from that of fast-twitch fibers.
- Muscles vary in their number and organization of motor units.
- Physiologically, the near-simultaneous activation of individual muscle fibers in a motor unit produces a summed MUAP.

Techniques for Modulating Muscular Force

One response to the requirement for additional force is to activate more motor units. If the nervous system didn't have a structured plan for determining which motor units to activate, it might take considerable effort and time to determine which motor units to use to meet the force demand. After all, even small intrinsic muscles in the hand have about 100 motor units. The task of selecting which motor units to activate for small or moderate-size forces could be quite arduous for the nervous system. Fortunately, there is a well-described organizational sequence for determining which motor units to activate. Almost invariably, motor units are recruited by increasing size—the smallest motor units are recruited first, and larger units are recruited according to increasing force demands. Larger motor units are generally composed of large motoneurons whose axons have fast CV. Their twitch forces are larger than those of small motor units. Elwood Henneman was among the first to describe this, in the 1960s (Henneman et al. 1965), and we now refer to this scheme as the *Henneman size principle.*

A corollary is that motor units are also deactivated in an orderly fashion, with the largest motor units turned off first as force decreases. The number of motor units recruited has a large impact on the EMG, since more motor units means that EMG amplitude increases. The process through which we increase the number of active motor units is termed *recruitment.* The use of motor unit recruitment varies with different muscles. Large muscles tend to avail themselves of recruitment strategies, continuing to recruit motor units up to 80% of maximal effort or higher. Small muscles rely less on motor unit recruitment to grade muscular force (Seki and Narusawa 1996).

An alternative means of grading muscular force is to increase the frequency with which motor units are active. This is termed *rate coding* or changing the *discharge rate* (or *firing rate*). Motor units have a minimal firing rate of about 5 to 10 impulses per second, though this varies for different muscles (Freund et al. 1975; Phanachet et al. 2004; Tanji and Kato 1973). As the demand for more force increases, firing rate increases. Maximal firing rates can exceed 60 impulses per second in some muscles (Kamen et al. 1995), and small muscles tend to rely more on firing rate modulation than do larger muscles (Seki and Narusawa 1996). Since each MUAP contributes to the EMG signal, the greater the frequency of motoneuron activation (and hence the larger the firing rate), the greater the amplitude of the electromyogram.

The pattern of muscular activation can also affect muscular force, and sometimes these patterns affect activity in the electromyogram. Burke and colleagues (1970) were among the first to show how these activation patterns can be expressed. As seen in figure 1.9, a single extra pulse can cause a prolonged increase in muscular force, while a single missed firing can cause a prolonged depression in muscular force.

Sometimes a motor unit will fire with two discharges separated by a very short interpulse interval. These *doublets* are more prevalent at the onset of muscular contraction and are more prevalent in young adults than in older adults (Christie and Kamen 2006). Doublets seem to be important for the rapid rise in muscular force, particularly during rapid or forceful contractions (Garland and Griffin 1999). Naturally, the production of a doublet can affect the electromyogram, particularly if many motor units discharge using doublets in a short time interval.

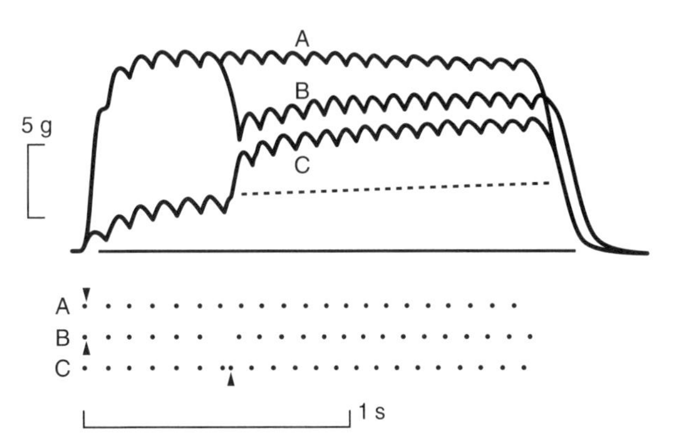

FIGURE 1.9 The effect of altering the pattern of stimulation can be studied in an isolated animal preparation. The solid lines indicate the force produced using different patterns of stimulation. In A and B, the muscle is first stimulated by two double pulses separated by a short interpulse interval. Deleting a single additional pulse (as in B) can result in a loss of force, while the insertion of a single additional pulse into a stimulus train (as in C) produces a large enhancement of muscular force.

Reprinted from SCIENCE. R.E. Burke, P. Rudomin, and F.E. Zajac III, 1970, "Catch property in single mammalian motor units," *Science* 168: 122-124.

Pairs of motor units sometimes fire simultaneously more often than one would expect by chance occurrence alone, and this can also affect the EMG pattern. This simultaneous-firing pattern (termed *synchronization*) has been reported in virtually every muscle group in which it has been investigated (Kamen and Roy 2000). It has been suggested that the frequency of synchronization might be greater with exercise training or during fatigue. However, these ideas have yet to be systematically investigated. Nevertheless, the synchronous discharge of many motor units would likely affect the electromyogram by increasing the amplitude of the signal, and *possibly* decreasing the frequency characteristics, though considerably more research is needed to resolve this issue.

These patterns of motor unit activation can be summarized through examination of figure 1.10. According to the size principle, the smallest motor units are recruited

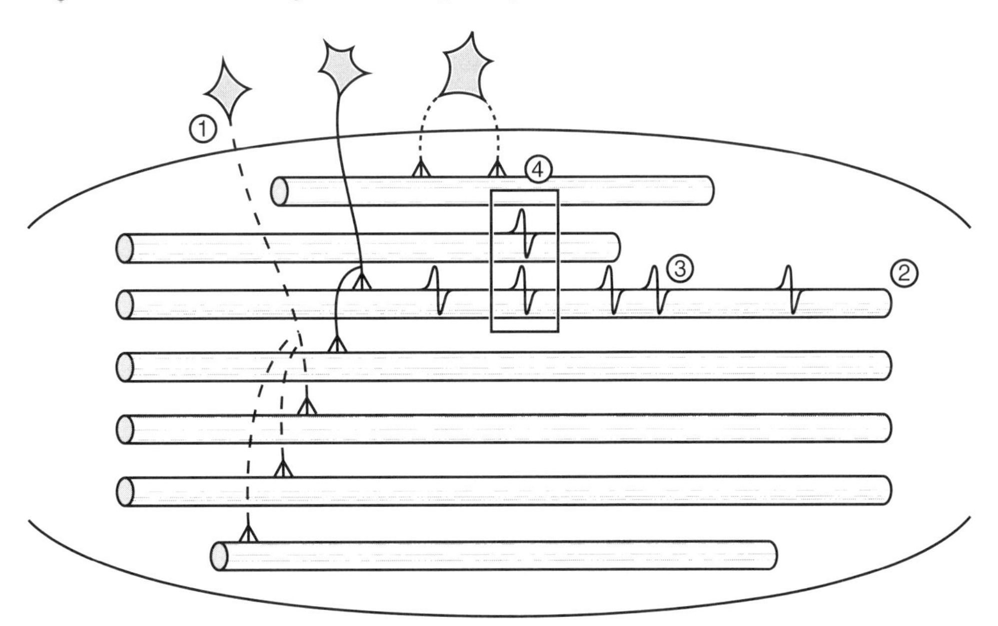

FIGURE 1.10 Muscular force can be graded by numerous mechanisms. Motor unit recruitment involves the activation of motor units of increasing size (1). The firing rate of each motor unit can be controlled (2). Doublet firing (3) or motor unit synchronization (4) may produce large forces. The activation of antagonist and synergistic muscles also serves to grade muscular force.

first (1). These smaller motor units may be located in the deeper portions of the muscle. Increasing motor unit firing rate (2) can increase the level of muscular force. Patterns of motor unit activation may also alter muscular force, and these changing patterns include double firing (3) and motor unit synchronization (4).

KEY POINTS

The difficulty of determining which motor units should be involved in a specific movement requires that the nervous system activate motor units in a logical and organized manner.

- The order of motor unit activation is determined by the size principle, according to which small motor units are activated at low forces and larger motor units are activated as force requirements increase.
- The process of electrically activating motor units is termed recruitment, and motor units are deactivated (or derecruited) in the opposite order—the smallest motor units remaining active at the lowest muscular forces.
- Motor units vary the frequency of action potentials (firing rate) in a process termed rate coding.
- Nonlinear firing patterns such as doublet firing can effect large changes in muscular force. Pairs of motor units may also fire simultaneously, and this is termed motor unit synchronization.

Other Physiological Influences on the Electromyogram

We have already mentioned the effect of CV, motor unit recruitment, firing rate, and motor unit control properties on the electromyogram. Here are some other physiological variables that affect the EMG recording:

- **Muscle length.** The EMG amplitude does vary with muscle length (Babault et al. 2003), although the available evidence indicates that this is due to changes in moment arm at different muscle lengths (Nourbakhsh and Kukulka 2004). With passive stretch, the amplitude of the MFAP decreases (Libet and Feinstein 1951). The frequency characteristics of the MUAP depend on muscle fiber length, tending toward lower frequencies with increasing muscle length (Bazzy et al. 1986; Okada 1987). This occurs apparently because muscle stretch increases conduction distance. The latency between MFAPs lengthens, causing a slowing of the MUAP. It is also possible that at longer lengths, input from muscle spindles synchronizes motor unit activity (Mori and Ishida 1976), producing lower EMG frequencies.
- **Tissue filtering.** Multielectrode recording has demonstrated that the amplitude of the action potential decays with increasing distance from the muscle fiber (Buchthal et al. 1957). When the surface electromyogram is recorded, the amplitude and frequency characteristics of the signal are affected by the intervening tissue between the electrodes and the muscle fiber. This tissue creates a low-pass filter effect, and the effects of the low-pass filter increase with increasing distance (De la Barrera and Milner 1994; Gath and Stålberg 1977; Lindström and Petersén 1983). This tissue filtering effect is one of several reasons why the surface electromyogram is biased toward the recording of muscle fibers that are closer to the electrode.
- **Fiber length.** As already discussed, fibers within a muscle volume can differ in length, sometimes as much as fivefold. The fiber length has an influence on the shape

of the MUAP. Short fibers exhibit a tendency to higher spectral frequency (Dimitrova et al. 1991; Inbar et al. 1987).

- **Muscle temperature.** As noted earlier, muscle temperature has been shown to affect the characteristics of action potentials in excitable tissues. Cold temperatures tend to depress excitability and the speed of conduction, while slightly warmer temperatures enhance CV (Kimura 2001; Rutkove 2001). Temperature has other effects as well:
  - The duration of the action potential propagating along the muscle fiber increases with lower temperatures (Buchthal et al. 1954). Consequently, for the most part, cold temperatures tend to result in lower EMG spectral frequencies (Petrofsky and Lind 1980; Winkel and Jørgensen 1991) and produce effects that are similar to those observed with muscle fatigue.
  - The amplitude of the EMG signal increases as muscle temperature is reduced (Winkel and Jørgensen 1991), although some studies have shown that root-mean-square (RMS) amplitude does not seem to be affected by changes in muscle temperature (Holewijn and Heus 1992; Krause et al. 2001). Consequently, it is prudent to attempt to maintain a constant temperature in the laboratory environment during EMG recordings.

KEY POINTS

- Ongoing changes in muscle length can affect EMG amplitude.
- The low-pass tissue filtering effect can bias the EMG signal toward muscle fibers that are closer to the electrode.
- Muscle fiber temperature can affect the CV and frequency characteristics of the action potential and ultimately influence the frequency characteristics of the EMG signal.

FOR FURTHER READING

Brown, W.F. 1984. *The physiological and technical basis of electromyography.* Boston: Butterworths.

Cram, J.R., G.S. Kasman, and J. Holtz. 1998. *Introduction to surface electromyography.* Gaithersburg, MD: Aspen.

Dumitru, D. 2000. Physiologic basis of potentials recorded in electromyography. *Muscle & Nerve* 23: 1667-1685.

Gans, C. 1982. Fiber architecture and muscle function. *Exercise and Sport Sciences Reviews* 10: 160-207.

Kimura, J. 2001. *Electrodiagnosis in diseases of nerve and muscle: principles and practice.* 3rd ed. New York: Oxford.

Loeb, G.E., and C. Gans. 1986. *Electromyography for experimentalists.* Chicago: University of Chicago Press.

MacIntosh, B.R., P.F. Gardiner, and A.J. McComas. 2006. *Skeletal muscle: form and function.* 2nd ed. Champaign, IL: Human Kinetics.

McComas, A.J. 1977. *Neuromuscular function and disorders.* Boston: Butterworths.

Sumner, A.J. 1980. *The physiology of peripheral nerve disease.* Philadelphia: Saunders.

chapter 2

Bioelectricity

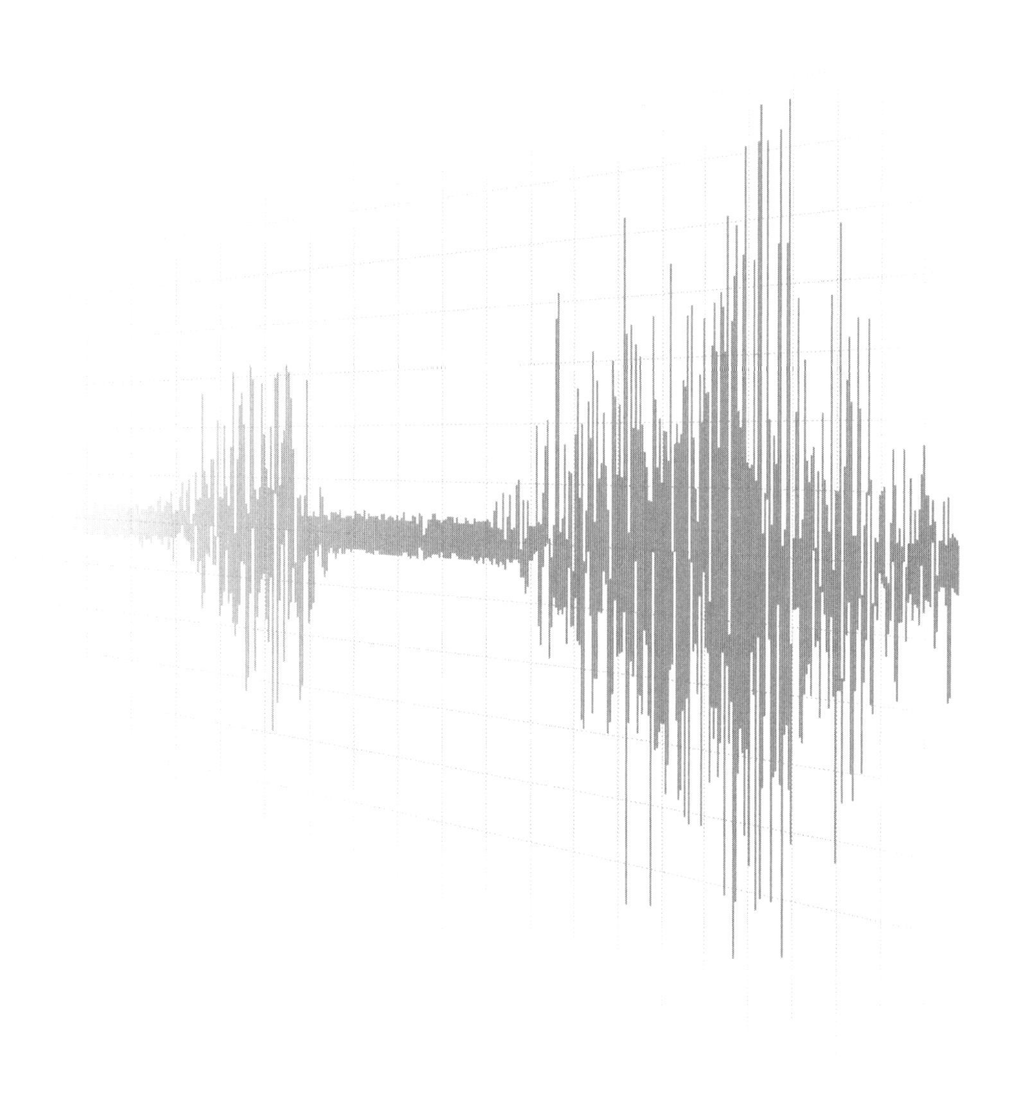

This chapter is designed to move from basic to intermediate concepts in electricity. It is important to present electrical concepts in some degree of depth because these provide the language for understanding EMG, instrumentation, and the methods used to process the resulting signal. We will start with a traditional approach wherein concepts are defined and developed from the physical sciences. We then apply the information to a biophysical understanding of the electromyographic signal. Examples in appendixes are provided to help explain the more difficult concepts.

Forces in Electricity

The basic unit of measurement in EMG is the *volt.* In the following sections we lay the foundation for a deeper understanding of the origin of the volt as the basic unit of measurement. When two charges are within proximity of each other, there is a force between them. The magnitude of the force between the two charges is proportional to the distance between them. Determining the forces between two charges as a function of distance, by moving one relative to the other at different points in space, is equivalent to mapping the electric field around the stationary charge. Because there are forces between the two charges, moving a charge within the electric field requires work (force applied over a distance). Electric potential energy is a position-dependent ability to perform work on the charge. The relationship between electric potential energy and work gives rise to the definition of the volt as a unit of measurement, but the force between two charges is the basis of the definition.

Electric Charge

The **coulomb (C)** unit is a specific number of elementary charges. Electrons have an elementary charge of -1.6×10^{-19} C whereas protons have an elementary charge of $+1.6 \times 10^{-19}$ C. One coulomb of negative charge (−1 C) is equivalent to 6.25×10^{18} electrons. Likewise, one coulomb of positive charge (+1 C) represents 6.25×10^{18} protons (figure 2.1). A common theme in electricity that can be difficult to grasp is that net charge is relative. When electrons (negative charge) move in one direction, they leave behind a relative positivity. Thus, +1 C means that there are 6.25×10^{18} fewer electrons in the object than before movement occurred.

There is a force exerted between two charged particles, which acts to move two particles with the same sign away from each other and two particles with opposite

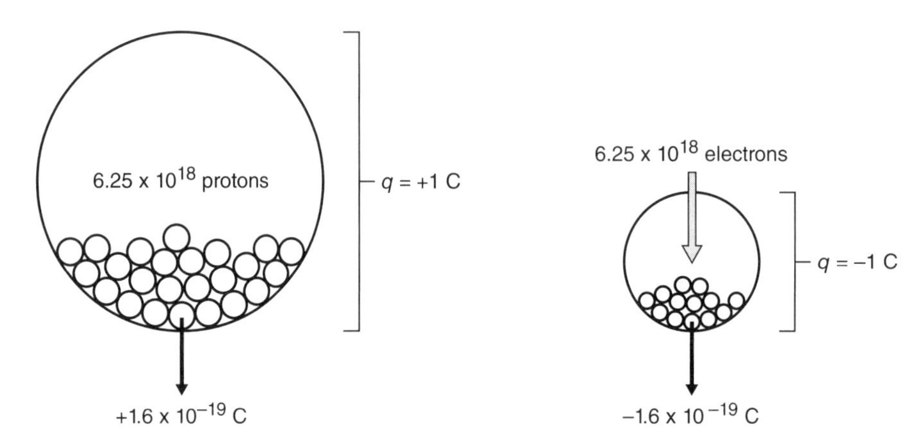

FIGURE 2.1 Fundamental quantities of electrons and protons assembled together to compose the basic unit of charge, the coulomb. The difference in size between plus and minus one coulomb illustrates that protons are physically larger than electrons.

signs toward each other. This force is described quantitatively by Coulomb's law, taking into account the magnitude of the two charges and the radial distance between them, in a way that allows conversion between electrical and mechanical units and concepts. The force between two positive charges ($+Q$), each one coulomb in magnitude, placed one meter apart is

$$F = k\frac{Q_1 Q_2}{r^2}$$

where (r) is the radial distance between the two charges and k ($k = 9.0 \times 10^9\,\mathrm{Nm}^2/\mathrm{C}^2$) is a proportionality constant so that electrostatics can be expressed in the more familiar unit of newtons:

$$F = 9.0 \times 10^9 \frac{\mathrm{Nm}^2}{\mathrm{C}^2} \times \frac{1\,\mathrm{C} \times 1\,\mathrm{C}}{1\,\mathrm{m}^2}$$

$$F = 9.0 \times 10^2\,\mathrm{N}$$

The force associated with Coulomb's law is enormous in magnitude because the size of the two charges is actually quite large, but it serves as a frame of reference with which other charge systems may be compared. For example, the magnitudes of charges involved in EMG are on the order of nanocoulombs (nC = 10^{-9} C), resulting in only tiny forces. Coulomb's law is also important because it forms the basic building block for other definitions in electricity.

KEY POINT

The coulomb unit is a specific number of elementary charges. One coulomb of negative charge is equivalent to 6.25×10^{18} electrons. Likewise, one coulomb of positive charge represents 6.25×10^{18} protons.

Electric Fields

When an electric charge ($+Q$) is placed at some point in space, it creates a state of electric stress within its general vicinity called an **electric field (*E*).** If another, much smaller charge ($+q_0$) is placed in the electric field, the first charge ($+Q$) exerts an electrostatic force on the second ($+q_0$) as a result of the field. The electric field surrounding the charge ($+Q$) can be mapped using the second charge ($+q_0$) to create a vector field. To map the electric field with vectors, the force exerted by $+Q$ on ($+q_0$) is calculated at each point in space using Coulomb's law. Coulomb's law predicts that the force will be the same between the two charges if the radial distance remains unchanged (positions a, b, and c in figure 2.2). As the radial distance increases, the electrostatic force decreases (positions d and e in figure 2.2).

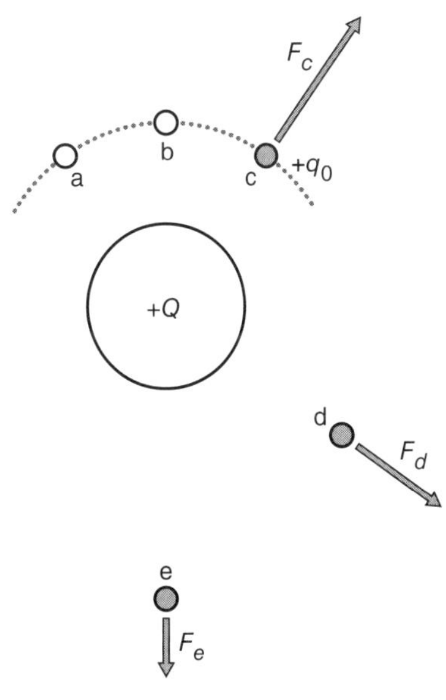

FIGURE 2.2 The force (F) exerted by $+Q$ on a small test charge $+q_0$ placed at points a through e. The size of the vector is equal to the magnitude of the force.

Remember that force is a vector that has both magnitude and direction. If the resulting force (F) at each point is divided by the magnitude of the charge (q_0), the vectors are said to be "normalized" with respect to the magnitude of q_0. Normalization makes the electric field independent of the magnitude of q_0 and describes only the effects of the charge (Q) creating the field at that point in space. A map of the electric field therefore consists of small normalized vectors in space that point in the direction of the resulting force. The magnitude of the electric field is then the force per unit charge (newtons per coulomb) at that particular point: $E = F / q_0$. Calculation of the electric field at a particular point due to two charges is presented in appendix 2.1. The appendix also illustrates how electric field lines are generated based on the sample calculation.

KEY POINTS

- When an electric charge is placed at some point in space, it creates a state of electric stress within its general vicinity called an electric field.
- If another, much smaller charge is placed in the electric field, the first charge exerts an electrostatic force on the second as a result of the field. The magnitude of the electric field (E) is then the force per unit charge at that particular point, and is highly dependent on the radial distance between the two charges.

Electric Potential Energy

Coulomb's law introduced the concept of electrostatic forces that was used to determine the electric field surrounding a charge ($+Q$). The electric field acts on other charges (i.e., q_0) within the vicinity; they may be attracted or repelled depending on their polarity (+ or −). The charge (q_0) can also have different amounts of electric potential energy depending on its position within the electric field. The amount of work (W) required to move a charge between two points within the electric field is then related to the difference in electric potential energy between the two points. As will be defined later, the difference in electric potential energy between two points is measured in volts (V)—the volt being the fundamental unit of amplitude measurement in EMG.

The mechanical analogy is often invoked to obtain a more intuitive understanding of the origin and meaning of volts because we are more familiar with gravitational fields. First, consider the work (W) done in translating an object across a horizontal surface: It is the product of force (F) multiplied by displacement (d): $W = Fd$. One joule is the work done by one newton of force acting through a distance of one meter. The unit of joules is used to avoid confusion with torque, which has the same basic units (N·m). If force is not acting in the same direction as displacement, then trigonometry must be used to find the vector component of force that is in the same direction as displacement: $W = Fd\cos(\theta)$. This is critical to understanding the sign convention in relation to electric potential.

The following example, depicted in figure 2.3, illustrates how work and energy can be defined in terms of either gravitational or electrical fields. First, consider an object resting at ground level in position A (figure 2.3*a*). If a force is applied beneath the object that is infinitesimally greater than the weight, the object will move very slowly toward position B. The goal is to move the object upward but not give it any additional energy by applying a large force. The object moves so slowly that velocity and kinetic energy are essentially zero. In this case, we are able to consider only

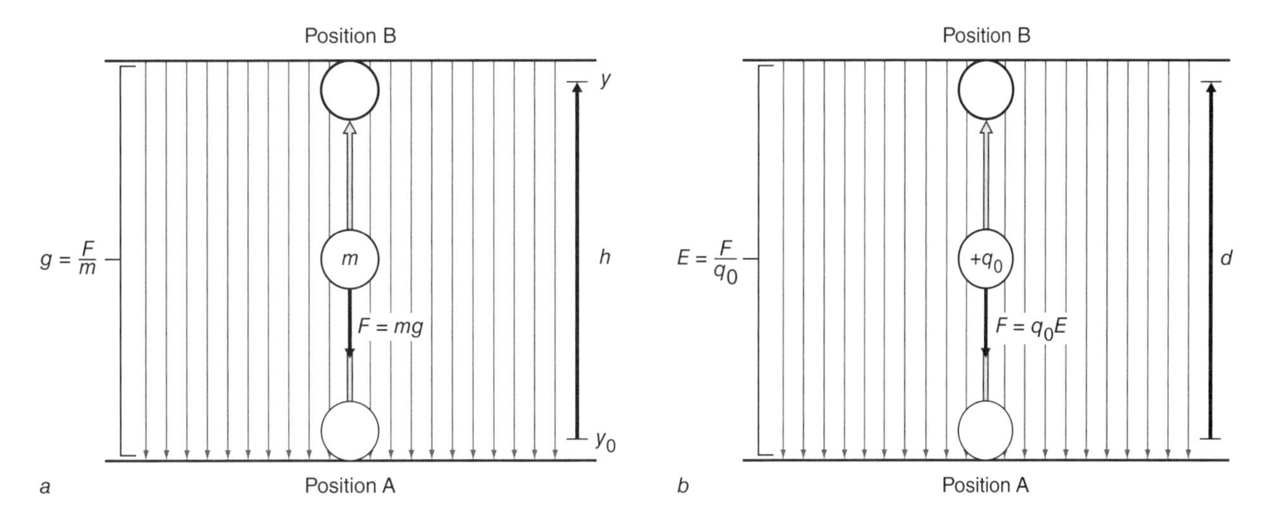

FIGURE 2.3 Mechanical and electrical analogs of *(a)* gravitational and *(b)* electrical fields.

the effects of the gravitational field. Gravitational force (F_g) is equal to the weight of the object (mg). The vertical displacement of the object is given by $h = (y - y_0)$. However, because movement of the object is in a direction that is opposite (i.e., 180°) to the gravitational force, the work done by the gravitational field is actually negative. This arises because of the sign convention:

$$W_g = F_g h \cos(180°) = mgh(-1) = -mgh$$

The minus sign means that the gravitational field transfers potential energy (U_g) to the object as it rises above the ground. The magnitude of the potential energy at position B is equal to the negative value of the work done by the gravitational field: $U_g = -W_g$.

The fact that potential energy at a specific point is always negative should not be of any concern. Only the difference in potential energy (ΔU_g) between two points can be physically measured. The difference in potential energy can be either positive or negative, depending on whether the distance from the reference point increases or decreases. The difference in the potential energy between two points (ΔU_g) is therefore equal to the difference in negative work: $\Delta U_g = -\Delta W_g$. Where height is given by displacement in the y-direction, the general expression for potential energy in this example is

$$U_b - U_a = mg(y - y_0).$$

By convention, position A is at $y_0 = 0$ to serve as a reference point wherein the object has a potential energy of $U_a = 0$. The resulting change in potential energy in moving the object from position A to B is a positive quantity: $U_b = mgy$. If the object were to move from position B to A, the work done by the gravitational field would be positive. However, the change in potential energy would then be negative because the initial position (y_0) at B is greater than the final position (y) at A.

To translate mechanical to electric potential energy, consider a small positive charge (q_0) in a uniform electric field between two oppositely charged plates (figure 2.3*b*). The electric field lines project from the positive toward the negatively charged plate. Electric force is derived from the electric field: $F_e = q_0 E$, where q_0 and E are the electric analogs of mass (m) and gravity (g), respectively. In this case, displacement of q_0 is given by (d). The work done on charge (q_0) by the electric field as it is moved from position A to B is negative:

$$W_e = F_e d\cos(180^\circ) = q_0 Ed(-1) = -q_0 Ed$$

The magnitude of the electric potential energy at position B is therefore equal to the negative value of the work done by the electric field: $U_e = -W_e$. We stated earlier that only a difference in potential energy between two points is physically measurable. This definition is therefore extended so that the difference in electric potential energy between two points is equal to the difference in negative work: $\Delta U_e = -\Delta W_e$.

Recall that the electric field strength $(E = F / q_0)$ at a particular point is expressed in terms of force (N) per unit charge (C). Similarly, the electric potential energy at a particular point may be expressed as potential energy per unit charge: $V_a = U_a / q_0$. This normalized quantity is called the electric potential (V) at point A. It is important to keep in mind that the electric potential at a particular point still depends on its location relative to a reference position. The potential difference, or, simply, the electric potential between points A and B, is equal to the difference in negative work $(-\Delta W_{ba})$ in moving the charge from point A to point B: $\Delta U_{ba} = U_b - U_a = -\Delta W_{ba}$. If each term in the equation is divided by q_0, the result reduces to

$$V_{ba} = V_b - V_a = -\frac{\Delta W_{ba}}{q_0}$$

where the normalized potential difference (V_{ba}) between points A and B is equal to the negative value of work per unit charge required to take the charge from point A to point B. The normalized definition of electric potential in terms of joules per coulomb gives rise to a new unit, **volt (V).** In the International System of Units (SI), one joule of work is required to move one coulomb of charge through a potential difference of one volt.

Electromyographic recordings require a minimum of two electrodes because we are measuring the difference in electric potential between two points. As will be described in later chapters, the most basic electrode configuration is one "recording" electrode on the muscle and the other "reference" electrode over an electrically neutral tissue area, such as its tendon. Since the charges associated with the muscle action potential are on the order of nanocoulombs (nC), the resulting difference in electric potential may be in microvolts (μV or 10^{-6} V) or millivolts (mV = 10^{-3} V). It is also more common to refer to the difference in electric potential between the two electrodes as muscle electrical activity, which is measured in microvolts or millivolts. The concepts developed thus far are utilized in appendix 2.2 to illustrate the calculation of the electric potential difference at an electrode due to a dipole, as occurs on a muscle fiber during the depolarization and repolarization phases of the action potential.

KEY POINTS

- The potential energy of a charge depends on its location within the electric field. There must always be a reference point wherein potential energy is zero.
- Work is done on the charge by the electric field as it is moved from position A to B within the electric field. The magnitude of the electric potential energy at the second position (position B) is equal to the negative value of the work done by the electric field. The normalized potential difference is equal to the negative value of work per unit charge required to take the charge between those two points. The normalized definition of electric potential difference in terms of joules per coulomb gives rise to a new unit, volt (V).

Volume-Conducted Potentials

Recording the muscle fiber action potential (MFAP) through a medium (i.e., extracellular fluid and tissues) is termed **volume conduction.** Volume conduction is one of the most fundamental topics in EMG because it explains how the size and shape of the resulting potential depends on the location of the recording electrode (Brown 1984). The concepts involved in volume conduction extend to recording motor unit action potentials (MUAPs), but the single fiber is used to understand the basic principles.

The MFAP travels at constant velocity along the muscle fiber, maintaining its shape as it propagates toward, passes underneath, and then travels away from the electrode. Because MFAP shape is constant, the depolarization and repolarization phases can be viewed as remaining stationary on the muscle fiber while the electrode is allowed to move relative to the MFAP. Changes in the shape of the MFAP due to its location relative to the electrode may then be studied in this way.

The next step is to understand that the depolarization and repolarization phases of the MFAP can be thought of as negative and positive charges, respectively, that sit adjacent to each other on the muscle fiber. Because the negative and positive charges are linked together through physiological events, they represent a **dipole** system. A coordinate system may then be placed between the negative and positive charges of the dipole to provide a reference for the location of the electrode. The physical system is depicted in figure 2.4. The two observation lines correspond to electrode location in the y-direction, near and far above the dipole. There are five different electrode positions in the x-direction, approaching the stationary dipole, passing over it, and then moving away from it (figure 2.4*a*).

The graph (figure 2.4*b*) shows the evolution of two MFAPs as the electrode is moved along both observation lines. The y-axis is the electric potential recorded at the electrode in microvolts and the x-axis is the x-position of the electrode relative to the dipole center, where 0 mm is directly between the two charges. The MFAP for the observation line close to the dipole is tall and narrow while the MFAP for the observation line far from the dipole is short and wide. The x-axis could easily be converted to time in milliseconds because we know that muscle fiber conduction velocity is approximately 4 m/s. Time on the x-axis would be appropriate if a stationary electrode recorded a MFAP propagating along the muscle fiber.

To understand how differences in MFAP shape arise, we must keep three facts in mind. First, the dipole consists of equal and opposite charges wherein the net potential is determined by the difference (Δr) between the radial distances (r_1 and r_2) between each charge and the electrode. The closest charge dominates the net potential. The net potential is recorded by the electrode and corresponds to the MFAP amplitude in microvolts. Second, the evolution of the MFAP is governed by changes in the difference (Δr) between the radial distances (r_1 and r_2) as the electrode moves along the observation line. A close-up view of the difference (Δr) in radial distances (r_1 and r_2) is shown in appendix 2.2, figure 2.2.2. Third, the geometric relationship between the dipole and electrode governs *how* the difference (Δr) between the radial distances (r_1 and r_2) changes as the electrode moves along the observation line. This relationship is different for the two observation lines. In the following paragraphs, we describe how the difference (Δr) between the radial distances (r_1 and r_2) changes as the electrode moves along the two observation lines, resulting in MFAPs of different size and shape.

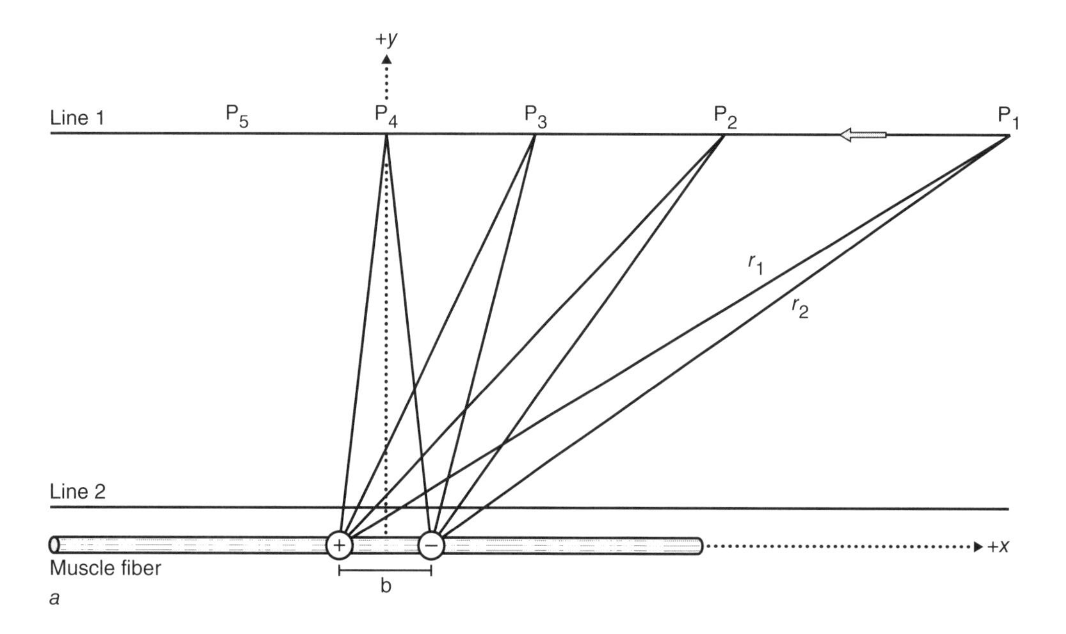

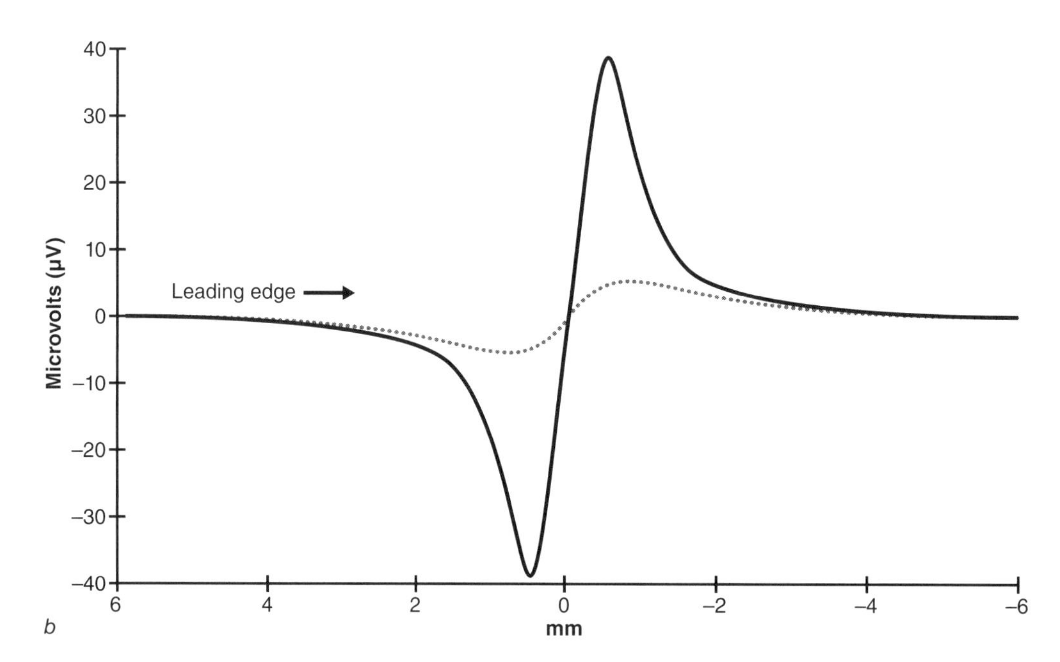

FIGURE 2.4 *(a)* Volume conduction of muscle fiber action potentials (MFAPs). The two lines represent near and far observation distances for electrode placement. The near distance is 1 mm above the muscle fiber while the far distance is approximately 10 mm away. *(b)* The evolution of the MFAPs associated with the two observation lines as the electrode is moved from left to right with respect to the stationary dipole. The dotted and solid lines correspond to MFAPs for the far and near distance, respectively.

The Far Observation Line

When the radial distance (r_1 and r_2) is much greater than the distance between charges in the dipole ($r \gg b$), the difference (Δr) is very small. The net potential for the first electrode position P_1 is zero because r_1 and r_2 are nearly equal. The change in electrode position from P_1 to P_3 is associated with a negative deflection because there is an increase in the difference (Δr) between r_1 and r_2 and the closest charge is negative. However, the great radial distance also results in a change in the difference (Δr) between r_1 and r_2 that is more gradual. There is a steady increase in negativity of the net potential as the electrode moves along the far observation line. The angle between r_1 and r_2 may be used to indicate the rate of change in the difference (Δr) between r_1 and r_2. It should be evident from figure 2.4 that there is a gradual increase in the angle between r_1 and r_2 as the electrode moves between P_1 and P_3.

Although the change in electrode position from P_1 to P_3 results in an increase in the difference (Δr) between r_1 and r_2, it is relatively small because the radial distance is still great, even though the electrode moves closer to the dipole. The net potential is small for the far observation line, so the MFAP that is recorded is small in amplitude. The gradual increase in the difference (Δr) between r_1 and r_2 for the far observation line achieves its maximum close to P_3, before reaching the negative charge. The net negative potential then steadily decreases toward zero at dipole center; r_1 and r_2 are equal in length, so neither charge can dominate the sum.

The geometry of the physical system is mirrored as the electrode moves past the dipole center and beyond. There is a gradual increase in the difference (Δr) between r_1 and r_2 as the electrode moves between P_4 and P_5. In this case, the positive charge is closest to the electrode and dominates the net potential, so the MFAP reverses polarity. Maximum positivity is achieved close to electrode position P_5, past the positive charge. Because the maxima are attained outside the dipole, the distance between the negative and positive peaks is greater than the distance (b) between the two charges. Because the resulting MFAP is low in amplitude with a gradual increase to maxima that occur outside the dipole distance (b), it has a wide appearance.

The Near Observation Line

Although the near observation line is close to the dipole, the radial distance at the first electrode position P_1 is still great enough that the difference (Δr) between r_1 and r_2 is negligible. The net potential at the first electrode position P_1 is therefore zero. Both r_1 and r_2 shorten as the electrode moves from P_1 to P_2. However, because the observation line is so close to the dipole, there is no appreciable change in difference (Δr) between r_1 and r_2. The net potential stays close to zero. Recall that the angle between r_1 and r_2 may be used to indicate the rate of change in the difference (Δr) between r_1 and r_2. The angle between r_1 and r_2 appears to remain constant between P_1 and P_2. Appreciable changes occur only as the electrode approaches close to the dipole. It is a useful exercise for readers to draw the example in order to be convinced of this fact.

The difference (Δr) between r_1 and r_2 starts to change slightly as the electrode moves past P_2. The net negative potential appears to increase slightly off of baseline at this point. As the electrode approaches P_3, the difference (Δr) between r_1 and r_2 starts to change abruptly, giving rise to a rapid increase in negativity. The magnitude of this change is much greater for the near observation line. The net negative potential is therefore much greater and the MFAP is larger in amplitude, due simply to geometry. The net potential achieves its maximum directly over the negative charge of the dipole. From this point on, there is a rapid decrease in the net negative potential toward zero at P_4—the same as for the far observation line.

The mirrored geometry past electrode position P_4 results in the same events, but in reverse order. That is, there is a rapid increase in net positivity of the potential until its maximum is achieved directly over the positive charge. The net positivity then rapidly decreases between the positive charge and electrode position P_5, coming close to zero thereafter. Because the negative and positive maxima are attained directly over their respective charges within the dipole, the distance between MFAP peaks equals the dipole distance, which is $b = 0.05$ mm. The resulting MFAP for the near observation line is therefore tall and narrow in appearance by comparison to the one generated for the far observation line.

Tripole Representation of the Muscle Fiber Action Potential

A more accurate physical representation of the MFAP is that of a **tripole** (+ − +) (Dumitru 2000; Loeb and Gans 1986). Figure 2.5 illustrates the electrochemical events involved in generating the MFAP. The electric potential associated with each electrochemical event is depicted immediately below the muscle fiber as it would be recorded by an extracellular electrode. All these events are occurring at the same time as the action potential propagates along the muscle fiber, from left to right toward the electrode. However, imagine that we can freeze these events in time to explore the electrode recordings further.

To focus first on the point of depolarization, when Na^+ ions rush into the muscle fiber they leave behind a relatively strong negativity in the extracellular space. This strong negativity is called a **current sink** because positive charges are drawn to it. If an electrode is placed directly over the depolarization event, a negative potential is recorded (position 1). However, the current sink is so strong that it attracts positive ions from the membrane area in front of the depolarization event. This forward membrane area is called a weak **current source** area because it provides the positive ions that are drawn to the current sink. An electrode placed in front of the depolarization event would record a slight positivity (position 2). As positive ions leave the

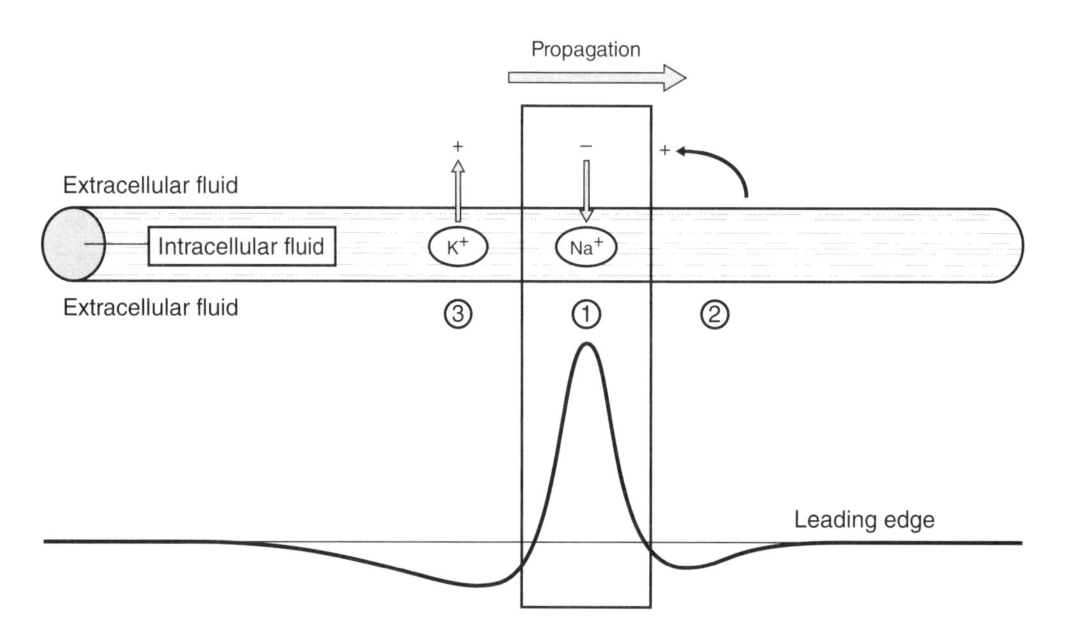

FIGURE 2.5 Electrochemical events involved in generating the muscle fiber action potential (MFAP). In standard recordings, the evolution of the MFAP is expressed as a function of time and viewed from left to right. For this example, the MFAP is aligned "spatially" with the electrochemical events in the muscle fiber above to allow understanding of the relationship between the MFAP and the muscle fiber. In electrophysiology, the convention is for the negative polarity to be above the horizontal line and the positive polarity to be below it.

forward membrane area, the charge difference across the membrane decreases, which leads to passive depolarization of the muscle fiber. The ion channel–mediated rush of positive ions (K^+) outside the muscle fiber gives rise to the repolarization event and is a strong current source. An electrode placed directly over the repolarization event would record a large positivity (position 3).

Thus, as the MFAP propagates along the muscle fiber from left to right toward the electrode, the leading edge (weak current source) is detected first, followed by the depolarization phase (current sink) and then the repolarization phase (strong current source). The MUAP is also triphasic because it is the linear sum of all its associated muscle fibers. Electrical stimulation of the peripheral nerve results in the activation of a large number of motor units simultaneously. The triphasic waveform is still apparent in the simultaneous depolarization and repolarization of the recruited motor units. The evoked potential is logically termed the **massed action potential** (or **M-wave**). It is also called the **compound muscle action potential (CMAP)** due to the linear summation of all the constituent MUAPs. The large number of muscle fibers involved in the evoked response results in an electric potential that is several millivolts in magnitude.

KEY POINTS

- The dipole consists of equal and opposite charges wherein the net potential is determined by the difference (Δr) between the radial distances (r_1 and r_2) between each charge and the electrode.
- The evolution of the MFAP is governed by changes in the difference (Δr).
- The geometric relationship between the dipole and electrode governs how the difference (Δr) changes. As the MFAP propagates along the muscle fiber from left to right toward an electrode, the leading edge (+) is detected first, followed by the depolarization phase (–) and then the repolarization phase (+).

Essentials of Electric Circuits

The electromyographic signal is ultimately fed to an amplifier that magnifies the relatively small voltage to a level that can be measured. An electric circuit then alters the frequency content of the incoming electromyographic signal to minimize (filter out) noise from the surrounding environment or other sources before it is actually stored on the computer for later analysis. The first principles of understanding filtering are based on the electric circuit and how resistors and capacitors are arranged within the circuit to accomplish this task. The electric circuit is also important for understanding the physical properties of nerves and muscle fibers as related to the flow of current and electric potential.

Capacitance

Any conductive material can be viewed as a reservoir or source of electric charge. If a conducting wire is connected to the reservoir, electric charge will then flow from it. A device, apparatus, or material that can store electric charge is called a **capacitor.** Charging a conductor is analogous to pumping air into a flat tire. As more air is pumped into the tire, the pressure opposing the flow of additional air becomes greater. Likewise, as more charges of the same sign are transferred to the conductor, the electric potential of the conductor becomes higher, making it increasingly more difficult to transfer additional charge.

The most basic capacitive system involves the use of two parallel conductive plates placed a short distance (*d*) from each other (see figure 2.3*b*). The insulator between them is air. If the plates are connected to a battery, they acquire the charge associated with the terminal to which they are connected. The charges are then mutually attracted to the inside of the oppositely charged plate facing them. Each electric field line from a positive charge projects to a negative charge on the opposite plate, so now the plates contain an equal number of opposite charges. The negatively charged plate lowers the net potential at the surface of the positively charged plate, allowing more charges to be added to the positive plate. Without the negatively charged plate, the analogy of pumping air into a tire would continue. That is, it would get progressively harder to add positive charges until the positively charged plate spontaneously discharged and ionized the air around it, much as when a tire is overinflated until it "blows."

The amount of charge Q acquired by each plate is proportional to the potential difference between the two plates (V_{ba}): $Q \propto V_{ba}$. The constant of proportionality that equates Q and V_{ba} is the capacitance (C) of the system: $Q = CV_{ba}$. Rearranging, the units of capacitance is coulombs per volt (Q/V_{ba}), which is defined as a farad (F). A capacitance of one farad means that one coulomb of charge has been transferred to the conductor, increasing its potential by one volt. Recall that a coulomb is a huge amount of charge, so the magnitude of capacitance usually encountered is on the order of microfarads (μF; 10^{-6} F) or picofarads (pF; 10^{-12} F). Capacitance (C) depends on the size and shape of the two conductors, their relative position (*d*), and the characteristics of the medium (insulating material) that separates them. In this case, the plates are assumed to be separated by air. The distance (*d*) between the two plates is also assumed to be very small compared to the area (*A*) of each plate. Because the plates are equal and oppositely charged, it is appropriate to refer to the charge as (Q) without a "+" or "–" sign.

Although circuits are typically represented by wires attached to a battery, a circuit can be generally thought of as any path that allows for the flow of charge. For this reason, nerves and muscles are often represented by electrical circuits to allow better understanding of their functional properties. Concepts in instrumentation and signal processing are also represented by electrical circuits. The electrical circuit is presented here to contribute to the depth of understanding of EMG.

The first example is for a single capacitor linked by a conductive wire to a battery (figure 2.6). A battery is an electrochemical device that maintains a potential difference (V) between two terminals. The negative terminal **(cathode)** is the low-potential sink, and the positive terminal **(anode)** is the high-potential source. When the switch is closed and the circuit is complete, the potential difference between the two terminals results in an electric field within the conductive wire that causes the flow of electrons. The magnitude of this potential difference is referred to as the electromotive force of the battery; the unit of measurement is still volts.

The electric field causes the flow of electrons from the top plate of the capacitor toward the positive terminal of the battery. The top plate loses electrons and becomes positively charged. The electric field simultaneously causes the flow of an equal

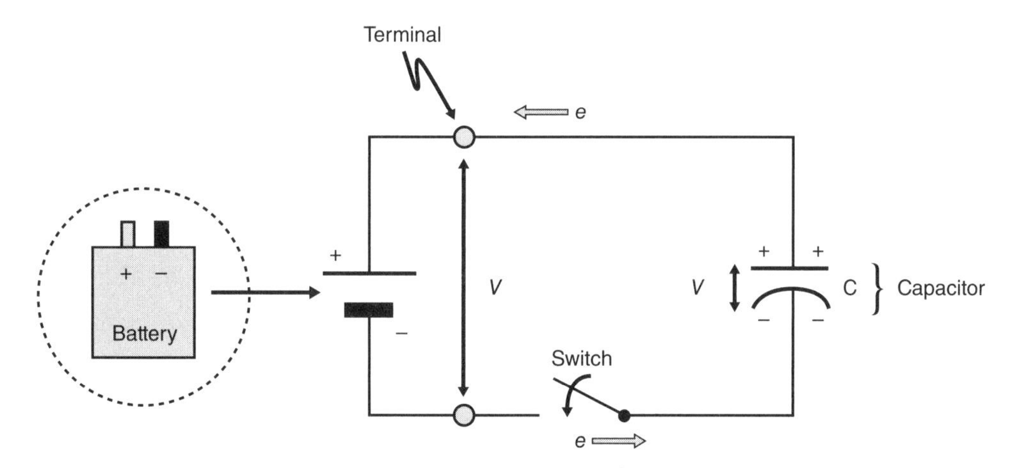

FIGURE 2.6 The basic elements of an electric circuit. In this example, a capacitor is linked in series by a conductive wire to a battery.

number of electrons from the negative terminal of the battery toward the bottom plate of the capacitor. The bottom plate then gains electrons and becomes negatively charged. The process continues until the two plates become equally and oppositely charged and the potential difference between the plates matches that of the battery. When the potential between the battery terminals and their associated plates is equal, the electric field within the wire falls to zero, and the flow of electrons ceases. The capacitor is now fully charged.

Charging capacitors connected in series (figure 2.7*a*) follows a slightly different process than that just outlined. When the switch is closed, electrons flow from the top plate of the first capacitor (C_1) to the positive terminal of the battery, and also from the negative terminal of the battery to the bottom plate of the third capacitor (C_3). There is no direct connection between the battery and the remaining plates (those outlined in the box). The flow of charge between the top plate of C_1 and the bottom plate of C_3 must therefore be induced by the electric field within the wire. Specifically, the flow of electrons to the bottom plate of C_3 repels electrons in the plate just above it. The bottom plate becomes negative and the top plate becomes positive as electrons are driven away from it. Electrons continue to move and charge the bottom and top plates of the second (C_2) and first (C_1) capacitors in succession in exactly the same way. The inductive process between the negative terminal and the top plate of C_1 is simultaneously matched by the polar-opposite sequence of events between the positive terminal and the bottom plate of C_3. When the potential between the battery terminals equals that of the three capacitors, the electric field within the wire falls to zero, and the flow of electrons ceases. The three capacitors are now fully charged.

Because the charge (Q) between capacitors connected in series must be induced, each plate has the same charge (Q) on it: $Q = Q_1 = Q_2 = Q_3$. If charge remains the same but the area over which it is distributed is increased, the overall electric potential will decrease. For a series arrangement, each capacitor represents additional surface area over which to distribute the same charge. Consequently, there is a successive decrease in electric potential associated with each capacitor in the series. The sum of

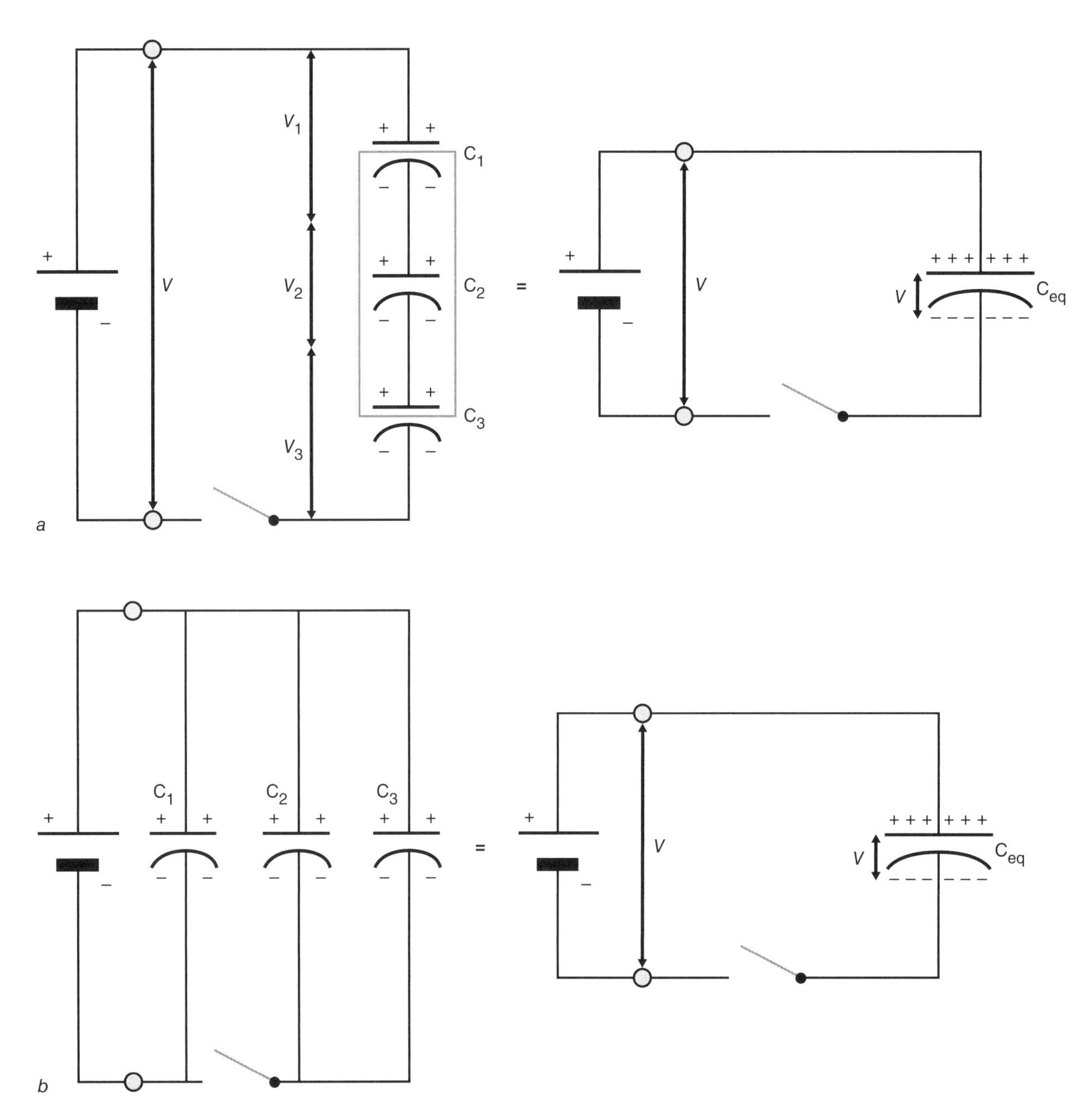

FIGURE 2.7 Electrical circuits composed of capacitors arranged *(a)* in series and *(b)* in parallel.

the potential differences across the three capacitors is equal to the potential difference supplied by the battery: $V = V_1 + V_2 + V_3$. The relationship for capacitance ($C = Q/V$) can be used to substitute for V:

$$\frac{Q}{C} = \frac{Q_1}{C_1} + \frac{Q_2}{C_2} + \frac{Q_3}{C_3}$$

Dividing through by Q because all charges are equivalent,

$$\frac{1}{C} = \frac{1}{C_1} + \frac{1}{C_2} + \frac{1}{C_3}.$$

The three capacitors in series can be replaced by a single equivalent capacitor (C or C_{eq}) using the preceding equation. Notice that the equivalent capacitor is depicted as a larger surface area for holding charges (see figure 2.7*a*).

Capacitors connected in parallel receive a direct charge (Q) from the battery so that the total charge stored by the capacitors is $Q = Q_1 + Q_2 + Q_3$. The situation is identical to the first example wherein both plates of a single capacitor are directly connected to a battery (figure 2.7*b*). The potential difference (V) across all three capacitors when they are fully charged is equal to that of the battery: $V = V_1 = V_2 = V_3$. The relationship for capacitance ($C = Q/V$) can be used to substitute for Q:

$$CV = C_1V_1 + C_2V_2 + C_3V_3$$

Factoring out V because the potential difference is the same for the battery and capacitors,

$$C = C_1 + C_2 + C_3.$$

The three capacitors in parallel can be replaced by a single equivalent capacitor (C or C_{eq}) using the preceding equation. Expressions for finding the equivalent capacitor for parallel and series arrangements are important because they allow us to reduce and solve more complicated circuits to determine their physical characteristics. A worked example for capacitors within an electric circuit is provided in appendix 2.3.

KEY POINTS

- A battery is an electrochemical device that maintains a potential difference (V) between two terminals. The negative terminal (cathode) is the low-potential sink, and the positive terminal (anode) is the high-potential source. The potential difference between the two terminals results in an electric field within the conductive wire that causes the flow of electrons when the switch is closed to complete the circuit. The potential difference across the terminal is the electromotive force of the battery. The unit of measurement is still volts.
- Because the charge between capacitors connected in series must be induced, each plate has the same charge on it. If charge remains the same but the area over which it is distributed is increased, the overall electric potential will decrease.
- For a series arrangement, each capacitor represents additional surface area over which to distribute the same charge. Consequently, there is a successive decrease in electric potential associated with each capacitor in the series. The sum of the potential differences across the three capacitors is equal to the potential difference supplied by the battery.
- Capacitors connected in parallel receive a direct charge from the battery so that it equals the total charge stored by the capacitors. The potential difference across all three capacitors when they are fully charged is equal to that of the battery.

Electric Current

Electric current (*i*) is the flow of like charges through a defined surface area (figure 2.8*a*). Negative charges moving to the left are equivalent to positive charges moving to the right, in the direction of the electric field (E). Remember that when electrons move away from an area they leave behind a relative positivity. By convention, the direction of flow of electric current is designated as the direction in which positive charges are free to move. If charges are moving perpendicular to a surface area (*S*), current is the rate at which charges flow through this surface area. The amount of charge (ΔQ) that passes through the area (*S*) in a given time interval (Δt) is the average current: $i_{ave} = \Delta Q / \Delta t$. However, the rate at which charge flows varies in time, so instantaneous current must be defined as the limit of the differential:

$$i = \lim_{\Delta t \to 0} \frac{\Delta Q}{\Delta t} = \frac{dQ}{dt}$$

The SI unit of current is the **ampere (A),** defined as one coulomb of charge passing through the area (*S*) in one second (C/s). The usual magnitudes involved in EMG are milliamperes (1 mA or 10^{-3} A) or microamperes (1 µA or 10^{-6} A).

Within the context of conductive wire for electrical circuits, the movement of electrons (negative charges) generates current flow. Metal atoms within the crystal lattice release their outermost electrons, and these electrons move randomly within the lattice, colliding sporadically with the metal atoms. These collisions represent an internal resistance to the movement of electrons. The movement of electrons resembles that of gas molecules because the repeated collisions result in a very erratic motion path (figure 2.8*b*). In the absence of an electric field, the number of electrons moving in opposite directions is equal within any given period time, so there is no net change in charge within any particular region. Thus, there is no current.

Consider electron flow inside a conductive segment within an electric circuit (similar to figure 2.6). When the switch is closed, a potential difference is applied across the conductor, and an electrical field is set up within the circuit. The force exerted on the charges due to the electric field $(F = qE)$ decreases the randomness of the movement. The electric field then accelerates the electrons as reflected by more parabolic paths between collisions. The electrons then transfer this additional energy to the metal atoms during the collisions, increasing their vibrational energy and the temperature of the conductor. The movement remains erratic but there is a net displacement toward the positive terminal, creating a current. Because electrons are still moving back and forth within the electric field, the net displacement (Δx) within a given period of time (Δt) is referred to as drift velocity (v_d).

The concept of current is better described with respect to a section of the conductor. The number of charge carriers (electrons in this case) in a given section of the conductor is $nA\Delta x$, where (n) is the number of charge carriers per unit volume (charge density), and the volume is given by $A\Delta x$. If the magnitude of an individual charge is represented by (q), then the charge in a given section is $\Delta Q = (nA\Delta x)q$. Knowing how long (Δt) an object traveled at a specific velocity (v_d) enables the determination of displacement: $\Delta x = v_d \Delta t$. The following steps show that substituting for Δx, then dividing both sides by Δt, yields the current within a given section of the conductor:

$$Q = nqAv_d \Delta t$$

$$i_{ave} = \frac{\Delta Q}{\Delta t} = nqAv_d$$

$$i = \frac{dQ}{dt} = nqAv_d$$

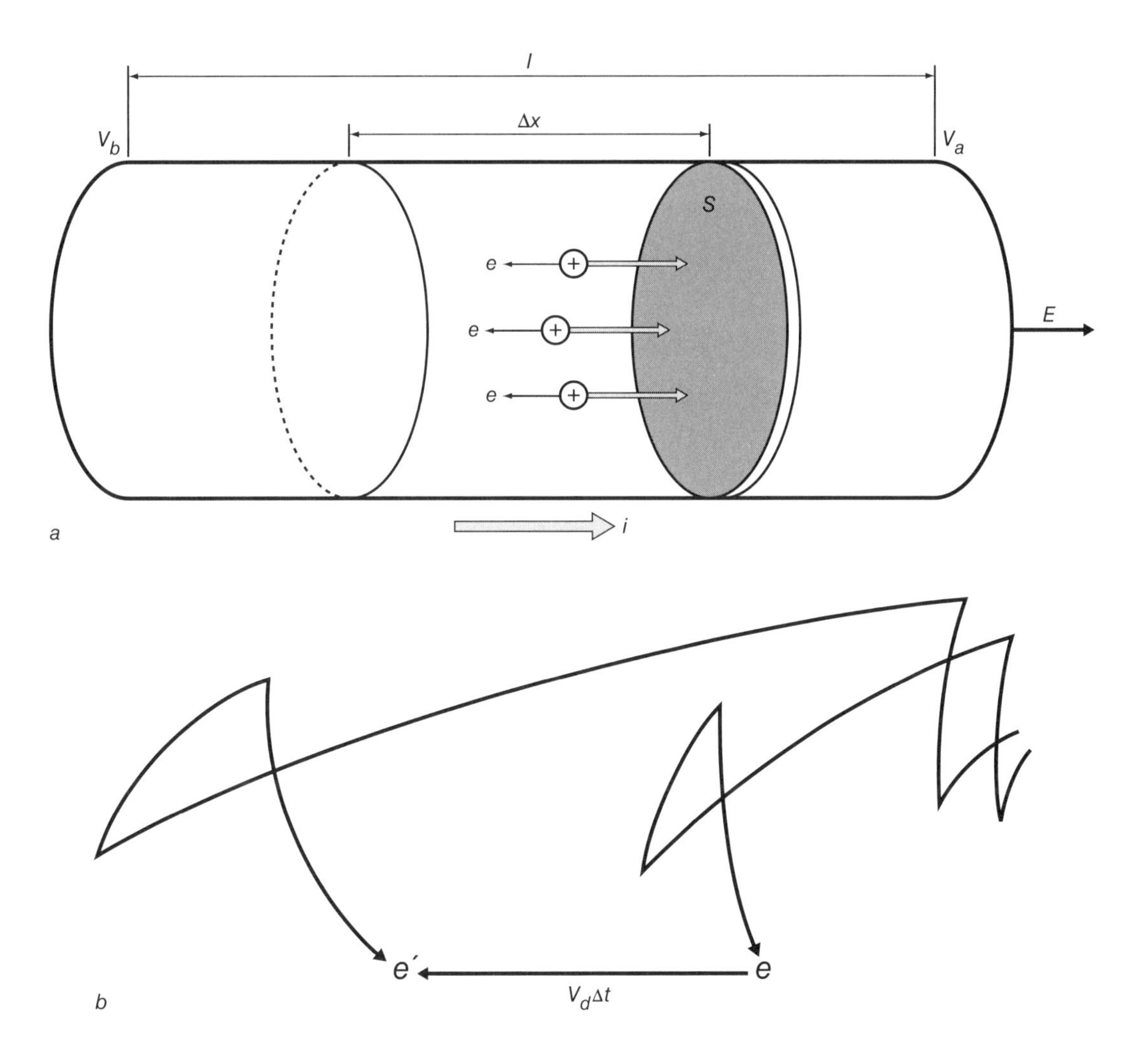

FIGURE 2.8 *(a)* Idealized movement of electrons within a section of conducting wire and *(b)* a more realistic zigzag pattern for electron movement. The two zigzag patterns reflect motion of the same electron in the absence (*e*) and the presence (*e*´) of an electric field. The electric field results in a "net" displacement ($\Delta x = v_d \Delta t$) within a section of the conductor.

KEY POINT

The amount of charge (ΔQ) that passes through the area (*S*) in a given time interval (Δt) is the average current: $i_{ave} = \Delta Q / \Delta t$. However, the rate at which charge flows varies in time, so instantaneous current must be defined as the limit of the differential. The unit of current is the ampere (A), defined as one coulomb of charge passing through the area (*S*) in one second (C/s).

Resistance

The amount of charge flowing (i) per unit area (A) is referred to as *current density* (J). The relationship for current ($i = nqAv_d$) is based on the number of charges moving through a section of the conductor. If current is divided by cross-sectional area of the conductor, then the current density is

$$J = \frac{i}{A} = nqv_d.$$

The units of current density are amperes (A) per meter squared (m^2). It is always important to remember the following sequence of events: When the switch is closed in an electrical circuit, a potential difference is established across the conductor that produces an electric field, and the result is a flow of charges (i.e., current). When the potential difference is constant as supplied by a battery, current density is proportional to the electric field:

$$J \propto E$$

The constant of proportionality that equates the two variables is conductivity (σ) of the material:

$$J = \sigma E$$

Although the electric field generates the current density, the magnitude of current density is also determined by how well the material allows the charges to flow (σ). The reciprocal of conductivity is the resistivity (ρ) of the material to the flow of charges:

$$\rho = \frac{1}{\sigma}$$

If the potential difference and resulting electric field are large but the conductivity of the material is low, the current density will be small. Thus, a low conductivity is equivalent to high resistivity. Material that is a good electrical conductor has low resistivity (high conductivity). If the material is a good insulator, then it has high resistivity (low conductivity). The use of either conductivity or resistivity depends on the context. Conductivity is used in electrophysiology because it is often easier to describe processes related to action potential generation as an increase in conductivity. However, resistance is often used to explain the material properties of muscles and nerves as an electrical circuit.

The relationship between potential difference and electric field can be used to quantify the flow of charges (i.e., current) within a section of wire with respect to resistivity of the material. First, recall that the work per unit charge required to move a charge from one point to another within the electric field is $\Delta V = -\Delta W_{ba} / q_0$. However, the definition for work done on charge by the electric field as it is moved between two points is $\Delta W_{ba} = -q_0 Ed$. Making the appropriate substitution and canceling for $(-q_0)$, the relationship between potential difference and electric field is $\Delta V = Ed$. In the case of the potential difference established over a segment of wire, the electric field over a distance can be replaced: $\Delta V = El$.

The next step is to link potential difference and current density. We may accomplish this by isolating for the E in current density $(J = \sigma E)$ and substituting into the new formula for potential difference: $\Delta V = J(l / \sigma)$. Substituting the alternate relationship for current density $(J = i / A)$ reveals the different factors governing the flow of charges:

$$\Delta V = i\left(\frac{l}{\sigma A}\right)$$

The reader should recognize that resistivity $(1 / \sigma)$ is embedded in the equation. However, it is not the only factor governing the flow of charges (i.e., current, i). The length (l) and cross-sectional area (A) of the wires are also important. All three factors together describe the **resistance** (R) of the conductor:

$$R = \left(\frac{l}{\sigma A}\right)$$

The relationship indicates that a greater conductor length (l) provides a greater resistance to the flow of charges, while a larger cross-sectional area (A) offers less resistance to the flow of charges (figure 2.9). An analogy is often made to the relationship between the length and diameter of an ordinary garden hose and the flow of water. The reciprocal of resistance is **conductance** ($1/R$).

Substituting for resistance (R) into the potential difference gives the more familiar relationship associated with Ohm's law: $\Delta V = iR$. If resistance is constant, then the flow of charges is directly proportional to the potential difference applied to the conductor. The resistance of a conductor is given by

$$R = \frac{\Delta V}{i}.$$

The SI unit of resistance is volts per ampere: One volt per one ampere is defined as one ohm (Ω). It is important to emphasize that resistance ($l / \sigma A$) takes into account the physical properties of the conductor while resistivity ($1 / \sigma$) is a property of the material from which the conductor is made. Substituting resistivity into the equation for resistance and rearranging,

$$\rho = \frac{RA}{l} = \frac{\Omega \cdot \text{m}^2}{\text{m}}.$$

The SI unit for resistivity is ohm-meters ($\Omega \cdot m$). Thus, once the physical dimensions (i.e., l and A) are known along with the material's ability to conduct charges (ρ), the resistance (R) can be calculated:

$$R = \rho\left(\frac{l}{A}\right) = \frac{\Omega \cdot \text{m} \times \text{m}}{\text{m}^2}$$

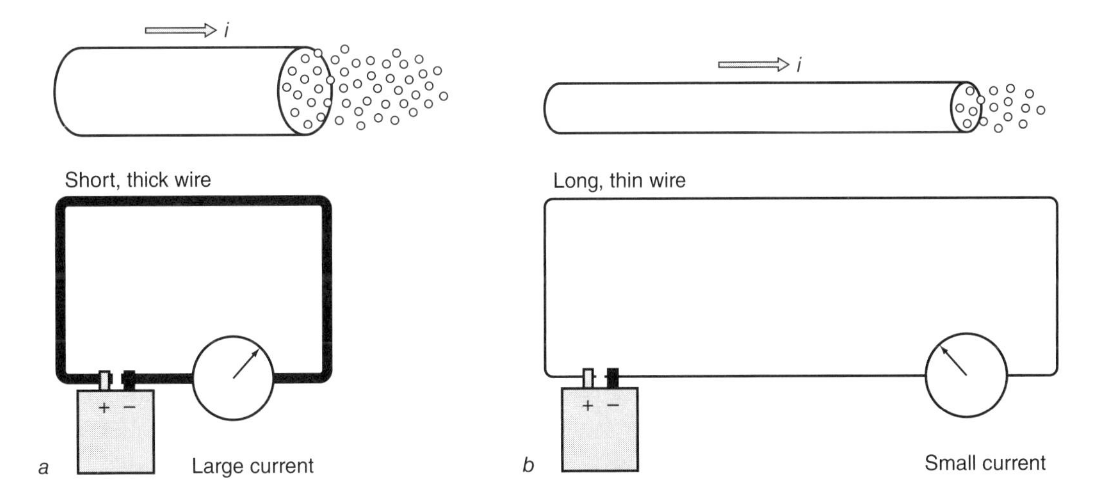

FIGURE 2.9 The length (l) and cross-sectional area (A) of the conductor affect the flow of charges in a way that is analogous to the way the dimensions of a pipe affect the amount of water that can flow through it. A shorter wire with thicker cross-sectional area can conduct a larger current *(a)*, whereas a longer and narrower wire with greater resistance conducts a smaller current *(b)*.

KEY POINTS

- The amount of charge flowing (i) per unit area (S) is referred to as current density (J). If current is divided by cross-sectional area of the conductor, then the current density is $J = i / A$, where the units are amperes (A) per meter squared (m^2).
- When the potential difference is constant as supplied by a battery, current density is proportional to the electric field: $J \propto E$. The constant of proportionality that equates the two variables is conductivity (σ) of the material: $J = \sigma E$. The reciprocal of conductivity is the resistivity (ρ) of the material to the flow of charges.
- The relationship between potential difference and current density reveals the different factors governing the flow of charges: $\Delta V = i(l/\sigma A)$, where $(1/\sigma)$ is resistivity of the material to the flow of charges (i.e., current, i), which is also regulated by the length (l) and cross-sectional area (A) of the wires. All three factors together describe the resistance (R) of the conductor: $R = (l/\sigma A)$.
- Substituting for resistance (R) into the potential difference gives the more familiar relationship associated with Ohm's law: $\Delta V = iR$. If resistance is constant, then the conductor is said to obey Ohm's law wherein the flow of charges is directly proportional to the potential difference applied to the conductor. The resistance of a conductor is given by $R = \Delta V/i$. The unit of resistance is volts per ampere: One volt per one ampere is defined as one ohm (Ω).

Electrical Energy

An electric circuit that includes a piece of highly resistive material (resistor) is illustrated in figure 2.10. The negative terminal of the battery (low potential) will be used as an arbitrary starting point. When the switch is closed, an electrochemical reaction within the battery transfers energy to the charges to create a potential difference between the two terminals (V_{ba}). The electric potential energy gained by each charge is equal to the amount of chemical energy lost by the battery (conservation of energy).

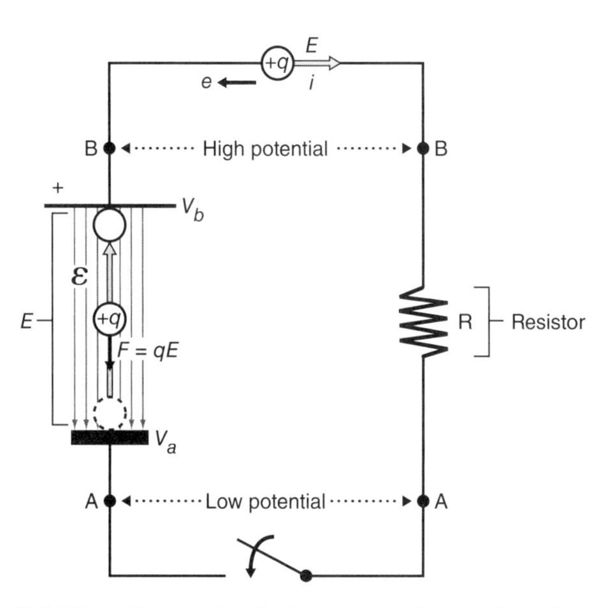

FIGURE 2.10 An electric circuit with a single resistor. The source (battery) is enlarged to illustrate the basic concepts underlying electromotive force (ε).

The energy transferred to each charge (joules per coulomb) is called the **electromotive force ($\mathcal{E}$)** of the battery. The electromotive force is not really a force in the literal sense, but it does perform work on the charge to bring it from a lower to a higher energy potential as if it were an external force.

Recall that the potential difference between two points (V_{ba}) is equal to the negative of the work ($-W_{ba}$) required to move the charge the distance between the two points: $V_{ba} = -W_{ba} / q$. The work done by the electromotive force is $-W_{ba} = qV_{ba}$, where the negative sign indicates that energy is supplied by an external source to move the charge against the electric field between the two terminals. The electromotive force ($\mathcal{E}$) must therefore equal the potential difference across the terminals. For example, a battery with an electromotive force of 12 V can perform 12 J of work on every coulomb of charge to give it an electric potential energy of 12 V as it travels between the negative and positive terminals.

Consistent with the concept of electric potential energy developed for charged plates, the positive charges naturally flow clockwise from high to low potential energy toward the negative terminal. Resistance to the flow of charges offered by the conductive wire is negligible compared to that of the resistive material.

As the charges enter the resistor, they collide with the atoms within the resistor. The collisions transfer electric potential energy from the charges and increase the kinetic energy of the atoms within the resistor, which is expressed as heat (conservation of energy). The charges therefore perform work on the resistor, transferring energy to it. In doing so, the charges experience a potential drop between B and A as given by $V_{ba} = iR$. The charges once again arrive at the negative terminal with low potential. To maintain a constant flow of current (i), the rate at which the battery performs work on the charges to increase their potential energy across the terminals must equal the rate at which the charges perform work on the resistor and lose potential energy. Where $W_{ba} = qV_{ba}$ and $i = dq / dt$, a general expression for the rate of transfer of electric potential energy is

$$\frac{dW_{ba}}{dt} = \frac{dqV_{ba}}{dt};$$

$$\frac{dW_{ba}}{dt} = iV_{ba}.$$

Since the rate of work is called power, the rate at which electric energy is delivered to the circuit by the source (battery) is called **electric power:** $P = iV_{ba}$. The unit of power is joules per second, or watts (W):

$$i\left(\frac{\text{coulomb}}{\text{second}}\right) \times V_{ba}\left(\frac{\text{joules}}{\text{coulomb}}\right) = P\left(\frac{\text{joules}}{\text{second}}\right)$$

Another very useful expression for electrical power that has many practical applications in EMG is obtained by substituting for V_{ba}:

$$P = i(iR) = i^2R$$

since

$$i = \frac{V_{ba}}{R};$$

$$P = \left(\frac{V_{ba}}{R}\right)^2 R;$$

$$P = \frac{V_{ba}^2}{R}.$$

Any device that is included in the electric circuit provides resistance to the flow of charges, and may be considered as just another type of resistor. As the electric charges flow through a lightbulb, for example, they excite the atoms of the filament, which generates both light and heat. All electric devices include a power rating in watts for a 120 V source. This rating reflects energy consumption by the device in joules per second, also termed i^2R losses.

In the previous section, we showed that whenever a charge passes through a resistor, it experiences a potential drop given by $V_{ba} = iR$. If several resistors are connected in series, the charges experience a potential drop across each resistor (figure 2.11*a*). The sum of each potential drop $(V_1 + V_2 + V_3 + \cdots + V_n)$ must equal the total voltage (V_{ba}) driving the charges across the resistor (conservation of energy). One can solve problems involving resistors within an electrical circuit by finding the equivalent resistor, in a manner similar to that outlined for capacitors. For two resistors connected in series, $V_{ba} = V_1 + V_2$. Since the same amount of current (i) must pass through each resistor in succession, $iR_e = iR_1 + iR_2$, where R_e is the equivalent resistor.

When resistors are connected in parallel, the same potential difference is applied across each resistor (figure 2.11*b*). The potential difference across resistors in parallel is therefore the same as that applied by the battery: $V_{ba} = V_1 = V_2$. However, the current from the source splits into different branches. The split results in less current (i) entering each individual resistor than leaving the battery. Because the charge must be conserved, the current that enters the branch must be the same as that leaving the branch: $i = i_1 + i_2$. The relationship for current $(i = V / R)$ can be used to substitute for i:

$$\frac{V_{ba}}{R_e} = \frac{V_1}{R_1} + \frac{V_2}{R_2}$$

Dividing through by V because the potential differences are equivalent,

$$\frac{1}{R_e} = \frac{1}{R_1} + \frac{1}{R_2}.$$

The difference between resistors in series and in parallel is illustrated by the original formulation for resistance:

$$R = \rho\left(\frac{l}{A}\right)$$

When resistors are connected in series, they have the same effect upon current flow as increasing the length of the conductor does. Resistors in series allow less

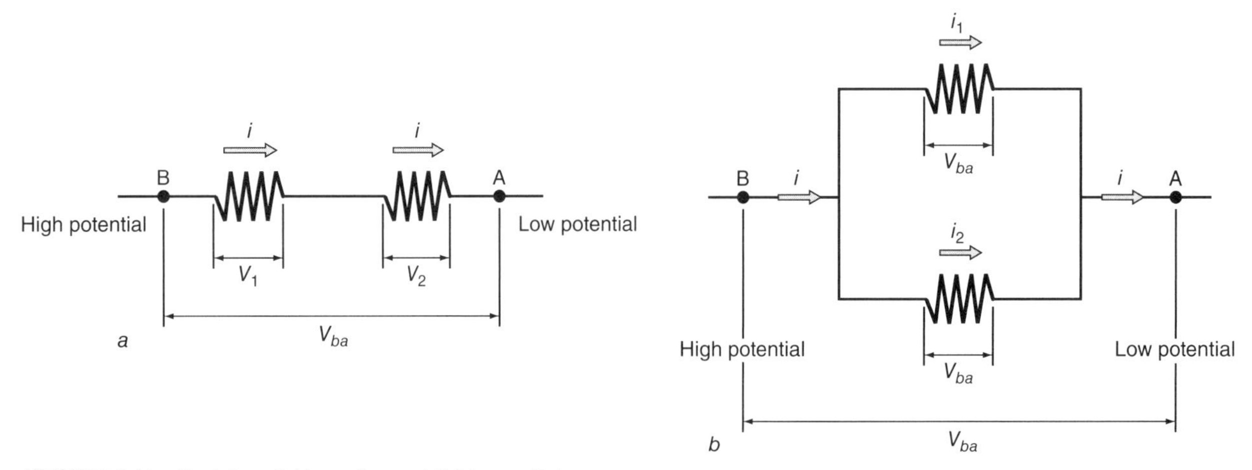

FIGURE 2.11 Resistors *(a)* in series and *(b)* in parallel.

current to flow than does any individual resistor alone. Resistors connected in parallel have the same effect as increasing the cross-sectional area of the conductor. Resistors in parallel allow more current to flow through the circuit than does any one individual resistor alone. An analogous situation is the consumption of a milk shake through two straws attached end to end (in series) versus two straws attached side by side (in parallel). A worked example for resistors in an electric circuit is provided in appendix 2.3.

 **KEY POINTS**

- To maintain a constant flow of current (*i*), the rate at which the battery performs work on the charges to increase their potential energy across the terminals must equal the rate at which the charges perform work on the resistor and lose potential energy. The rate of work is called power, and the rate at which electric energy is delivered to the circuit by the source (battery) is also called power. The unit of power is watts (W).
- If several resistors are connected in series, the charges experience a potential drop across each resistor. The sum of each potential drop must equal the total voltage driving the charges across the resistor (conservation of energy).
- Since the same amount of current must also pass through each resistor in succession, when resistors are connected in parallel, the same potential difference is applied across each resistor.
- The potential difference across resistors in parallel is therefore the same as that applied by the battery. However, the current from the source splits into different branches. The split results in less current entering each individual resistor than leaving the battery. Because the charge must be conserved, the current that enters the branch must be the same as that leaving the branch.

Resistors and Capacitors in a Circuit

The incorporation of a resistor and capacitor into the same circuit has both physiological and physical applications important to EMG. For example, the relationship between electromotive force, the flow of current through the resistor to the capacitor, and the resulting potential across the plates forms the basis for understanding the physical properties of muscle fibers. A resistor and capacitor in a circuit is also important for concepts in signal processing related to filtering the electromyographic signal.

Charging a Capacitor Through a Resistor

An electric circuit that includes a resistor and capacitor is termed an **RC circuit** (figure 2.12*a*, inset). When the switch is closed on an RC circuit (position A), the electromotive force ($\mathcal{E}$) establishes a potential difference across the circuit that drives electrons through the resistor (R) toward the positive terminal of the battery. A positive charge (q) is left behind on the top capacitor plate. At the same time, electrons are driven from the negative terminal of the battery toward the bottom capacitor plate. The accumulation of charges on each plate establishes a potential difference across the capacitor (V_C). Current flow then decreases as the potential difference across the

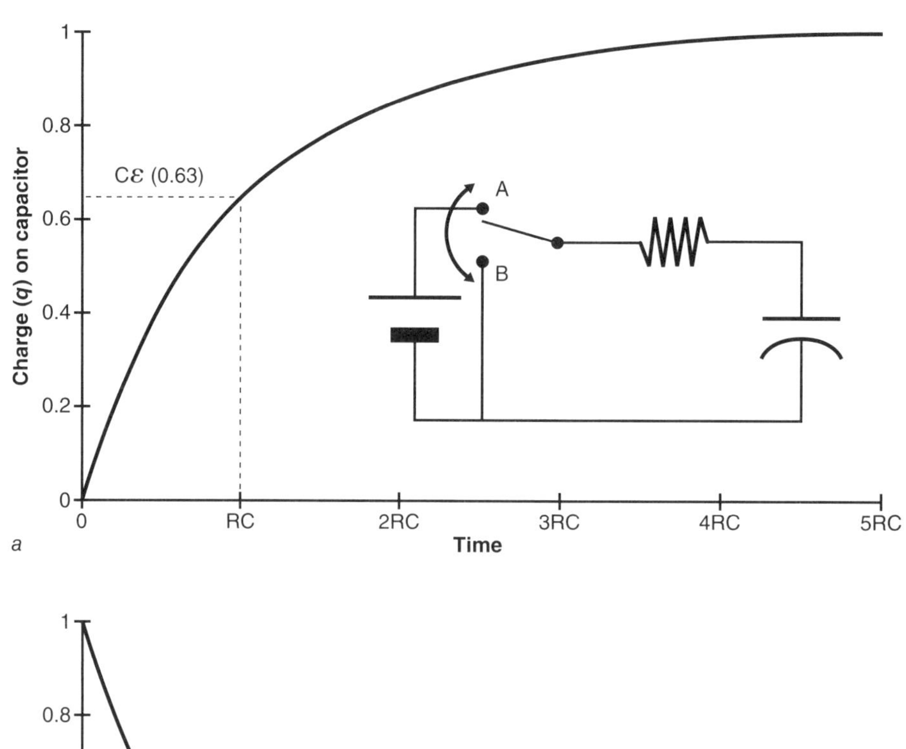

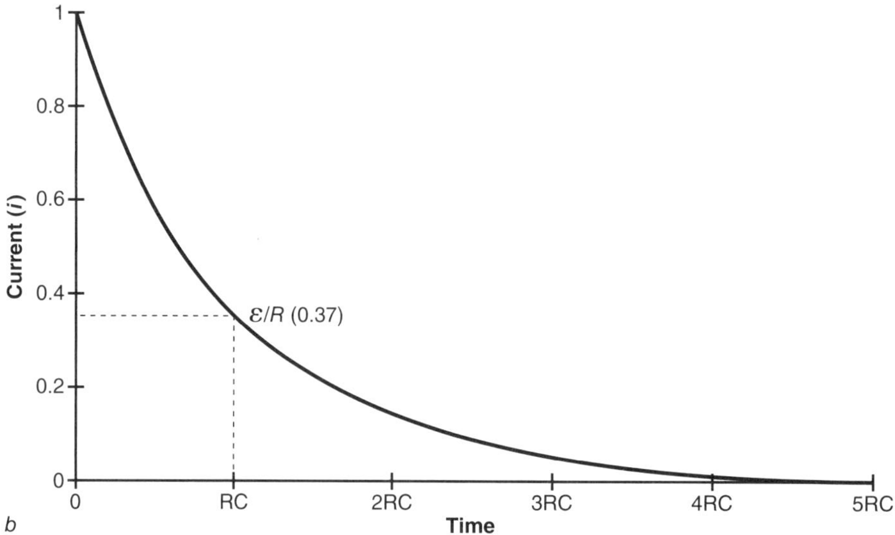

FIGURE 2.12 *(a)* The normalized charge–time graph for the resistor–capacitor (RC) circuit along with *(b)* the normalized current–time graph.

capacitor increases. When the potential difference across the capacitor equals the electromotive force ($\mathcal{E} = V_C$), current flow decreases toward zero. This is logical since the charges are now in static equilibrium: The potential difference created by the battery to drive charge into the circuit is matched by the potential difference at the capacitor (in the circuit), which can drive charge in the opposite direction.

Appendix 2.4 details derivation of the equations that describe the amount of charge $q(t)$ and current $i(t)$ on the capacitor plates as a function of time when the switch is closed in position A. The following equation is based on achieving static equilibrium between the potential difference across the battery and capacitor ($\mathcal{E} = V_C$):

$$q(t) = C\mathcal{E}\left(1 - e^{-\frac{t}{RC}}\right)$$

The exponential function means that the charge (q) on the capacitor takes a long time to reach its maximum value ($C\mathcal{E}$) wherein static equilibrium is achieved. The units for the RC quantity reveal that it is a time constant:

$$\Omega\left(\frac{V}{A}\right)\times F\left(\frac{C}{V}\right)=\frac{C}{A}$$

$$\frac{C}{A}=\frac{C}{\frac{C}{S}}=C\frac{S}{C}=\text{seconds}$$

When $t\,/\,RC = 1$, the quantity inside the parentheses represents a time multiple at which the capacitor is charged to a percent of the maximum value:

$$q = C\mathcal{E}(1-e^{-1})$$

$$q = C\mathcal{E}(0.63)$$

This result means that the capacitor achieves 63% of its maximum charge within the first time multiple (figure 2.12*a*). The capacitor is charged at 86% and 95% of maximum when $t\,/\,RC$ equals 2 and 3, respectively. The time multiple is termed the RC time constant (τ) and is given by $\tau = RC$.

The time course of current $i(t)$ can be obtained by differentiating with respect to the charge:

$$i(t)=\frac{\mathcal{E}}{R}e^{-\frac{t}{RC}}$$

Initial conditions ($t = 0$) occur at the instant the switch closes at position A (figure 2.12*b*). Current is maximal at this moment because $e^0 = 1$ and $i = \mathcal{E}\,/\,R$. As charge builds up on the capacitor plates and the potential difference increases (V_C), current decreases exponentially with a time constant $\tau = RC$. At the first time multiple, the flow of current has decreased to 37% of its maximum value:

$$i=\frac{\mathcal{E}}{R}e^{-1}$$

$$i=\frac{\mathcal{E}}{R}(0.37)$$

Discharging a Capacitor Through a Resistor

With the switch in position A and the capacitor fully charged, there is no current flow because the system is in equilibrium. If the switch is moved to position B, the circuit consists of a fully charged (Q) capacitor and a resistor (figure 2.12*a,* inset). In this situation the capacitor serves as a *nonrenewable* source. That is, the fully charged capacitor is similar to a battery, but it is without an electromotive force to maintain a potential difference and flow of current across the circuit. When the switch closes at ($t = 0$), positive charges move off the upper plate (high potential) through the resistor toward the bottom plate (low potential) of the capacitor. The potential drop across the capacitor is gained by the resistor as collisions transfer the electric potential energy from the charges and increase the kinetic energy of the atoms within the resistor, which is expressed as heat (conservation of energy). The charges therefore perform work on the resistor, transferring energy to it. However, when the capacitor is fully discharged, the flow of current ceases.

Appendix 2.4 also details derivation of the equations that describe both charge $q(t)$ and current $i(t)$ on the capacitor plates as a function of time when the switch is closed to position B. The following equation is based on the absence of a battery ($\mathcal{E} = 0$):

$$q(t) = Qe^{-\frac{t}{RC}}$$

Notice that the equation for $q(t)$ for a discharging capacitor is similar to that for $i(t)$ for a charging capacitor. In both cases, there is an exponential decrease in a quantity that occurs at a time constant $\tau = RC$. The same is also true for $i(t)$ during charging the capacitor. As before, the relationship for $i(t)$ may be obtained by differentiating with respect to $q(t)$:

$$i(t) = -\frac{Q}{RC}e^{-\frac{t}{RC}}$$

where the quantity Q / RC is actually the initial current at the moment the switch closes to position B. The negative sign $(-Q / RC)$ indicates that the direction of current flow associated with a discharging capacitor is now reversed.

The Muscle Fiber as a Resistor–Capacitor Circuit

In electrophysiology and biophysics, muscle fibers are represented as electrical cables with porous insulation such that current is able to leak to the surrounding area. Thus, the muscle fiber is basically a long cylindrical tube of conducting fluid (myoplasm) that is surrounded by a membrane. Current travels axially along the length of the muscle fiber through the myoplasm, but a portion also leaks through the cell membrane (figure 2.13). The electric potential depends on resistance to the flow of current in the axial and radial directions along the muscle fiber. Resistance (R) to the axial flow of current (i_{fiber}) depends on the resistivity of the myoplasm (ρ_m). Resistance to the radial flow of leakage current (i_{leak}) depends on the resistance per unit area of the membrane (R_m).

The membrane also has a capacitive function because charges of opposite sign exist on the two sides of the membrane, negative inside and positive outside. Analogous to the situation with the plate capacitor, the charge per unit area divided by the potential difference is capacitance of the membrane per unit area (C_m). Appendix 2.5 presents a worked example in which the axial and radial resistances for the muscle fiber are calculated. The appendix also illustrates how the two resistances relate to muscle fiber conduction velocity.

KEY POINTS

- When the switch is closed on an RC circuit, the amount of charge $q(t)$ and current $i(t)$ on the capacitor plates as a function of time is based on achieving static equilibrium between the potential difference across the battery and capacitor ($\mathcal{E} = V_C$).
- If a fully charged capacitor is then discharged through a resistor, the capacitor serves as a nonrenewable source. The equations that describe both charge $q(t)$ and

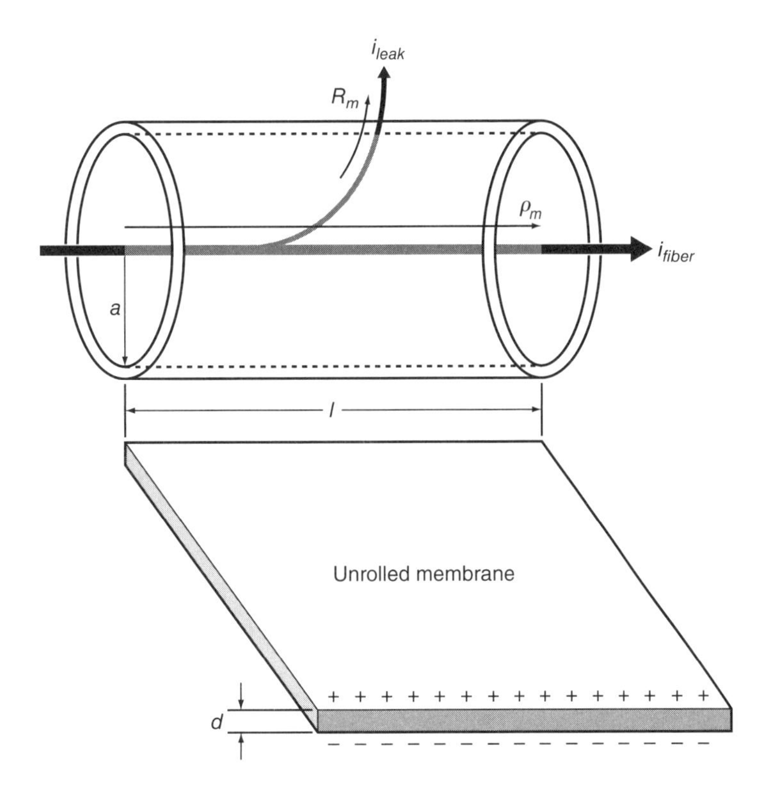

FIGURE 2.13 The muscle fiber is modeled as a simple cable with radius (*a*), axial resistance to the flow of current (ρ_m) through the myoplasm, and radial resistance to the flow of current through the membrane (R_m). If the muscle fiber membrane were unrolled, it would have a thickness (*d*) that separates the two charges, analogously to the two plates of a capacitor.

current *i*(*t*) on the capacitor plates as a function of time are based on the absence of a battery ($\mathcal{E} = 0$).

- The muscle fiber is basically a long cylindrical tube of conducting fluid (myoplasm) that is surrounded by a membrane. The electric potential depends on resistance to the flow of current in the axial and radial directions along the muscle fiber.
- Resistance (*R*) to the axial flow of current (i_{fiber}) depends on the resistivity of the myoplasm (ρ_m). Resistance to the radial flow of leakage current (i_{leak}) depends on the resistance per unit area of the membrane (R_m).
- The membrane also has a capacitive function because charges of opposite sign exit on the two sides of the membrane, negative inside and positive outside. Analogous to the situation with the plate capacitor, the charge per unit area divided by the potential difference is capacitance of the membrane per unit area (C_m).

Essentials of Alternating Current

The electromotive force ($\mathcal{E}$) provided by a battery is said to result in **direct current (DC)** because it is nonoscillating. Coiled wire rotating within a magnetic field results in an electromotive force ($\mathcal{E}$) that oscillates in sinusoidal fashion. A sine wave is the simplest form of alternating voltage and **alternating current (AC).** The electromyographic signal also consists of an alternating voltage, and many of the conventions associated with the sinusoidal function are applied to it.

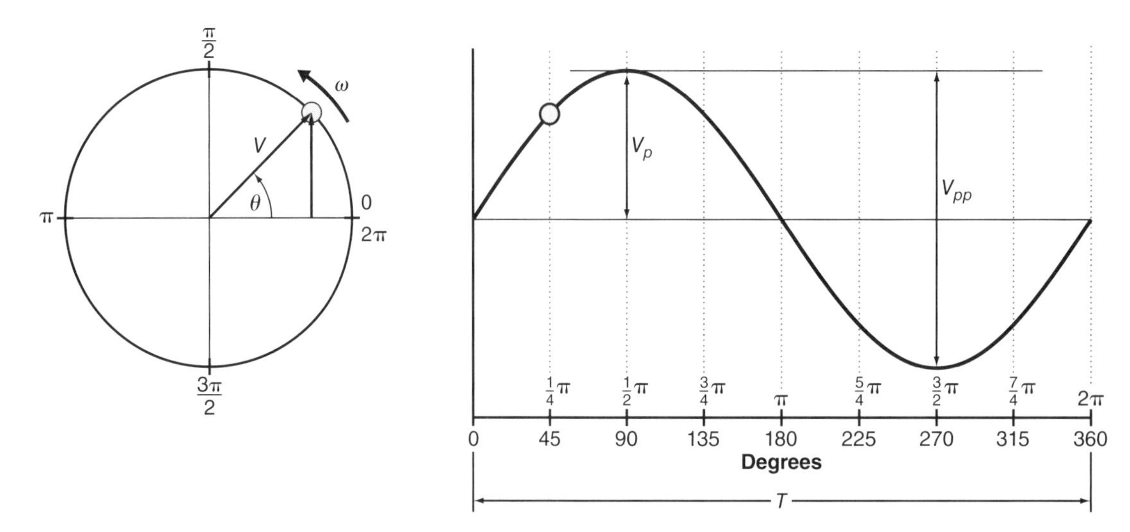

FIGURE 2.14 The vector rotates at a constant angular velocity (ω), and the vertical component gives the amplitude of the sinusoid at that particular point in time, as it rotates through one complete cycle (360° or 2π rad). Shown on the sinusoid are the peak (V_p) and peak-to-peak (V_{pp}) amplitudes.

Conventions of an Alternating Signal

The height of a sinusoidal waveform at any point in time is related to the vertical component of a vector (V) rotating counterclockwise at a constant angular velocity within a unit circle. One revolution of the circle is 2π radians completed in a time period *(T)* (figure 2.14). One cycle of the sinusoid is also completed in the same time period *(T)*. The number of cycles per second is the frequency (f) of the sinusoid where $f = 1/T$ and the unit is hertz (Hz). Because the sinusoid has been mapped onto the unit circle, it may also have an angular frequency (ω) as the number of revolutions per second in radians:

$$\omega = \frac{2\pi}{T} = 2\pi f$$

The length of the vector (V) represents the magnitude of the peak voltage (V_p). The peak voltage (V_p) is measured from the zero isoelectric baseline to the peak of the waveform. The voltage value at any time (t) is then given by the y-component (vertical) of the vector (V) obtained from basic trigonometry: $V = V_p \sin(\omega t)$. The vertical component oscillates between $+V_p$ and $-V_p$ with an angular frequency (ω). The absolute difference between $+V_p$ and $-V_p$ is defined as the peak-to-peak amplitude (V_{pp}). The expression $V = V_p \sin(\theta)$ is more familiar, but the two expressions (or equations) for V are equivalent because ($\omega \times t = 0$).

If two sinusoids have the same frequency but one is delayed in time with respect to the other, they pass through zero at different times and are said to be out of phase (figure 2.15). The angle between the two rotating vectors is called the **phase angle** (ϕ). The voltage V_1 "leads" V_2 because it reaches its peak first. Another way to express the temporal relationship is to say that V_2 "lags" behind V_1. Figure 2.15 shows that the phase angle between the two peaks is $\phi = 45°$ or $1/4\pi$ rad. The two peaks can be aligned in the following way. First, imagine that V_1 is advancing forward from right to left, toward the y-axis. Subtracting $\phi = 45°$ from the current position of V_1

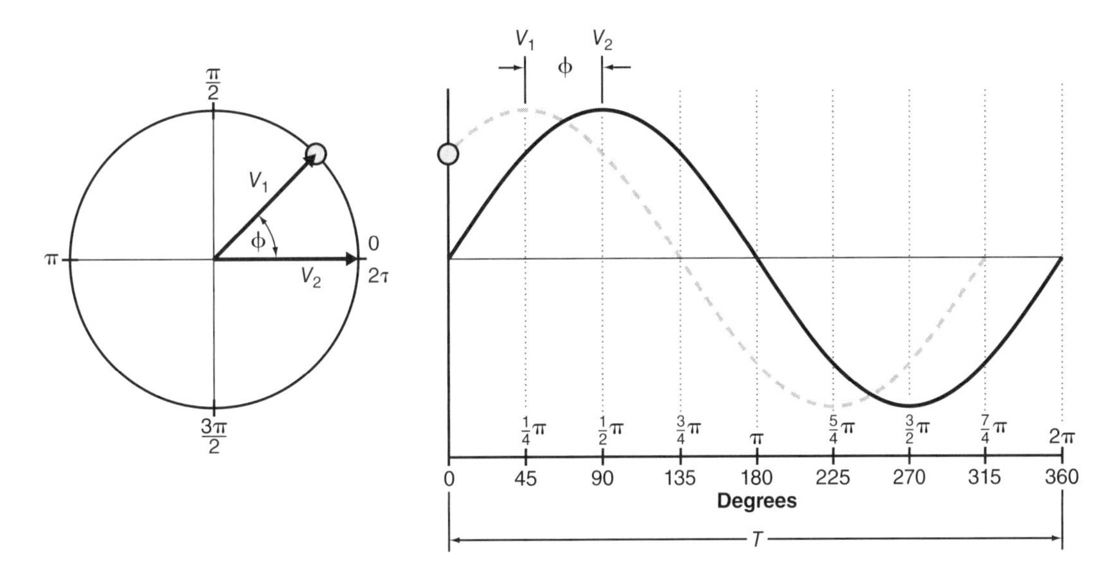

FIGURE 2.15 Two voltage sinusoids V_1 (dashed line) and V_2 (solid line) with a 45° phase lag (ϕ). Their associated vectors are represented in a unit circle in the left panel. The magnitudes of the two vectors are equal to the maximum voltage (V_p).

will move it backward, away from the y-axis to superimpose on V_2. The equation that shifts V_1 toward the right along the x-axis is $V_1 = V_p \sin(\omega t - 45°)$. In contrast, adding $\phi = 45°$ to the current position of V_2 will shift it to the left toward the y-axis to superimpose over V_1. The equation that shifts V_2 so that it superimposes over V_1 is $V_2 = V_p \sin(\omega t + 45°)$. A more general expression for the voltage waveforms is $V = V_p \sin(\omega t \pm \phi)$, where the term $\pm\phi$ shifts the waveform right or left along the x-axis.

KEY POINTS

- One revolution of the circle is 2π radians completed in a time period *(T)*. One cycle of the sinusoid is also completed in the same time period *(T)*. The number of cycles per second is the frequency *(f)* of the sinusoid where $f = 1/T$ and the unit is hertz (Hz). Because the sinusoid has been mapped onto the unit circle, it may also have an angular frequency (ω) as the number of revolutions per second in radians: $\omega = 2\pi/T = 2\pi f$.
- The length of the vector *(V)* rotating within a circle represents the magnitude of the peak voltage (V_p). The peak voltage (V_p) is measured from the zero isoelectric baseline to the peak of the waveform. The voltage value at any time *(t)* is then given by the y-component (vertical) of the vector *(V)* obtained from basic trigonometry: $V = V_p sin(\omega t \pm \phi)$, where $\omega t = \theta$ and ϕ is the phase lag or lead.

Effective Voltage and Current

This section provides the background for one of the most commonly used measures in EMG, the root-mean-square voltage of muscle activity. This measure actually has a functional meaning that is originally associated with calculating the power (*P*) delivered to the system by an electromotive force ($\mathcal{E}$). The "system" we are referring to is anything that impedes the flow of current. This may be a piece of resistive material or an electric device placed within the circuit; both have the same effect.

If an electric circuit contains an alternating voltage source and a resistor, both the voltage and current across the resistor will be in phase and vary in a sinusoidal fashion. The peak amplitude of the voltage across the resistor (V_p) will also have the same magnitude as that of the electromotive force ($\mathcal{E}$). Voltage across the resistor can then be expressed as

$$V = V_p \sin(\omega t).$$

Keeping in mind that R is a constant value and substituting V / R for i to obtain an expression for current as a function of time,

$$i = \frac{V}{R} = \frac{V_p \sin(\omega t)}{R};$$

$$i = i_p \sin(\omega t).$$

Electrons flow in one direction when the current is positive and in the opposite direction when the current is negative. If all the positive and negative values are summed to determine the average current over time, it should be apparent that the mean value is zero. However, electrons still move back and forth within the resistor, generating heat (i^2R losses), so power is still delivered to the circuit. To determine the average (mean) power delivered to the circuit, substitute the time-varying quantity of current:

$$P = i^2 R$$

$$P = \left(i_p^2 \sin^2(\omega t)\right) R$$

If the mean power is based on the average of squared values for $\sin(\omega t)$, the average will be a nonzero number:

$$\overline{i^2} = i_p^2 \overline{\sin^2(\omega t)}$$

Sine and cosine waves that have the same amplitude and frequency still sum to zero over one complete cycle. Likewise, the mean of the squared values will also be the same. If the mean square value for the following trigonometric identity is

$$\overline{\sin^2(\omega t)} + \overline{\cos^2(\omega t)} = 1,$$

and the mean square values of the sine and cosine functions are the same, then

$$\overline{\sin^2(\omega t)} = \frac{1}{2}.$$

Substituting back into the expression for the mean square value for current,

$$\overline{i^2} = i_p^2 \overline{\sin^2(\omega t)} = i_p^2 \frac{1}{2},$$

the average power for AC is

$$\overline{P} = \overline{i^2} R = \frac{1}{2} i_p^2 R.$$

A constant (DC) electromotive force ($\mathcal{E}$) results in constant values for voltage, current, and power. This is the same thing as the mean value of the AC quantities, over a period of time. The mean square values for AC quantities must be returned to scale by taking the square root:

$$\overline{i^2} = i_p^2 \frac{1}{2}$$

$$i_{rms} = \frac{i_p}{\sqrt{2}} = (0.7071)i_p$$

The root-mean-square current (i_{rms}) is also termed the effective current. One effective ampere is that amount of AC that will develop the same power as one ampere of DC. Thus, the equivalent relationship between AC and DC power:

$$\overline{P} = i_{rms}^2 R$$

If the identical steps are taken for alternating voltage, a similar result is obtained:

$$\overline{V^2} = V_p^2 \frac{1}{2}$$

$$V_{rms} = \frac{V_p}{\sqrt{2}} = (0.7071)V_p$$

The root-mean-square voltage (V_{rms}) is termed effective voltage. One effective volt is that alternating voltage that will produce an effective current (i_{rms}) of one ampere through a resistance of one ohm. An alternative and often used expression for average power is obtained by substituting for effective current:

$$\overline{P} = i_{rms}^2 R$$

$$\overline{P} = \left(\frac{V_{rms}}{R}\right)^2 R$$

$$\overline{P} = \frac{V_{rms}^2}{R}$$

KEY POINTS

- The mean value of alternating voltage and current is zero; however, they do have an effect within the circuit. The effective voltage and current is based on the root-mean-square value of the sinusoid. Each value is first squared to generate a nonzero mean. The square-root operation is then performed to return to the original scale.
- The root-mean-square current is i_{rms} = (0.7071)i_p, where (i_p) is the peak current. If the same steps are taken for alternating voltage, a similar result is obtained: V_{rms} = (0.7071)V_p, where (V_p) is the peak voltage. One effective volt is that alternating voltage that will produce an effective current (i_{rms}) of one ampere through a resistance of one ohm.

Capacitance in an AC Circuit

The function of analog filters cannot be understood unless capacitance in an AC circuit is already grasped. The next chapter, on instrumentation, describes analog filters in detail. When the switch is closed on a DC circuit that contains only a capacitor, charges build up on the capacitor plates and start to oppose the additional charges of like sign. Current flow then starts to decrease. This increase in charge continues until the capacitor is fully charged and the potential difference across the capacitor plates equals the electromotive force ($V_C = \mathcal{E}$). The system is in electrostatic equilibrium and the flow of current ceases.

In contrast, a capacitor in an AC circuit is alternately charged and discharged as the voltage and current reverse direction every half-cycle (figure 2.16). By necessity, an arbitrary starting point must be taken to understand the time course of a sequence of cyclical events. In this case, $t = 0$ is the instant a switch closes on the AC circuit. Voltage across the capacitor is zero and current flow is at a maximum (0°). The reason is that the plates are empty of charge and there is nothing to oppose the flow. Positive and negative charges then accumulate on the top and bottom plates of the capacitor, respectively. The buildup of charges results in a concomitant decrease in the flow of current. Current goes toward zero as the plates become fully charged at 90° ($\pi/2$ rad), one-quarter through the cycle.

At this point, the electromotive force ($\mathcal{E}$) starts decreasing, the capacitor discharges, and current flows in the opposite direction. The negative sign indicates a reversal in the direction of current flow. Voltage across the capacitor is zero, and current reaches a maximum at 180° (π rad), halfway through the cycle. Current is once again maximum at this instant because there are no charges on the capacitor to oppose the flow. This is the same moment when the electromotive force ($\mathcal{E}$) changes direction and starts charging the capacitor again. In this case, the polarities of charges building up on the capacitor plates are opposite to those in the first half of the cycle: The top plate is negative and the bottom plate is positive. The voltage across the capacitor achieves a

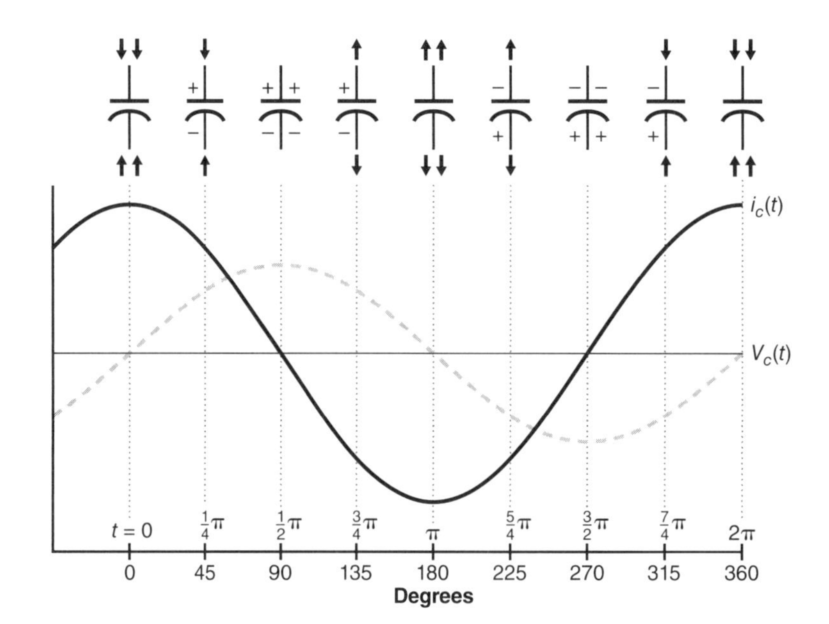

FIGURE 2.16 A circuit consisting of an alternating current source and capacitor. The series of capacitors illustrate the alternating charging and discharging of the capacitor. The current (solid line) and voltage (dashed line) are given below. Current and voltage are out of phase ($\phi = 90°$ *or* $\pi/2$ *rad*).

second maximum in the opposite direction at 270° (3π/2 rad). The maximum voltage prevents the flow of additional charges, and current ceases for a second time, three-quarters through the cycle. The cycle is complete when the decreasing electromotive force ($\mathcal{E}$) discharges the capacitor toward zero, and current flow achieves a maximum at 360° (2π rad). The system is now in the same state that it was at ($t = 0$) and the cycle can begin once again. Notice that through the entire sequence of events, current actually *leads* voltage by 90° (figure 2.16).

The potential difference across the capacitor is expected to have the same peak amplitude (V_p) and frequency as the alternating electromotive force ($\mathcal{E}$). Proceeding as before, the potential difference across the capacitor is $V = V_p \sin(\omega t)$. The goal here is to obtain an expression for current that also involves voltage. We accomplish this by using the relationship $V = q / C$ and solving for charge: $q = CV_p \sin(\omega t)$. The flow of current as a function of time may be obtained by differentiating charge (q) with respect to time:

$$i = \frac{d\left(CV_p \sin(\omega t)\right)}{dt} = \omega CV_p \cos(\omega t)$$

The fact that the resulting formula contains $\cos(\omega t)$ confirms the earlier observation that current leads voltage on the capacitor by 90° because $\cos(\omega t) = \sin(\omega t + 90°)$. Further, the maximum current (i_p) is reached when $\cos(\omega t) = 1$, so the expression reduces to

$$i_p = \omega CV_p.$$

Knowing that $V = iR$ and rearranging to isolate V_p:

$$V_p = \frac{i_p}{\omega C}$$

For reasons of symmetry of notation:

$$R = \frac{1}{\omega C}$$

A resistor impedes the flow of current and in the process dissipates energy (i^2R losses) as heat. A capacitor also impedes the flow of current as charges build up on its plates, but it does not dissipate energy. Rather, it *reacts* against changes in voltage by alternately absorbing and releasing charges (energy) into the circuit. A new term, **reactive capacitance** (X_C), is therefore introduced because the capacitor not only resists the flow of current but also regulates the flow of charges:

$$X_C = \frac{1}{\omega C}$$

Reactive capacitance has the same units as resistance (Ω), and it is inversely proportional to both the frequency (ω) of the voltage source and the magnitude of the capacitance (C). When the frequency (ω) of the voltage source increases, there is less time per cycle for charge to build up on the plates and oppose the flow of current. Thus, a capacitor tends to let high frequencies pass through the circuit. In contrast, as the frequency decreases toward $\omega = 0$, it starts to resemble a DC source wherein charges have enough time build up on the capacitor plates until the flow of current eventually ceases. In this case, low frequencies are cut from the circuit. An increase in capacitance (C) then allows a greater number of charges to accumulate on the plates. More current flows through the circuit before being opposed by charges on the plates.

Substituting reactive capacitance (X_C) for resistance (R), Ohm's law for AC can be rewritten:

$$V = iX_C$$

where V and i may be either peak or root-mean-square values. This relationship is valid only for peak and root-mean-square amplitudes because voltage (V) and current (i) are time-varying quantities in an AC circuit that are out of phase at any one instant.

KEY POINT

A capacitor impedes the flow of AC current as charges build up on its plates, but it does not dissipate energy. Rather, it *reacts* against changes in voltage by alternately absorbing and releasing charges (energy) into the circuit. The term reactive capacitance (X_c) is introduced because not only does the capacitor resist the flow of current, but there is also a frequency-dependent regulation of the flow of charges: $X_C = 1/\omega C$. Reactive capacitance has the same units as resistance (Ω), and it is inversely proportional to both the frequency (ω) of the voltage source and the magnitude of the capacitance (C).

Impedance

Both resistors and capacitors impede current flow, but in different ways. That is, resistance is one type of impedance, and reactive capacitance is another. Together, they represent the total or net opposition to the flow of current, termed **impedance (Z).** When the switch closes on an AC circuit that includes a resistor and capacitor, the flow of current will be associated with voltage drops across both elements (figure 2.17*a*). Where the peak voltage and current across the resistor have the same time-varying profile as the electromotive force ($\mathcal{E}$),

$$V_R = iR;$$

$$V_R = Ri_p\sin(\omega t).$$

Ohm's law for the voltage drop across the capacitor is more complicated because voltage and current are out of phase by $\phi = 90°$. Thus, the original formula is applicable only to peak and root-mean-square values. Fortunately, since the phase angle between current and voltage is known, the time-varying quantities can be aligned in the following way. Current leads voltage by 90°, so the lead can be subtracted from the time-varying current profile to align it with the voltage:

$$V_C = \frac{1}{\omega C} i_p \sin(\omega t - 90°)$$

$$V_C = X_C i_p \sin(\omega t - 90°)$$

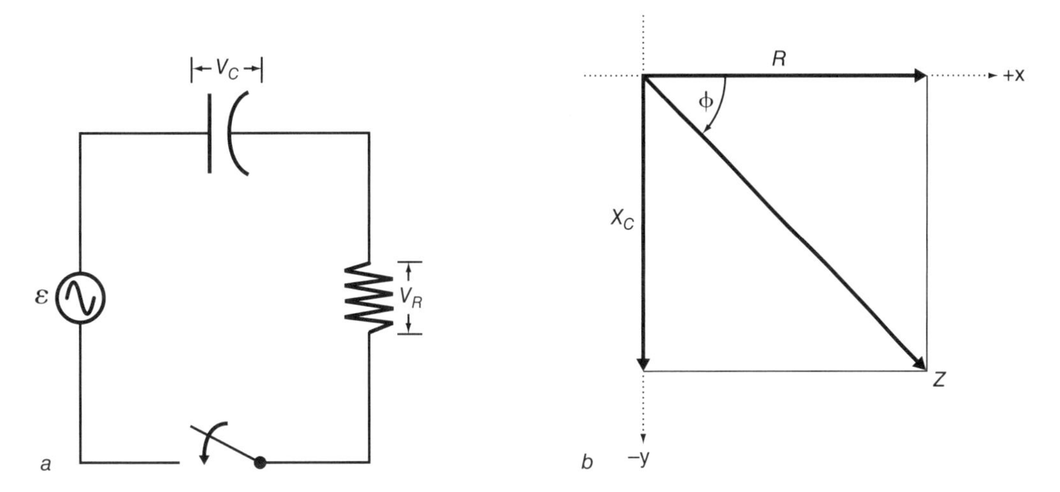

FIGURE 2.17 *(a)* A circuit consisting of an alternating current source, resistor, and capacitor. *(b)* Circuit impedance (Z) is the vector sum of resistance (*R*) and reactive capacitance (X_C).

The total voltage drop across the circuit is

$$V = V_R + V_C = Ri_p \sin(\omega t) + X_C i_p \sin(\omega t - 90°).$$

There is a geometrical way to determine the total or net opposition to the flow of current. Resistance and capacitive reactance may be represented as vectors on an x-y coordinate system (figure 2.17*b*). The magnitude determines vector length, and the phase angle with respect to current determines the direction. Voltage and current are in phase across a resistor, so the phase angle is 0°. The vector for resistance therefore lies along the x-axis. However, voltage lags (90°) behind current at the capacitor plates, so the phase angle is −90°. Using the sign convention that a clockwise rotation is negative, the reactive capacitance is plotted directly along the negative y-axis. The effective impedance of a resistor in series with a capacitor is then given by the Pythagorean theorem:

$$Z = \sqrt{R^2 + X_C^2}$$

The angle (ϕ) is the phase constant between the applied voltage ($\mathcal{E}$) and current within the circuit:

$$\phi = \tan^{-1}\left(\frac{X_C}{R}\right)$$

Cutoff Frequency for an Alternating Current Circuit

The net opposition to current flow (or impedance) is a lumped parameter that is sometimes referred to as AC resistance. The peak or root-mean-square values may be used to rewrite Ohm's law ($V = iR$) for AC circuits as

$$V_p = i_p Z.$$

However, the relationship is valid only for peak and root-mean-square values because the time-varying quantities of voltage and current in an AC circuit are out of phase. These quantities can, however, be adjusted for the difference in phase to obtain a time-varying relationship. Current leads voltage in the circuit by 90° due to the capacitor. The phase lead can be subtracted out from the expression for current to obtain voltage across the circuit:

$$V = Zi_p \sin(\omega t - 90°)$$

Recall that the expression for voltage across the resistor is

$$V_R = Ri_p \sin(\omega t).$$

Substituting for $i_p = V_p / Z$ into the voltage across the resistor,

$$V_R = R\frac{V_p}{\sqrt{R^2 + X_C^2}}\sin(\omega t).$$

If the impedance of the capacitor is small in comparison to that of the resistor, nearly all of the voltage in the circuit is across the resistor, and the phase shift is negligible. However, as capacitance increases, there is less voltage across the resistor and the phase delay increases. There is a characteristic frequency at which the reactive capacitance and resistance are equal. To determine the voltage at this frequency, simplify the expression for V_R by dividing through by R and knowing that $X_C = R$:

$$V_R = \frac{R}{\sqrt{R^2 + X_C^2}} V_p \sin(\omega t)$$

$$V_R = \frac{1}{\sqrt{2}} V_p \sin(\omega t)$$

The voltage across the resistor has decreased to

$$V_R = 0.707 \times V_p \sin(\omega t).$$

The phase delay has increased to

$$\phi = \tan^{-1}\left(\frac{X_C}{R}\right) = \tan^{-1}(1) = 45°.$$

The frequency at which $X_C = R$ is called the cutoff frequency (f_c) and is given by

$$R = X_C = \frac{1}{2\pi f_c C};$$

$$f_c = \frac{1}{2\pi RC}.$$

Notice that the frequency has been converted to hertz. The significance of the 0.707 reduction in voltage amplitude is based on the cutoff frequency and will reappear in later discussions in signal processing.

KEY POINT

The cutoff frequency for an AC circuit is the characteristic frequency at which the reactive capacitance and resistance are equal ($X_C = R$). The voltage across the resistor at this frequency has been reduced to $V_R = 0.707 \times V_p \sin(\omega t)$. The cutoff frequency ($f_c$) is given by $f_c = 1/2\pi RC$, where the frequency (f_c) has been converted from radians to hertz.

FOR FURTHER READING

Brown, W.F. 1984. *The physiological and technical basis of electromyography.* Boston: Butterworths.

Dumitru, D. 2000. Physiologic basis of potentials recorded in electromyography. *Muscle & Nerve* 23: 1667-1685.

Geddes, L.A., and L.E. Baker. 1968. *Principles of biomedical instrumentation.* New York: Wiley.

Katz, B. 1966. *Nerve, muscle, and synapse.* New York: McGraw-Hill.

Loeb, G.E., and C. Gans. 1986. *Electromyography for experimentalists.* Chicago: University of Chicago Press.

chapter 3

EMG Instrumentation

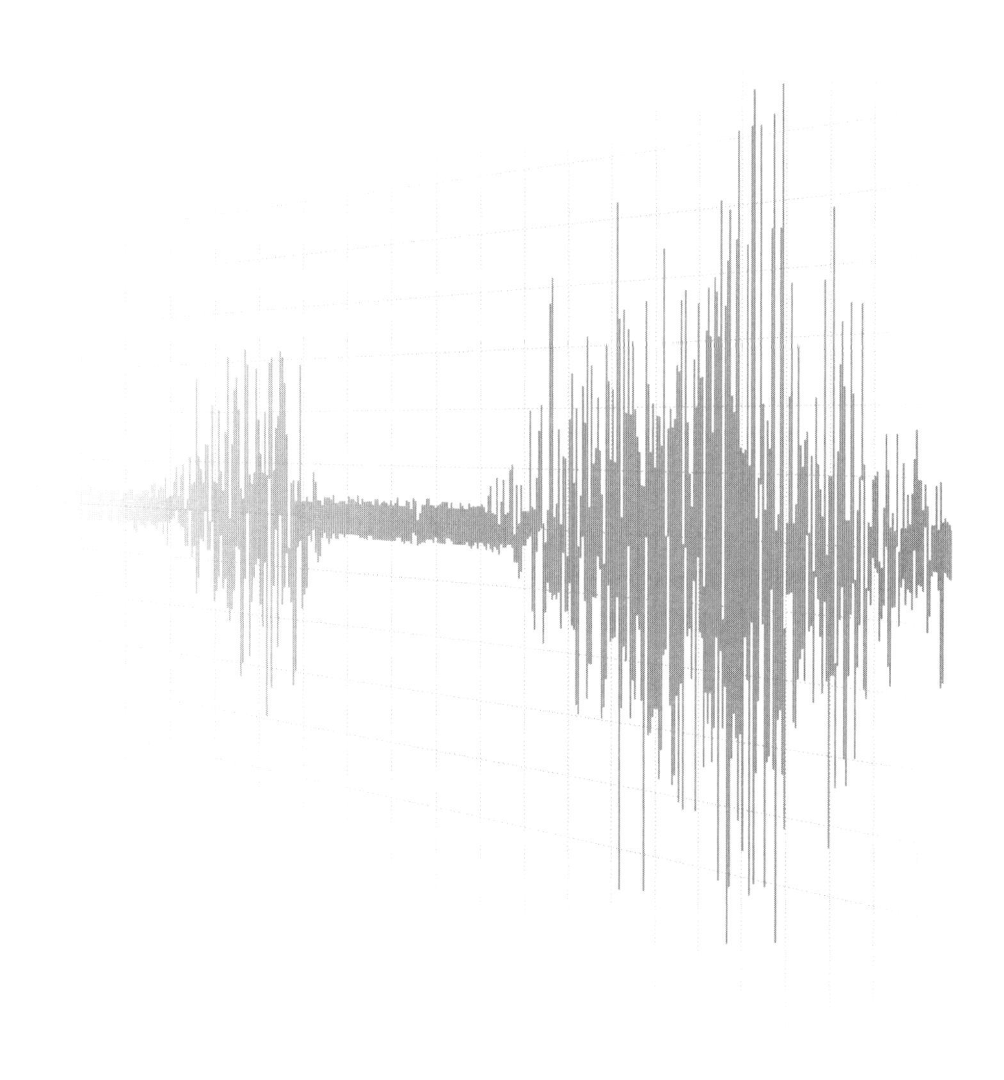

The significant time invested in laying a strong foundation in electricity will be rewarded by a deeper understanding of EMG instrumentation, which is best understood from the perspective of electrical circuits. A typical EMG experimental setup is illustrated in figure 3.1. The EMG signals from the surface electrodes are sent to an amplifier that increases the magnitude of the signal so that it may be digitized with high fidelity by the analog-to-digital conversion board residing in a computer. Additional mechanical measurements may or may not be obtained concurrently. This chapter reviews the basic principles of each of these EMG instrumentation components.

Electrodes

There are two basic types of electrodes: surface and indwelling. Surface electrodes are placed on top of the skin directly over the muscle whereas indwelling electrodes are inserted through the skin directly into the muscle. Both types of electrodes are made of conductive metals, and they perform the same function.

Electrodes convert the electric potential generated by the muscle into an electric signal that is conducted through wires to the amplifier, a process termed **signal transduction.** Most surface electrodes require the use of an electrolyte gel before being applied to the skin. The muscle fiber action potential generates extracellular currents that extend from the membrane to the electrode at the skin surface. As the dipole propagates along the muscle fiber, currents flow through the extracellular fluids

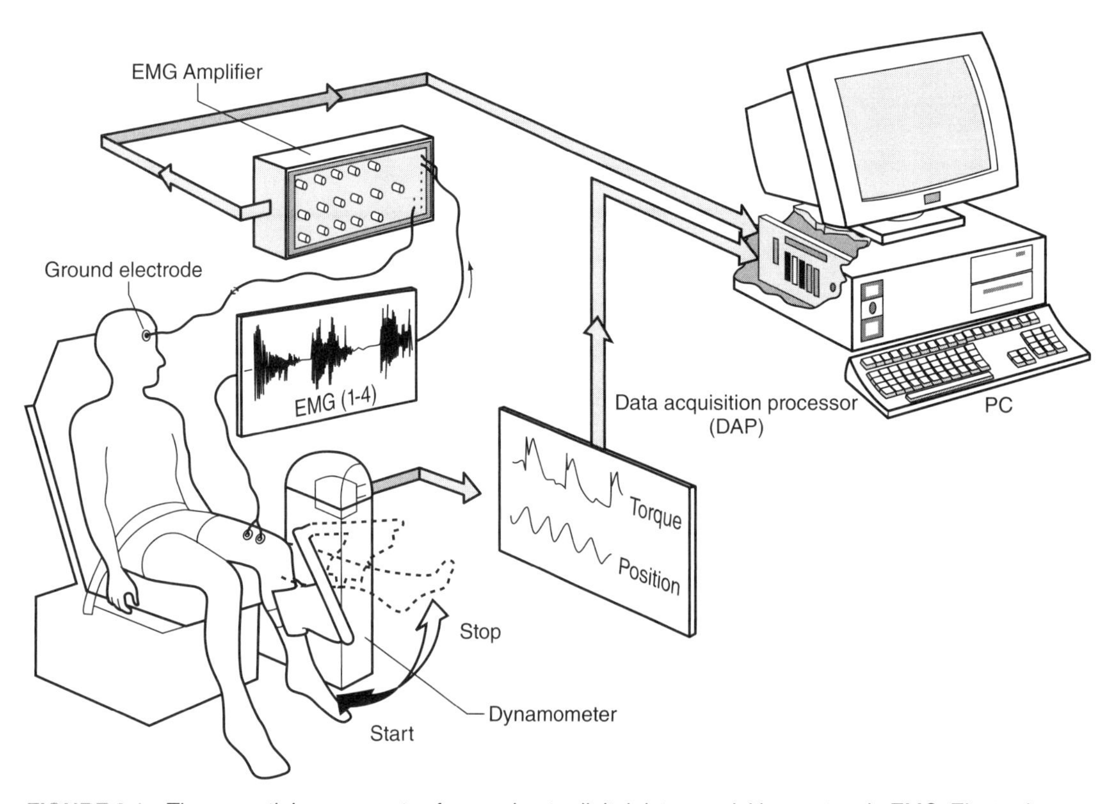

FIGURE 3.1 The essential components of an analog-to-digital data acquisition system in EMG. Electrodes are placed either on the surface of the muscle or into the muscle for indwelling recordings. The EMG signal is fed to an amplifier, which both filters and increases the magnitude of the signal before it is sent to the analog-to-digital conversion board residing in a computer. A mechanical signal, such as force or position, is often recorded concurrently.

Reprinted from *Journal of Electromyography and Kinesiology* 4(3), S. Karlsson, B.E. Erlandson, and B. Gerdle, "A personal computer-based system for real-time analysis of surface EMG signals during static and dynamic contractions," p. 11, copyright 1994, with permission from Elsevier.

and give rise to the potential gradients (this concept was introduced in the previous chapter). The changing potential gradients associated with the propagating dipole result in electrical currents in the electrode leads by capacitive conductance across the *metal–electrolyte interface* at the electrode contacts. The tiny currents in the electrode leads are then detected by the amplifier and increased to a magnitude that is large enough to be recorded. Therefore, the electrode is a device that converts ionic potentials generated by the muscles into electronic potentials that can be measured by an amplifier (Loeb and Gans 1986).

The Electrode-Electrolyte Interface

Surface electrodes are made from conductive materials that can range from plated precious metals (i.e., gold or silver) to simple stainless steel. Before the electrode is applied to the skin, the skin is lightly abraded to remove oils and layers of dead skin that contain only low levels of electrolytes necessary for conduction. An electrolyte gel is then applied to the electrode surface and rubbed into the skin so that it is absorbed into the stratum mucosum to make contact with the derma, where it can serve to decrease the recording resistance (R_s) through the skin (Tam and Webster 1977) (figure 3.2). This is termed the *electrode–electrolyte interface.*

When the metal comes into contact with the electrolyte gel, there are two critical electrochemical events that govern the recording properties of surface electrodes. First, the metal itself attracts ions from the electrolyte gel. The type of ion (positive or negative) that is attracted depends on the electrochemistry specific to the metal and the electrolyte gel. The result is a localized increase in the concentration of one type of ion at the electrode surface. Oppositely charged ions then align themselves relative to the electrode surface but slightly farther away, so that there is a small space within the electrolyte gel, near the electrode surface, that has a neutral charge. Second, there is a tendency for the metal to actually discharge ions into the electrolyte gel, leaving behind an excess of free electrons in the metal. This is the same process as corrosion. The type of ion that is discharged into the

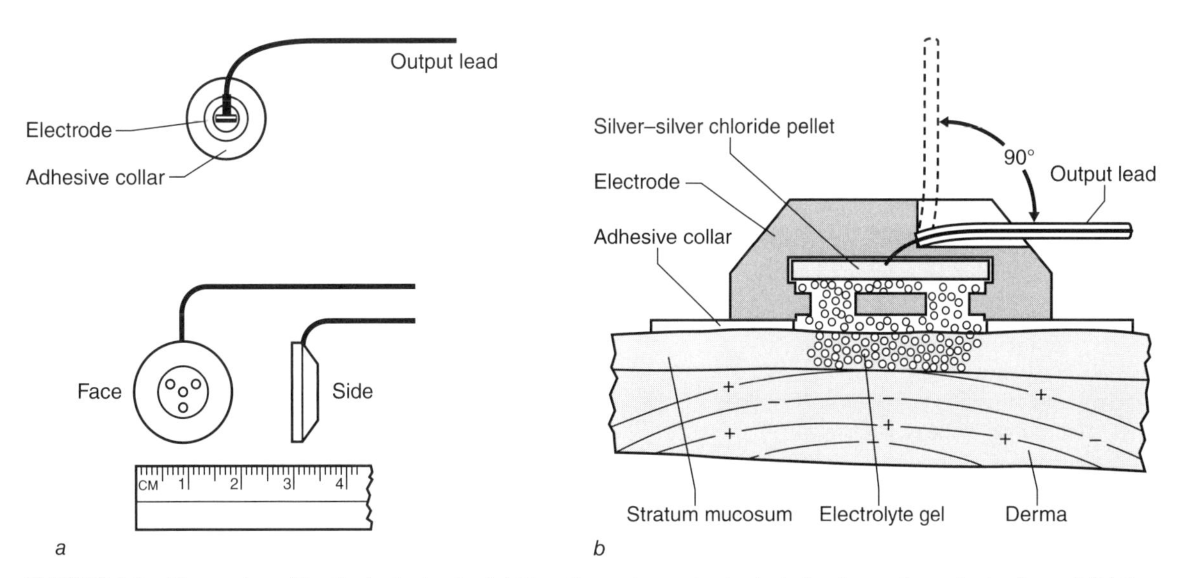

FIGURE 3.2 The surface (floating) electrode. *(a)* The dimensions of a typical circular surface electrode and *(b)* the skin–electrode interface.

electrolyte is the source of the original attraction from the electrolyte to the metal surface. Together, the two electrochemical interactions give rise to a *dipole layer* of charge at the electrode–electrolyte interface that behaves like a capacitor (*C*). The dipole layer is the source of EMG signal input impedance from the muscle to the electrode. It is important to understand that these same events occur with indwelling electrodes. In the case of indwelling electrodes, the electrolyte solution is the tissue fluid (Cooper 1963; Misulis 1989).

The skin, gel, and electrode interfaces function as a complex physical system that alters the EMG signal in deterministic ways. The recording properties for a single electrode can be modeled quite adequately as an equivalent circuit (figure 3.3). The bulk resistance of the electrolyte gel (R_s) is in series with the capacitive effects of the electrolyte dipole layer at the electrode surface (C_e). There is an additional resistor in parallel to denote the resistance of the chemical reaction (activation energies) that moves the charge at the interface (R_f) (Cooper 1963; Misulis 1989). The simplest application of the circuit equivalent can be seen through changes in electrode surface area. A decrease in electrode surface area will result in an increase in resistance (R_s) but a decrease in capacitance (C_e). The overall result is an increase in electrode impedance. The opposite is true for an increase in electrode surface area.

The physical properties of electrodes have serious consequences for the detection of muscle activity because they can induce a frequency-dependent voltage drop, which means that they can change the amplitude and frequency content of the EMG signal. Thus, the electrodes can also act as a filter (Geddes et al. 1967). One can better appreciate the filtering properties of electrodes by understanding that their circuit equivalent is very similar to the analog filters described in "Amplifier Characteristics" (p. 72). Due to the number of commercially available surface electrodes, there is no standardized recording surface geometry. It is good practice to report the electrode geometry from which the reader may calculate the recording surface area, as it has an impact on the observed EMG signal. As long as the surface area is equivalent, there is no functional difference in the electrical recording characteristics between square and circular electrodes (Jonas et al. 1999).

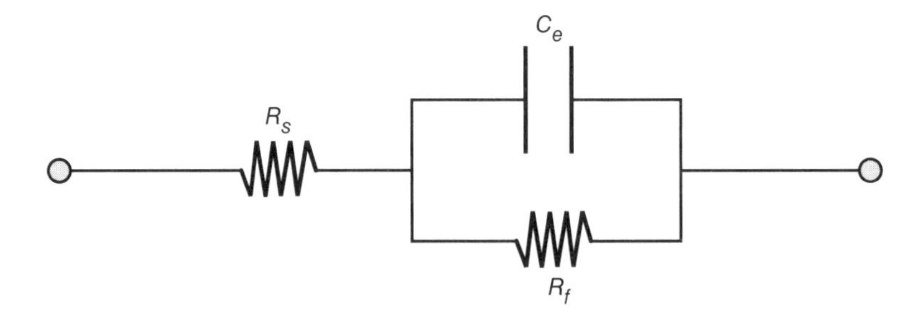

FIGURE 3.3 The equivalent circuit for a single electrode. The equivalent circuit is composed of the bulk resistance of the electrolyte gel (R_s), the capacitive effects of the electrolyte dipole layer at the electrode surface (C_e), and the resistance of the chemical reaction (activation energies) that moves the charge at the interface (R_f). The higher the activation energy required for the chemical reaction to take place, the greater the resistance to signal transduction.

KEY POINT

Electrodes convert the electric potential generated by the muscle into an electric signal that is conducted through wires to the amplifier, a process termed signal transduction. The changing potential gradients associated with the propagating dipole result in electrical currents in the electrode leads by capacitive conductance across the metal–electrolyte interface at the electrode contacts.

Half-Cell Potential

When the electrochemical reaction between the metal and the electrolyte stabilizes (i.e., reaches equilibrium), a potential difference is established by the dipole layer. The electrolyte gel just outside the electrode surface achieves a different potential than the rest of the surrounding medium. The potential difference between the electrolyte at the electrode surface and the surrounding medium is termed the **half-cell potential** (Cooper 1963; Misulis 1989).

The half-cell potential of a single electrode results in a direct current (DC) offset in the biological signal. However, this DC offset should be part of the **common mode signal** and will cancel if two electrodes are used. Anything that destabilizes the electrochemical reaction between the metal and the electrolyte can result in erratic changes in the half-cell potential (Huigen et al. 2002). This would be observed as potential changes originating from the electrode itself (noise) in the absence of any biological signal. Surface electrodes used in EMG are frequently plated with silver–silver chloride (Ag-AgCl). These electrodes are typically used with an electrolyte gel that contains either sodium chloride or potassium chloride (NaCl or KCl, respectively). The electrochemistry between Ag-AgCl metal surface and the electrolyte is highly stable (Cooper 1963; Misulis 1989). There are exceptions: (1) Repeated muscle contractions may change electrolyte gel ion concentrations due to sweating, and (2) electrolyte gel temperature may increase because of muscle metabolic heat production. The electrochemistry between the Ag-AgCl and the electrolyte gel then becomes unstable. These are important factors to consider when interpreting changes in the EMG signal associated with repeated contractions (Bell 1993).

KEY POINT

The electrochemical reaction between the electrolyte gel and the metal surface of the electrode creates a dipole layer wherein the gel just outside the surface of the electrode achieves a potential difference with respect to the surrounding medium.

Electrode Types

Surface and indwelling recordings are the two basic ways to record muscle electrical activity (figure 3.4). The two methodologies are associated with different types of recording electrodes, each with its own advantages and disadvantages. Muscle electrical activity may be recorded using either a monopolar or bipolar electrode configuration for both surface and indwelling methods.

Surface Electrodes

Early surface electrodes were constructed of a simple square or circular conductive metal plate (figure 3.4*a*). A thin layer of electrolyte gel was applied to the plate, and the plate was secured to the skin with tape. Plate electrodes were susceptible to *motion artifact.* Motion artifact is a mechanical disturbance that changes the thickness of the thin film of the electrolyte between the metal plate and the skin during muscle contraction. The charge distribution within the electrolyte gel is momentarily altered until both the half-cell potential and input impedance regain equilibrium (Ödman and Öberg 1982). Simple plate electrodes may be more problematic than other designs, but they can still be used effectively if the electrolyte gel is applied sparingly and then secured properly to the skin surface with adhesive tape.

A more effective way of minimizing motion artifact is through eliminating direct contact between the metal surface and the skin. Thus, a common design feature of most commercial surface electrodes includes a recording surface that is recessed away from the skin. The second major function of electrolyte gel is then to maintain a conductive path between the metal surface and the skin by forming an *electrolyte bridge.* A change in the orientation of the metal surface relative to the skin is not a problem as long as the electrolyte bridge is maintained. This type of electrode is also referred to as a **floating electrode.** The metal recording surface is usually recessed within a plastic casing, and the whole unit is attached to the skin surface with two-sided adhesive collars (Geddes and Baker 1968) (figure 3.2, p. 57). The floating electrode belongs to the general class of so-called **passive electrodes** because there are no additional electronics associated with the unit itself. The electrolyte gel is the only signal transduction mechanism.

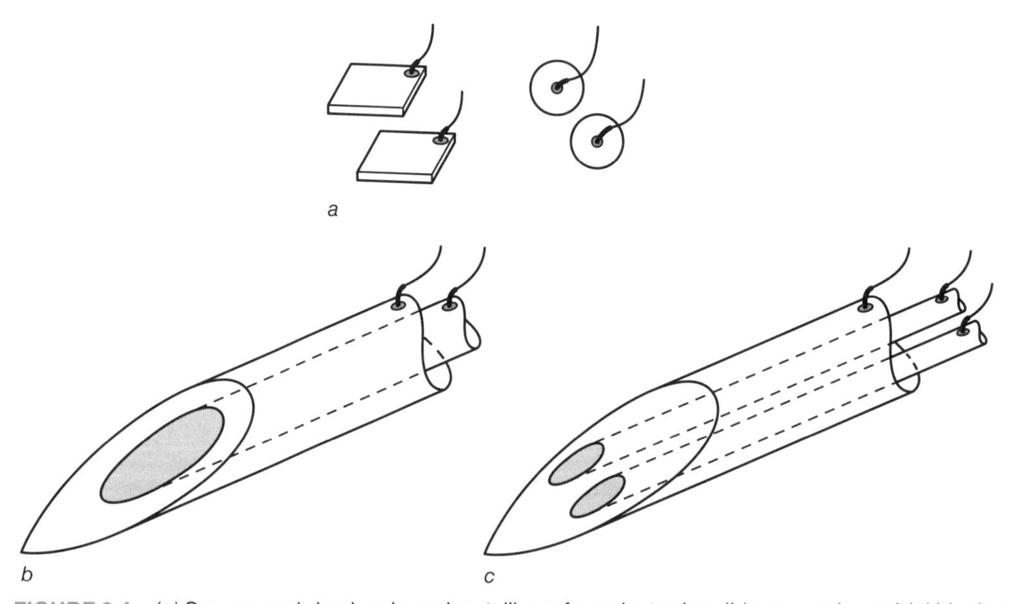

FIGURE 3.4 *(a)* Square- and circular-shaped metallic surface electrodes, *(b)* concentric, and *(c)* bipolar needle electrodes. The wires for the surface electrodes correspond to G1 and G2, with the ground not shown. The central wire for the concentric electrode corresponds to G1, while the wire connected to the cannula is G2, with the ground not shown. The bipolar needle electrode has direct G1 and G2 connections, with the ground connected to the cannula.

Reprinted from J. Goodgold and A. Eberstein, 1978, *Electrodiagnosis of neuromuscular disease,* 2nd ed. (Baltimore: Williams & Wilkins), 55. Used with permission of Dr. Goodgold.

Active electrodes incorporate a preamplifier within the small case that houses the metal recording surface. The metal recording surface then makes direct contact with the skin. The magnitude of the EMG signal is increased "at" the skin surface by a factor of 10 or more before it is transmitted through the electrode leads to the main amplifier unit. As long as the skin is thoroughly cleaned so that the natural electrolytes present in the derma can conduct the signal, electrolyte gel is not necessary to facilitate signal transduction. The complex electrochemical interaction between the metal recording surface and the electrolyte gel is eliminated (Roy et al. 2007). However, the additional advantage of active electrodes is that the resulting EMG signal strength is large in comparison to the surrounding environmental noise (Johnson et al. 1977). Both the size and configuration of the active electrodes are by necessity fixed to accommodate the physical dimensions of the preamplifier unit. Active electrodes are therefore more restrictive than passive electrodes with respect to the size and location of the muscle that can be recorded.

The general advantage of all surface electrodes is that they are noninvasive and easy to apply. Their use is, however, limited to superficial muscles that are large enough to support electrode mounting on the skin surface. It is difficult to isolate the activity of just one muscle using surface EMG detection. The entire limb may be viewed as a volume of conductive tissue. The electrical activity of muscles anywhere within the limb volume may be conducted through the intervening tissue to reach the electrode at some distance away on the skin surface (Dumitru and King 1992). Volume-conducted potentials from unrelated muscles that are "mixed in" with the signal of interest are referred to as **cross-talk** (Farina et al. 2002). Cross-talk is particularly problematic for smaller muscles within a complex mechanical arrangement, such as the forearm (Mogk and Keir 2003).

Indwelling Electrodes

Electrical activity may be recorded using either a single needle or two wires implanted within the muscle. The needle electrode is used within the clinical and research settings wherein patients or participants perform static muscle contractions and the activity of individual motor units is the focus of the electrophysiological study. Wire electrodes are also selective enough to record the activity of individual motor units. However, since the wires may be anchored within the muscle, they are generally used to record the interference pattern of deep muscles not normally accessible by surface electrodes, during dynamic contractions.

It is important to use sterile gloves and to ensure that the insertion site has been carefully sterilized with alcohol or iodine pads to minimize potential infection. It also is common practice to recommend that study participants refrain from aspirin intake 48 h prior to an indwelling electromyographic study to minimize the formation of a hematoma at the insertion site. There are commercially available disposable, sterile needle and fine-wire electrodes. If reusable needles or "homemade" fine-wire electrodes are preferred, sterilization methods must conform to any regional health and safety regulations. There may also be local regulations concerning the qualifications of individuals who perform any procedure with needles. It is important to become familiar with institutional guidelines for exposure to blood-borne pathogens through inadvertent needle piercing and for the safe disposal of used needles. A discussion of these matters is beyond the scope of this book.

Needle Electrodes Needle electrodes are used to detect motor unit action potentials (MUAPs). The size of the needle used (23-28 gauge) depends on the number of recording wires running down the center of the cannula. The wire (usually stainless steel, platinum, nichrome, or silver) is typically between 25 and 100 μm in diameter, and it emerges from the tip of the needle flush with the bevel (15-20°). The recording surface is isolated from the center of the cannula. Concentric electrodes are frequently used for clinical neurodiagnostic recordings (Daube 1991). The term *concentric needle* is used because a top view of the wire tip (recording surface) and the bevel of the needle shows concentric rings (figure 3.4*b*). If a single wire is used for a monopolar configuration, it is the active recording surface (G1), and the cannula is the reference (G2). A separate ground electrode must be placed on the skin and connected to the amplifier. Two wires placed side by side are used for a **bipolar configuration**, but now the cannula functions as the ground (figure 3.4*c*). A more specialized needle exists that incorporates a small side-port window on the cannula from which four wires *(quadrifilar)* are exposed (figure 3.5). The wires are connected to yield three sets of bipolar recordings with the cannula as the ground. Each bipolar recording corresponds to a unique view of the same MUAP to increase the accuracy of identifying individual MUAPs. Motor unit action potentials tend to look similar across one bipolar channel, but no two MUAPs look alike across all three channels. The quadrifilar electrode is an important technological advance because misidentified MUAPs contribute to erroneous results regarding the characteristics of motor unit discharge behavior (Mambrito and De Luca 1984; Kamen et al. 1995; Akaboshi et al. 2000).

The small surface area and short interelectrode distance (50-200 μm) makes needle electrodes ideal for detecting potentials from a very limited volume of tissue (Andreassen and Rosenfalck 1978; Nandedkar et al. 1985). However, the needle must be moved or withdrawn and reinserted multiple times into other compartments to obtain more representative activity of the whole muscle (Podnar 2004; Podnar and Mrkaić 2003).

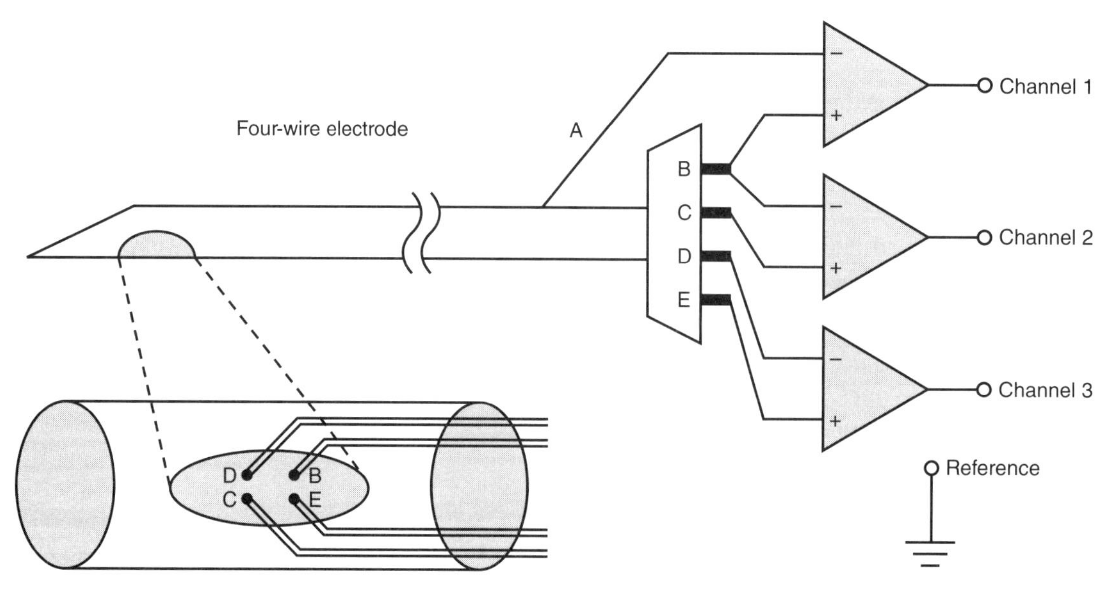

FIGURE 3.5 The quadrifilar needle electrode. Four wires running down the center of the cannula emerge from a side port and are arranged in a square array. The four wires yield three bipolar detection channels. The cannula also serves as the ground.

The needle is generally held by hand to maintain its position, though one may secure it by taping it to the skin. In either case, needle electrodes are susceptible to movement, and caution should be exercised when they are used for dynamic contractions.

Significantly more training is required for the proficient use of needle versus surface electrodes. Although they are ideal for recording the activity of deep muscles, correct placement requires a detailed knowledge of musculoskeletal anatomy. The needle electrode can be coupled with an audio monitor so that the investigator can hear the sounds associated with the electrode recordings. As the needle moves through the fascia there is little or no sound because the recording surfaces are too distant from any active muscle fibers. The needle then meets with slight resistance until a "pop" is felt as it passes through the fascia into the muscle. A light contraction of the target muscle at approximately 10% of maximum effort is helpful for determining the location of the needle relative to a motor unit (MU). The firing of distant motor units will produce a low "thudding" sound. Progression of the needle closer to a motor unit will result in a sharp ticking noise. If the strength of the contraction is increased while the needle remains in the same position, the rate of ticking will increase and distinct rhythmic patterns evolve that indicate the recruitment of additional motor units. The advantage of the needle electrode is that it may be repositioned closer to the motor unit to obtain the highest-quality recordings, with the help of audio feedback (Daube 1991; Barkhaus and Nandedkar 1996; Okajima et al. 2000).

Wire Electrodes Wire electrodes, sometimes called *fine-wire electrodes,* are typically used in a bipolar configuration. An insulated wire (50 μm) is threaded through a 27-gauge hypodermic needle so that a small loop is formed as it emerges from the bevel (figure 3.6). Approximately 4 mm of insulation at the middle of the wire is etched off with a sharp knife prior to its insertion into the needle, or it may be burned off with an alcohol flame or a match after the loop has been formed. Insulation must also be removed from the wire tips that exit the hub so that they can be connected to the amplifier. A spring assembly is quite useful for this purpose (Basmajian et al. 1966).

The loop is then cut and the wires are trimmed to leave a 2 mm recording surface at the most distal ends. Next the wires are staggered and folded back to form hooks that rest on the bevel. These hooks anchor the wires to the muscle when the needle is retracted. It is necessary to stagger the wires by a safety margin greater than the length of the exposed recording surface to avoid contact between the bared tips, which would short-circuit the electrode (Basmajian and Stecko 1963). There is a risk that the wires may still short after being inserted in the muscle. Consequently, each wire may be inserted separately so that they do not short by touching each other. A recording needle may also be coupled with an audio monitor to facilitate insertion of each wire to the correct depth (Gabriel et al. 2004). Following implantation, the wires are taped to the skin surface, allowing for strain relief so that they are not pulled out accidentally during the test session.

Compared to needle electrodes, fine-wire EMG electrodes have both advantages and disadvantages. Once the wires have been implanted, they cannot be reinserted into a different area of the muscle. The wires can be retracted to some degree to obtain a better recording location. Otherwise, a new fine-wire electrode must be inserted. On the other hand, because the hooked wire is embedded in the muscle, it is less susceptible to movement than the needle electrode, which makes it easier to detect muscle activity during dynamic contractions. Fine-wire electrodes are capable of detecting

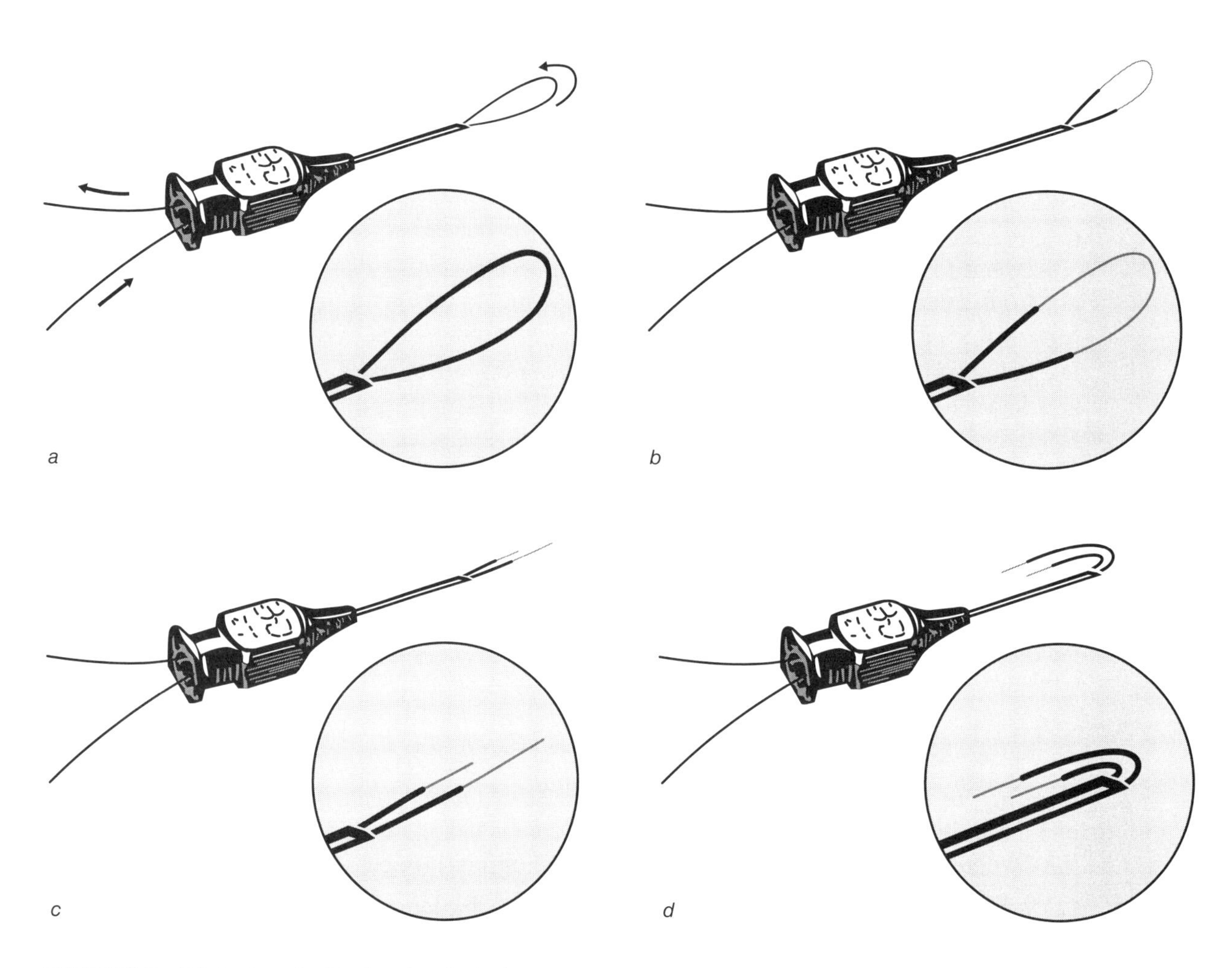

FIGURE 3.6 Fabrication of a bipolar wire electrode.

Reprinted, by permission, from J.V. Basmajian and G. Stecko, 1963, "A new bipolar electrode for electromyography," *Journal of Applied Physiology* 17: 849.

individual MUAPs but not to the same degree as electrodes with a smaller recording area. The larger surface area and greater interelectrode distance between the wires result in less selective recordings.

Very selective recordings allowing accurate identification of individual MUAPs can be obtained with wire electrodes that have a small recording area. One can make these selective electrodes by using a sharp knife to cut the wire at a 90° angle to the wire surface. Thus, the recording area is reduced to the cross-sectional area of the wire (Rich and Cafarelli 2000; Forsman et al. 2001; Westad et al. 2003). However, the most frequent and appropriate use of fine-wire EMG is for recording the interference pattern from deep musculature (An et al. 1983; Funk et al. 1987; Kaufman et al. 1991; Jacobsen et al. 1995). For both fine-wire and needle recordings, it is important to situate the electrode toward the middle of the compartment, as recordings near the border may be contaminated by cross-talk (English and Weeks 1989). The invasiveness of inserting a needle into the muscle, as well as the associated pain, is often cited as a major disadvantage of indwelling electrodes. If a needle is placed near a dense area of motor end plates, the subject will indeed complain of a painful aching feeling within the muscle. However, it is easy to alleviate the pain by redirecting the needle toward another area of the muscle. Discomfort can be minimized with small (0.5-1.0 mm) movements of the needle as it is redirected to obtain the best motor unit recordings (Strommen and Daube 2001).

KEY POINT

There are two types of surface electrodes: (1) Passive electrodes contain no additional electronics associated with the unit itself as the electrolyte gel is the only signal transduction mechanism, while (2) active electrodes also incorporate a preamplifier within the small case that houses the metal recording surface. Indwelling electrodes record muscle electrical activity using either a single needle or two wires implanted within the muscle.

Tissue Filtering

The propagation of electric currents through the muscle tissue is frequency dependent. As the frequency of the signal increases, a rapid decline in the amplitude is detected at the electrode recording surface. Because there is progressive amplitude attenuation of high-frequency signals, muscle tissue is categorized as a **low-pass filter.** That is, muscle tissue allows lower-frequency EMG signals to pass through it "relatively" unchanged, but distorts higher-frequency signals. The relationship between signal frequency and the amount of attenuation is also distance dependent. An electrode placed farther away from active fibers has a lower initial signal amplitude and greater high-frequency attenuation. Muscle tissue is also anisotropic: Impedance to the radial flow of current is roughly five times greater than that for the axial flow of current, in the direction of the muscle fibers. Muscle anisotropy arises primarily from the parallel arrangement of muscle fibers (Lindström and Magnusson 1977; Andreassen and Rosenfalck 1978; Nandedkar et al. 1984; Gielen et al. 1984).

Tissue filtering has practical significance because the amplitude and frequency content of surface EMG signals are greatly reduced compared to those with indwelling signals. However, tissue anisotropy means that the amplitude of indwelling EMG is highly dependent on whether the electrode is oriented parallel or perpendicular to the muscle fibers. Overall, the single greatest factor governing indwelling EMG amplitude is distance from the active fibers (Andreassen and Rosenfalck 1978).

KEY POINT

The signal recorded at the skin surface is lower in amplitude and frequency than that recorded with indwelling electrodes because muscle tissue has low-pass filtering properties.

Electrode Configuration

Electrode configuration for surface EMG refers to the number of recording surfaces and their arrangement relative to muscle, tendon, and bony surfaces. The two most common electrode configurations for surface and indwelling electrodes are monopolar and bipolar arrangements. In both cases, there are two detection surfaces and a ground electrode. More complicated electrode configurations can be viewed as a natural extension of the bipolar case.

Monopolar Recordings

The following three electrodes are used for a **monopolar configuration:** The first electrode is the active recording surface (G1); the second is the reference (G2), which is used to determine a potential difference; and the third is the **ground.** The active electrode (G1) is placed on the muscle; the reference (G2) is placed on an electrically neutral location such as a tendon; and the ground is placed on a bony surface distant to G1 and G2. If evoked potentials are being recorded, the ground is usually between the stimulator and G1. Electrodes in this configuration are referred to as monopolar because only one electrode is used to record muscle activity. The terms G1 and G2 are a holdover from the early days of electrophysiology when they referred to grids 1 and 2 of vacuum tube amplifiers that resulted in positive and negative outputs, respectively (Lagerlund 1996). However, the terms G1 and G2 are still used in clinical EMG laboratories and serve as a standard reference point for describing electrode placement and polarity (Calder et al. 2005).

The muscle contains areas with dense collections of motor end plates. These areas may be identified as focal points on the skin surface where the lowest possible electrical stimulation will produce a minimal muscle twitch. Each identified area is defined as a **motor point** (Walthard and Tchicaloff 1971). The motor point is often confused with the *innervation zone* that arises through the branching of motor end plates within a well-defined area. The innervation zone is a small region or band of muscle tissue wherein MUAPs originate and then propagate bidirectionally toward each tendon. The same is true for motor points, which may also be situated within the innervation zone (Masuda and Sadoyama 1987). Ideally, innervation zones should be electrically identified before the electrodes are applied. An alternative is to use an anatomic reference chart that describes the location of the motor points (Walthard and Tchicaloff 1971).

The compound muscle action potential (CMAP) is preferentially recorded using a monopolar configuration with the active electrode directly on the electrically identified motor point, for two reasons. If the CMAP latency is required for calculating motor nerve conduction velocity, then the motor point will be the first site of depolarization. The CMAP latency recorded at any position other than the motor point will include the time that it takes for action potentials to travel along the muscle fibers to reach the electrode. Measures of CMAP area, amplitude, and duration are also used to track the progression of neuromuscular disorders; and the true shape profile of the signal is altered as it travels down the muscle fibers to reach the electrode. There are two probable mechanisms: (1) a normally leaky muscle membrane and (2) a decrease in muscle fiber diameter toward the tendon (Kleinpenning et al. 1990). Monopolar recordings with the G1 electrode placed over the motor point allow for high-quality evoked potentials free of distortion. The main disadvantage of the monopolar configuration is that it does not take full advantage of the differential amplifier design to reduce unwanted noise in the EMG recordings. Great care must therefore be taken to ensure that the testing room is isolated and relatively free of unwanted electrical interference.

It is important to keep in mind when reading the literature that the polarity of the recorded phases of the CMAP depends on whether the G1 or the G2 electrode was placed over the innervation zone. If G1 is placed over the innervation zone, the negative (depolarization) phase will appear below the isoelectric (zero) baseline. The polarity will be reversed if G2 is placed on the innervation zone. Adding to the confusion, the conventional way to display waveforms in clinical electrophysiology is for the negative portion of the signal to be displayed above the isoelectric baseline.

Bipolar Recordings

Bipolar recordings are defined similarly for surface and indwelling EMG. A bipolar configuration has both G1 and G2 electrodes placed over the muscle. Signals from the G1 and G2 electrodes are fed to an amplifier that inverts the G2 input. The ground is placed on a neutral site such as a bony prominence, typically near G1 and G2. This basic setup takes full advantage of amplifier circuitry that is designed to minimize unwanted interference signals from electromagnetic fields in the surrounding environment. The amplifier achieves this by subtracting G2 from G1. The details are presented in the section on amplifiers. In this section we review the fundamental recording characteristics of bipolar electrodes.

 **KEY POINT**

Both surface and indwelling electrodes may be arranged in either a monopolar or bipolar configuration. There is one active detection surface for a monopolar configuration, and there are two active detection surfaces for a bipolar configuration.

Interelectrode Distance

The following discussion assumes that G1 and G2 are defined similarly for indwelling and surface electrodes. It is also assumed that the muscle fiber action potential (MFAP) and MUAP recorded by an indwelling electrode have the same basic shape as the CMAP recorded by a surface electrode, except that the latter is greater in amplitude and longer in duration. The interelectrode distance for bipolar indwelling and surface electrodes is an important consideration as it affects both the amplitude and frequency content of the EMG signal.

Figure 3.7 illustrates two different interelectrode distances and the resulting EMG signals. Recall that the MFAP, MUAP, or CMAP may be represented as a traveling dipole. The dipole passes beneath G1 first, and the negative (depolarization) phase

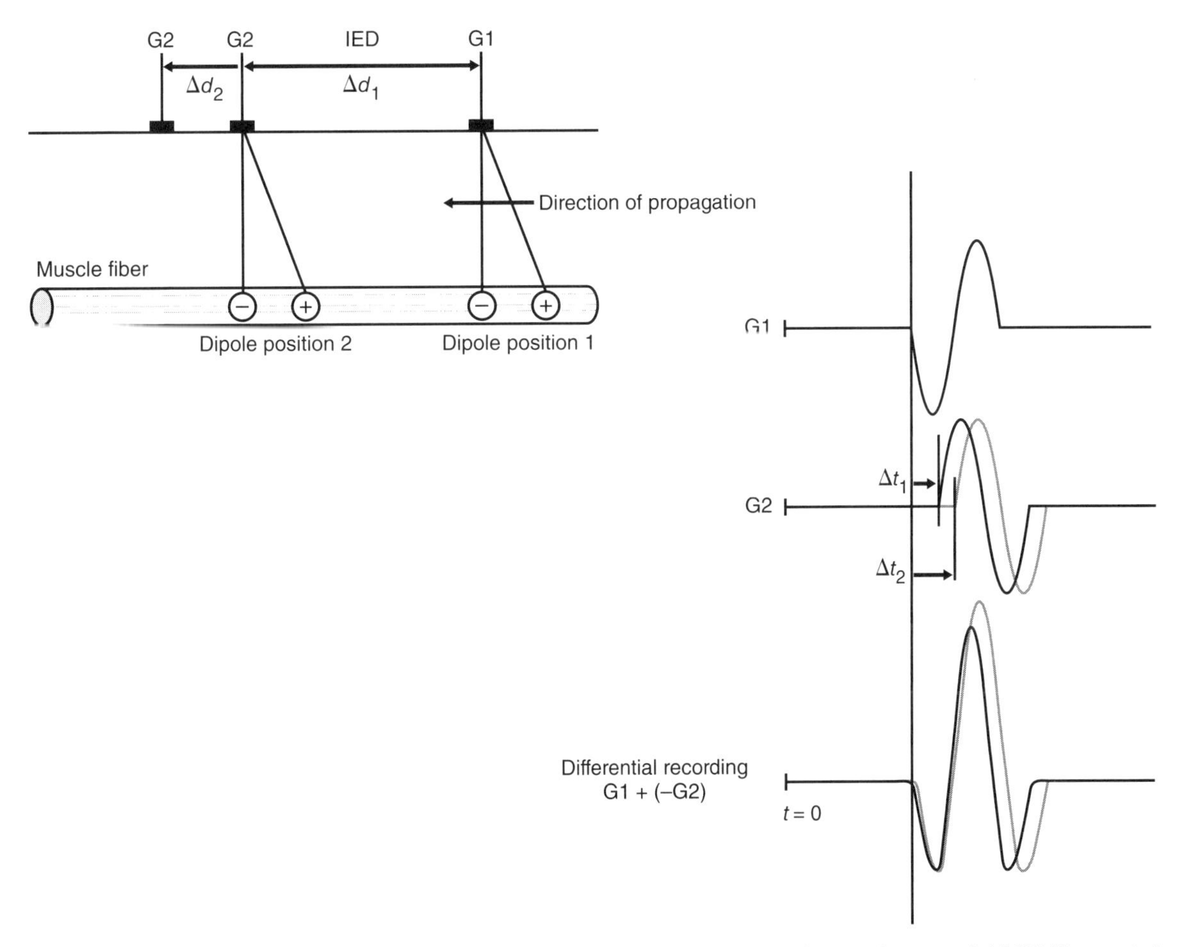

FIGURE 3.7 The effect of increased interelectrode distance (IED) on the muscle fiber action potential (MFAP) recorded by bipolar electrodes, G1 and G2. As the interelectrode distance increases, the time that it takes for the MFAP to travel from G1 to G2 increases. The resulting differential recording [G1 + (–G2)] is a MFAP that is longer in duration and greater in amplitude.

is below baseline. It then propagates down the muscle fiber toward G2 where the signal is inverted. The signal that passes beneath both electrodes is biphasic, but their summation [G1 + (–G2)] is triphasic. Bipolar recordings introduce additional phases that translate to higher-frequency components in the EMG signal than would be obtained with monopolar recordings. Repositioning G2 farther away with respect to G1 will force the dipole to travel a greater distance $(\Delta d_2 > \Delta d_1)$. Conduction velocity remains constant, so it will take longer for the dipole to arrive at G2 $(\Delta t_2 > \Delta t_1)$. The summation of potentials at G1 and G2 still has three phases, but now the EMG signal is longer in duration and greater in amplitude. Longer-duration potentials result in lower-frequency components in the EMG signal. It should be evident that the opposite is true for shorter interelectrode distances: They result in smaller-amplitude EMG signals with higher-frequency components. Because the amplitude and frequency content of EMG signals are altered by interelectrode distance, bipolar electrodes function as a **spatial filter** (Lynn et al. 1978). Evoked potentials are preferentially recorded using a monopolar configuration on the motor point to avoid distortion associated with spatial filtering due to bipolar electrodes (Tucker and Türker 2007).

Consider once again a propagating dipole (figure 3.8). It should be understood at this point that the propagating dipole applies to the MFAP, MUAP, and CMAP but that the dipole spacing is wider for the latter two. At a certain conduction velocity (v), two successive crests of the action potential, equal to one wavelength (λ), will be underneath both electrodes simultaneously (top fiber). The summation between G1 and G2 will result in wave cancellation. When the wavelength is equal to the interelectrode distance $(\lambda = d)$, the frequency (f) that is canceled in the EMG signal is $f = v / d$. Frequencies that involve a simple multiple $(n = 1, 2, \text{ and } 3)$ of the con-

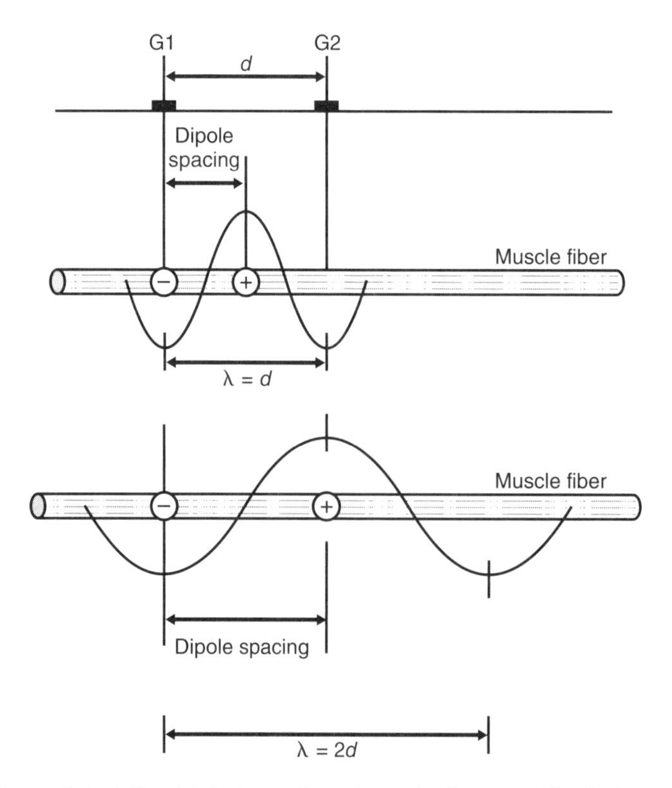

FIGURE 3.8 The spatial relationship between interelectrode distance (d), dipole spacing, and wavelength (λ) of the muscle fiber action potential.

duction velocity are also canceled. When the dipole spacing is equal to the interelectrode distance (λd), the negative and positive phases are centered underneath G1 and G2, respectively (bottom fiber). The result is perfect addition of the two phases. In this case, the wavelength is equal to twice the interelectrode distance ($\lambda = 2d$). The frequency passed by the electrodes is $f = v / 2d$. Frequencies that involve an odd multiple ($n = 1, 3, 5...$) of the conduction velocity are present in the EMG signal. The relationship $f = v / 2d$ always simplifies algebraically to $\lambda = d$ for frequencies that are an even multiple ($n = 2, 4, 6...$) of the conduction velocity, which results in cancellation. Noninteger values experience only partial attenuation.

The practical impact is that a bipolar detection system acts like a **comb filter** by allowing some frequencies in the EMG signal to pass, but not others. If the interelectrode distance and muscle conduction velocity are known, frequencies present in the EMG signal can be calculated (Lindström and Magnusson 1977). Given the important relationship between interelectrode distance and the amplitude and frequency content of the surface EMG signal, caution must be applied to its interpretation: "Successful interpretation of the physiological properties of the EMG signal depends on the separation of the physiological properties and filtering/contamination influences on the EMG signal" (Sinderby et al. 1996, p. 290).

Selectivity

Selectivity refers to the ability to record meaningful muscle activity from a local volume of tissue rather than cross-talk from neighboring muscle fibers. Surface electrodes are at an initial disadvantage, but the situation can be improved. The dipole spacing for muscle fibers is constant. However, the *effective* dipole spacing depends on the geometry of the radial distance between the muscle fiber and the detection surface. The dipole spacing for more distant fibers *appears* to be greater than that of closer fibers (see figure 2.4, p. 24). If the interelectrode distance is kept as short as possible, it will match the dipole spacing of muscle fibers closest to the detection surfaces, while the amplitude and frequency contribution from more distant fibers will be attenuated. Recall that the CMAP recorded from the skin surface is greater in magnitude and longer in duration. The CMAP duration is also related to its wavelength (λ) and can be used to determine the optimal interelectrode distance for surface electrodes, as it varies according to the size of the muscle. The interelectrode distance may range from 0.5 cm for the thenar to 1 cm for the biceps brachii. One way to check that the interelectrode distance is sufficient is to evoke a CMAP. If the classic triphasic CMAP shape appears free of distortion, other than the normal spatial filtering associated with bipolar recordings, then the interelectrode distance is sufficient.

At the other extreme, the typical interelectrode distance for needle electrodes (50-200 μm) ensures that selectivity is not a problem for this indwelling recording technique because it is appropriate for the much shorter wavelength (λ) of MUAPs (Andreassen and Rosenfalck 1978). It is very difficult to control for interelectrode distance of wire electrodes, which is the main reason they do not offer the same degree of selectivity as needle electrodes. There are more elaborate fabrication methods that involve braiding the wires and securing them with medical-grade epoxy so that the interelectrode distance is fixed at the wire radius (25-50 μm). However, it is not easy to cut the wires with a razor to expose a regularly shaped detection surface.

Interelectrode distance is the main factor affecting local selectivity for surface EMG recordings. Selectivity cannot be improved by decreasing electrode surface area, which only increases impedance and results in greater noise contamination in the signal. A general rule of thumb is that electrodes can detect "meaningful" electrical activity from a spherical volume of muscle tissue having a radius equal to the interelectrode distance (Lynn et al. 1978). This is referred to as the electrode **pickup area** or **detection volume.** For a fixed interelectrode distance, the detection volume is the same whether the muscle is large or small. However, for a smaller muscle, a given detection volume is a greater percentage of the total muscle volume. Thus, surface EMG recordings are more representative for smaller muscles than for larger ones.

It is tempting to increase the interelectrode distance to increase the detection volume for a larger muscle. An increase in the interelectrode distance from 2 to 4 cm can increase the amplitude and decrease the frequency content of the surface EMG signal (Beck et al. 2005). However, the difference observed in the EMG signal produced by an increase in interelectrode distance is consistent with the filtering function of bipolar electrodes. Increasing the interelectrode distance does not necessarily mean that the electrodes will record the electrical activity from deeper muscles (Fuglevand et al. 1992; Elfving et al. 2002). The detection volume of surface electrodes includes only the most superficial fibers of the muscle. These muscle fibers belong to larger, higher-threshold motor units (Knight and Kamen 2005). The surface EMG signal may therefore represent a biased view of the overall muscle activity.

Interelectrode distance is an issue for both surface and indwelling electrodes, as it affects the selectivity of the recordings as well as the amplitude and frequency content of the signal. The smaller the interelectrode distance, the more selective the recordings, because the electrodes record from a smaller volume of tissue.

Considerations for Electrode Placement

Interelectrode distance for surface recordings usually ranges between 5 and 20 mm depending on the size of the muscle. Shorter interelectrode distances are possible, but there is a practical limit associated with a greater risk that the electrolyte gel will form a **salt bridge** between the two recording surfaces across the skin. This will reduce the potential difference between the two electrodes, and the observed EMG amplitude will be much lower than expected. It is important to remove excess electrolyte gel.

In contrast to placement with the monopolar configuration, bipolar surface electrodes should *not* be placed within the innervation zone. Muscle fiber action potentials originating from the innervation zone travel bidirectionally, toward the tendon at either end of the muscle. If G1 and G2 were to straddle the innervation zone, both electrodes would "see" the same potential, and cancellation similar to that previously described would occur. Realistically, the cancellation would not be perfect. The signal would be lower in amplitude and would contain higher-frequency components because the summation between G1 and G2 would result in more irregular peaks.

For static contractions, the electrodes should be placed 20 mm away from the innervation zone to minimize the effects of **temporal dispersion.** Temporal dispersion is a function of (1) conduction velocity, which is dependent on fiber diameter; (2) the dispersion of motor end plates; and (3) variability in the timing of activation between fibers within a motor unit. These spatial differences between fibers within the same motor unit result in the dispersion of their individual potentials as they propagate toward the tendon. Since dispersion increases with conduction time, the effect is called temporal dispersion. The situation is analogous to the 100 m sprint wherein runners appear to be even initially, early off the blocks, then spread out during the progression of the race as the faster athletes pull ahead.

Placing the electrodes near—*but not over*—the innervation zone will summate the MFAPs when temporal dispersion is minimal and will produce the greatest EMG magnitude. The summation of potentials that are more temporally dispersed will result in a signal that is lower in amplitude and longer in duration, containing more low-frequency components. Temporal dispersion has low-pass filtering effects on the MUAP. In contrast, if the electrodes are placed too close to the tendon, the end effects of potentials near the tendon will contribute high-frequency components to the EMG signal (see appendix 3.1) (Lateva et al. 1996; Dimitrova et al. 2001).

The recommendation to keep the electrodes 20 mm from the innervation zone is based on the observation that this distance corresponds to the point at which the estimates of muscle fiber conduction velocity and the frequency content of the EMG signal become more stable (Li and Sakamoto 1996a, 1996b; Sakamoto and Li 1997). For dynamic contractions, the electrodes should be placed halfway between the innervation zone and the tendon. The midpoint location represents a balance between two competing factors: (1) the need to compensate for muscle shortening, which can bring the innervation zone closer to the electrode, and (2) the necessity to avoid the increased contribution from potentials generated by the muscle fiber–tendon end effects (see appendix 3.1) (Schulte et al. 2004; Martin and MacIsaac 2006).

Zipp (1982) detailed recommended electrode locations for various muscles based on anatomical landmarks. One should employ this method in order to standardize electrode positions across subjects participating in the same study. Cram and colleagues (1998) have presented detailed information about surface electrode placement, as well as behavioral tasks used to verify their location. The electrode placements are given for commonly studied muscles and for those of clinical interest. The authors also include potential sources of cross-talk and artifacts—information that is extremely valuable. More recent recommendations for best practices in EMG methodology are provided by the Surface EMG for the Non-Invasive Assessment of Muscles (SENIAM) project. The main results of the SENIAM project were published by Hermens and colleagues (2000) in the *Journal of Electromyography and Kinesiology*. The full report includes electrode placements for 27 muscles and can be obtained on CD-ROM through the SENIAM Web site (www.seniam.org). Perotto and colleagues (2005) have provided a text that is specific to indwelling recording and has the advantage of presenting detailed anatomical landmarks and distance referencing for precise electrode placement.

KEY POINTS

- A motor point is anatomically defined as an area of the muscle with a dense collection of motor end plates and may be determined electrically with very minimal electrical stimulation. Innervation zones are broader areas of the muscle associated with the branching of end plates away from the motor point.
- Compound muscle action potentials are preferentially recorded with a monopolar electrode configuration directly over the motor point. For kinesiological applications, bipolar electrodes should be placed approximately 2 cm away from the motor point.

Amplifier Characteristics

The range of EMG values reported in the literature can vary widely depending on the type of contraction, the size of the muscle, and other methodological and technical differences between studies. Maximal isometric contractions can generate peak-to-peak (P-P) amplitudes of 5 mV for surface EMG. Indwelling EMG is not attenuated by tissue filtering to the same degree as surface recordings and can reach a maximum of 10 mV. The largest P-P amplitudes (30 mV) are associated with evoked potentials because they do not have the wave cancellation associated with voluntary contractions. The main point is that these voltages are still relatively small and that special instrumentation is needed to record them (Winter 2005). The essential components of an amplifier that one needs to know about in order to understand its function are the (1) differential gain, (2) input impedance, (3) common mode rejection ratio, and (4) frequency response of the amplifier relative to the acquired signals.

Differential Gain

The basic function of the amplifier is to increase the magnitude of the signal so that it can be displayed on an oscilloscope or sent to a computer for analog-to-digital (A/D) conversion with a high level of fidelity. Amplifiers are more formally known as *operational amplifiers.* This term originates from a time when analog circuits were used to perform mathematical operations. Of particular interest is the *summing unit* depicted in figure 3.9*a*. In this figure, the G1 and G2 signal inputs are being summed. Input 1 (+) is termed the *noninverting input*, and input 2 (–) is the *inverting input*. The noninverting input is in phase with its output whereas a 180° phase shift exists for the inverting input. The voltage output is therefore proportional to the difference between the two input voltages:

$$V_o = A(G1 - G2),$$

where the multiplier (A) ranges from 10 to 10^6 times, depending on the original magnitude.

The idea of a *differential amplifier* might seem counterintuitive at first. If both electrodes (G1 and G2) are placed on the muscle and receive the same signals, the sum of the inputs should be zero. However, it is important to remember that action potentials must propagate along the muscle fibers. If G1 and G2 are placed in a bipolar configuration on the muscle, the signal that passes underneath G1 will appear about 2 ms later at G2. The exact latency depends on muscle fiber conduction velocity and interelectrode distance (IED). The point is that G1 and G2 do not detect the exact same biological signal at the exact same time. A signal that is present in both electrodes simultaneously is termed *common mode.* If a common mode signal is present, it is considered noise.

For example, the human body serves as an "antenna" for electromagnetic radiation present in the environment. This is due to the *capacitive coupling* between the amplifier and input leads and any electromagnetic radiation in the vicinity. *Electrostatic induction* of power-line energy into the body from nearby wires or testing equipment is the most prevalent source of electromagnetic radiation. This is the same phenomenon as the increase in strength of the radio signal as you approach the tuner. Power-line noise is present in both electrodes simultaneously, and it is easily observed as the line

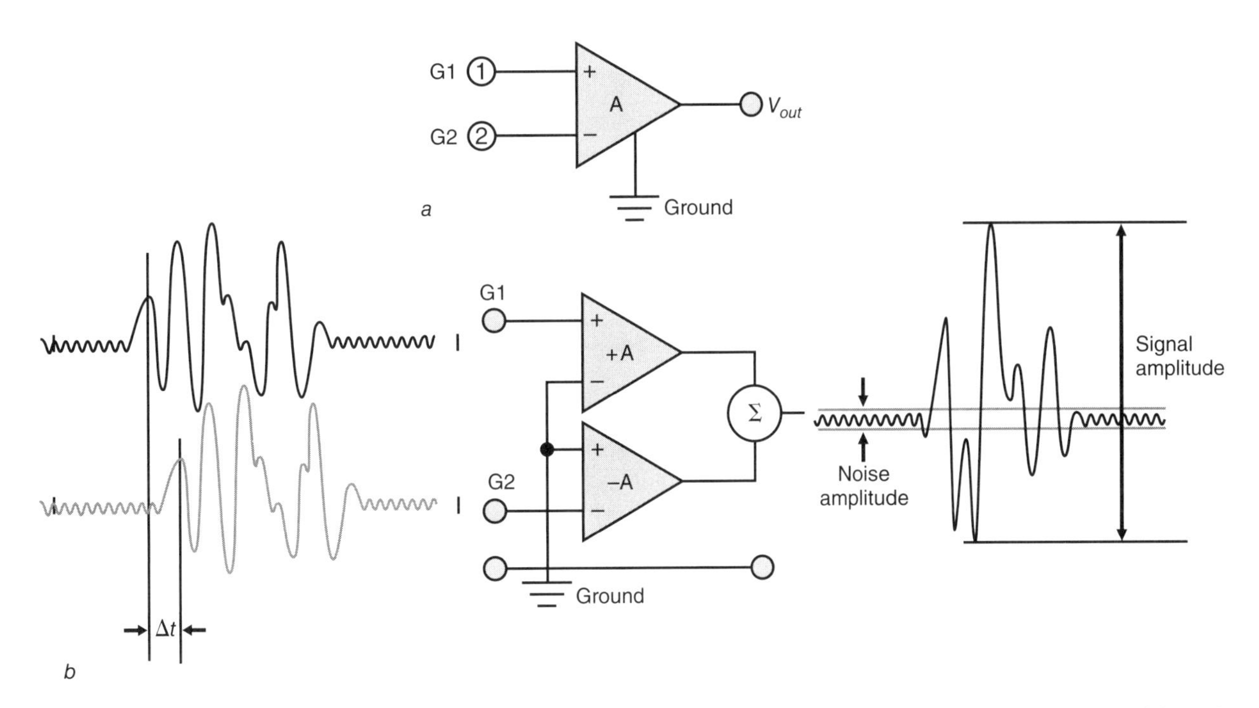

FIGURE 3.9 *(a)* The G1 and G2 wires are fed to a summing unit. Input 1 (+) is termed the noninverting unit, and input 2 (−) is the inverting unit. The letter "A" indicates the basic amplifier unit. *(b)* The basic amplifier is really two separate summing units linked to a common ground and output.

frequency component in the baseline of the surface EMG (sEMG) signal when the muscle is relaxed. In North America, this would be recorded as 60 Hz noise (Clancy et al. 2002).

The main purpose of the differential amplifier is to subtract out the common mode (noise signal) and amplify the difference (biological signal). The differential amplifier can be conceived of as two separate amplifiers linked to a common ground and output (figure 3.9*b*). First, notice that 60 Hz noise (common mode) is present in the baseline of both G1 and G2; it has the same magnitude and phase. The sEMG signal at G2 is delayed with respect to G1 by the time (Δt) taken for it to propagate along the muscle between the two recording surfaces. The G2 input is inverted so that the positive and negative components of the common mode cancel each other, leaving behind only the difference signal that is biological in origin. The difference signal is then multiplied by some magnitude set by the amplifier:

$$V_o = A\left[(G1 + \text{noise}) - (G2 + \text{noise})\right]$$

$$V_o = A(G1 - G2)$$

There is still 60 Hz power-line noise left in the baseline of the output, but its amplitude is greatly reduced. Because of natural differences in electrode–skin input impedances between G1 and G2, the common mode signal will not be identical at the two input terminals of the amplifier. Differences in the common mode also exist because it is impossible to build identical amplifiers. The resulting subtraction is never perfect in reality. A key performance criterion for differential amplifiers is how well they can actually subtract out the common mode signal; this is referred to as the *common mode rejection ratio* (CMRR). To determine the CMRR, a test signal is passed through only one of the two input terminals. The test signal will be amplified by some gain but without any subtraction. The increase in signal amplitude is called the *differential gain.* If the same test signal is passed through both amplifier inputs, it should be reduced in the output. The difference in amplitude is called the *common mode gain.* The CMRR is the ratio of the differential gain to the common mode gain. The minimum specifications for EMG amplifiers range from 10,000:1 to 100,000:1 (80-100 dB).

KEY POINTS

- The voltages involved in EMG are relatively small; and special instrumentation, the amplifier, is required to record them. The basic function of the amplifier is to increase the magnitude of the signal so that it can be displayed on an oscilloscope or sent to a computer for A/D conversion with a high level of fidelity.
- A single amplifier will receive the inputs from two recording electrodes, in a monopolar or bipolar configuration, plus a ground. Because the muscle action potential travels between electrodes, the amplifier magnifies the difference between the two recording surfaces.
- Electrical interference usually presents itself simultaneously in both electrodes and is termed common mode signal. The amplifier is designed to reject or minimize common mode signals to decrease noise interference.

Input Impedance

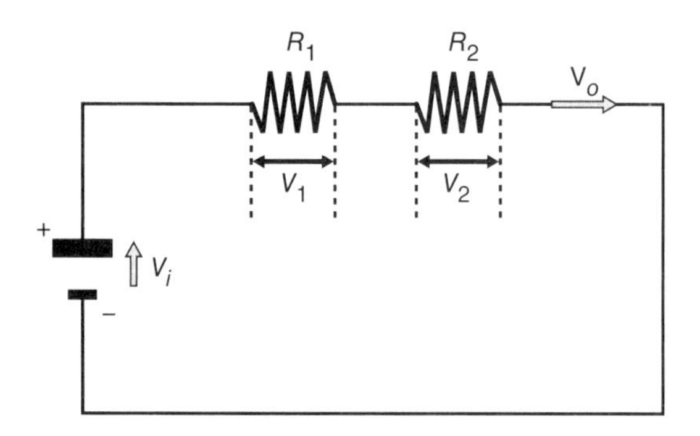

FIGURE 3.10 A circuit consisting of two resistors (R) in series illustrates the relationship between input voltage (V_i) and output voltage (V_o) using Kirchhoff's loop law.

Recall that impedance (Z) is a frequency-dependent form of resistance to the flow of alternating current. The total magnitude depends on both the resistive and capacitive elements within the circuit. Since the circuit under consideration in this section contains only resistors, "R" is still used to denote pure alternating current resistance.

High *input impedance* is another important characteristic for an amplifier. This also seems counterintuitive. If the amplifier is designed to measure very small signals, it would make sense that it should have as little resistance to their input as possible. The series circuit is fundamental for understanding why the opposite is true (figure 3.10). *Kirchhoff's loop law* states that the voltage drops around any closed loop circuit sum to zero. The voltage drop in any closed loop is therefore equal to the voltage rise in that same loop. It is the voltage drop that is actually measured:

$$V = V_1 + V_2$$

$$V = iR_1 + iR_2$$

Since the same current flows through R_1 and R_2,

$$V = i(R_1 + R_2);$$

$$i = \frac{V}{R_1 + R_2}.$$

Substituting for i into the formula $V_2 = iR_2$,

$$V_2 = R_2\left(\frac{V}{R_1 + R_2}\right);$$

$$V_2 = V\left(\frac{R_2}{R_1 + R_2}\right).$$

The same algebraic manipulations may be made for V_1. The voltage V_2 was, however, derived to illustrate that V represents the voltage going into the circuit (V_i) and V_2 is the voltage leaving the circuit (V_o). The following equation then shows that the decrease in output voltage is proportional to the voltage drop across each resistor:

$$V_o = V_i\left(\frac{R_2}{R_1 + R_2}\right)$$

We reviewed this circuit theory because the amplifier and muscle form a circuit when connected by electrodes and their associated wire connectors. The amplifier unfortunately draws current into the circuit by virtue of being connected to the two points across which the voltage is being measured. This decreases the potential difference between the recording electrodes, and ultimately the voltage recorded by the amplifier is less than the actual magnitude. The effect is formally known as *loading down* the circuit. Input impedance therefore refers to the resistance at the input terminals, which determines how much current the amplifier will draw from the voltage source. Consider an amplifier that has an input impedance (resistance) of 10 $k\Omega$ connected to a 1 mV source. Using *Ohm's relationship,* the amplifier draws

$$i = \frac{V}{R} = \frac{1\times10^{-3}\text{ V}}{1\times10^{4}\ \Omega} = 1\times10^{-7}\text{ A}.$$

If the input impedance is increased to 10 $M\Omega$ ($10^6\ \Omega$), the amplifier will draw only 1×10^{-9} A. High input impedance is therefore critical because the muscle as a voltage source has very little current capability.

To complicate matters further, *a certain amount of voltage is lost across the electrode due to its own intrinsic impedance properties.* The original magnitude of electrical activity generated by the muscle is therefore reduced even before it reaches the amplifier. Together, the electrode input impedance (R_e) and the amplifier input impedance (R_i) form a simple series circuit governed by Kirchhoff's law. The circuit equivalent is depicted in figure 3.11 to allow study of the situation. Remember that each electrode is essentially linked with its own amplifier subunit, so there are two series circuits, one for G1 (dotted box) and one for G2.

Consider a source of V = 2 mV from muscle electrical activity and a *skin–electrode input impedance* of $R_1 = 10\ k\Omega$ in series with an amplifier that has an input impedance of $R_2 = 10\ k\Omega$. From the circuit theory presented earlier, the amplifier will draw a current of

$$i = \frac{V}{R_1 + R_2};$$

$$i = \frac{2\times10^{-3}\text{ V}}{\left(1\times10^{4}\ \Omega\right)+\left(1\times10^{4}\ \Omega\right)} = 1\times10^{-7}\text{ A}.$$

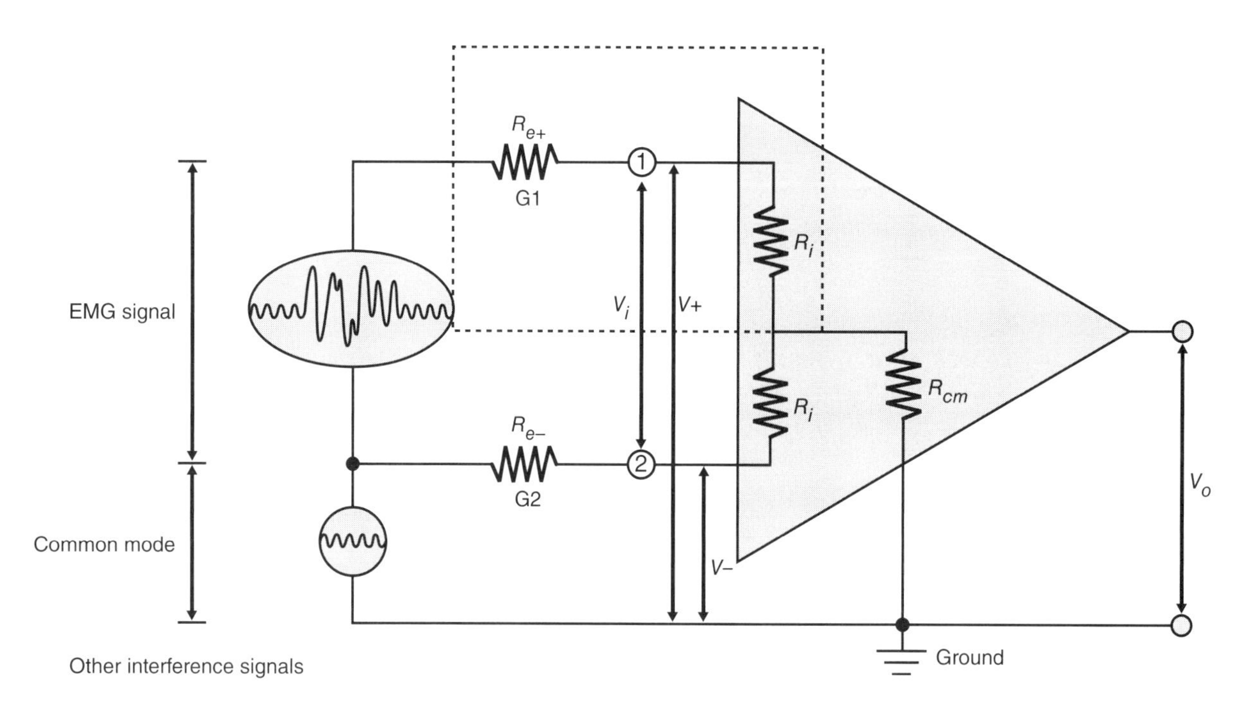

FIGURE 3.11 A schematic circuit of the differential amplifier is represented as two amplifiers with separate inputs (G1 and G2). The G1 and G2 electrodes are associated with skin–electrode input impedances R_{e+} and R_{e-}, respectively. The amplifiers' input impedance (R_i) is the internal resistance to each input. There is then the common mode resistance (R_{cm}) in contact with the ground. The dotted box outlines one of the two circuits that include R_e and R_i in series to illustrate how Kirchhoff's law works in impedance control.

The voltage drop across the amplifier is

$$V = iR_2;$$

$$V = \left(1\times10^{-7}\text{ A}\right)\times\left(1\times10^{4}\ \Omega\right) = 1\times10^{-3}\text{ V}.$$

Since the original signal is $2\text{ mV}(2\times10^{-3}\text{ V})$ and the voltage drop across the amplifier is $1\text{ mV}(1\times10^{-3}\text{ V})$, the remaining voltage drop across the electrodes is $2\text{ mV} - 1\text{ mV} = 1\text{ mV}$. The error introduced by use of an amplifier and electrodes with equal input impedances is

$$\frac{2\text{ mV} - 1\text{ mV}}{2\text{ mV}}\times 100 = 50\%.$$

Since the electrodes and amplifier form a series circuit, their input impedances must be considered *relative to each other*. The goal is to have most of the voltage drop occurring across the amplifier (where it is measured), relative to the electrodes. One can do this by decreasing the electrode input impedance through better skin preparation. Commercially available impedance meters create a circuit using the electrode wires. A sinusoidal voltage is then passed across the skin–electrode interface to test the impedance level. The acceptable standard is $10\text{ k}\Omega$ at 100 Hz. If skin–electrode input impedance is decreased to $R_1 = 1\text{ k}\Omega$, the voltage drop across the amplifier is increased to 1.82 mV and the error is only 9%. If the input impedance of the amplifier is then increased to $R_2 = 100\text{ k}\Omega$, the voltage drop across the amplifier increases to 1.98 mV and the error is only 1%. Impedance control is a matter of matching the source (electrodes) and the detector (amplifier) resistances. A high source impedance should be monitored by a detector that has a much greater input impedance. This is referred to as *impedance matching*. The rule of thumb is that the input impedance of the detector should be at least 100 times greater than that of the source.

We are now in a position to understand a more subtle aspect of amplifiers. Differential mode and common mode are actually two different functions. One can best appreciate this by remembering that the amplifier contains three inputs: two high-impedance signal input connections and one signal ground (figure 3.11). The amplifier increases the magnitude of the voltage difference between the signal input connections but attenuates the voltage difference between these two inputs and the ground reference connection. The two functions of the amplifier are further reinforced by the fact that they have separate impedances. If a voltage difference is applied *between* inputs 1 and 2, the impedance to the differential input (R_i) is measured with respect to these two inputs. If the same voltage is applied to *both* inputs 1 and 2, the impedance to the common mode (R_{cm}) is measured with respect to the ground. This may be intuitive since each electrode is an "antenna" for environmental noise, which is referenced to the ground (0 V).

If the electrodes' impedances (R_{e+} and R_{e-}) are not identical or if differences exist in the amplifier subunits, common mode and other interference signals at the noninverting (1) and inverting (2) inputs will be dissimilar. A less than ideal cancellation takes place. If the input impedance of the amplifier is very high, differences in the signal between inputs 1 and 2 due to such imperfections will be minimal by comparison. Thus, a high CMRR can be realized only if the input impedance is significantly greater than that of the source. The input impedance for indwelling electrodes is at least five times greater $\left(50 \times 10^3\,\Omega\right)$ than that for surface electrodes due to the smaller recording area. An input impedance of $10^9\,\Omega$ is more than sufficient to accommodate both surface and indwelling electrodes.

Bias Current

Bias current is the base current that must flow all the time to keep the electronics within the amplifier turned on. No current below the level of the amplifier bias can be detected. This current actually flows out of the amplifier and is injected into the leads. The result is a voltage drop across the electrodes in proportion to the impedance $(V = iR_e)$. Consider the typical input impedance for a surface electrode with electrolyte gel $(R_e = 50\ \text{k}\Omega)$ and a bias current $(i = 50\ \text{nA})$. The voltage drop across the electrodes is 2.5 mV, in the absence of any muscle activity. In reality, the bias current is normally so small $(1-2\ \text{pA})$ that it does not represent a safety hazard, but it may be sufficient to alter the recording properties of the electrodes by changing the electrochemistry of the metal–electrolyte interface with repeated use.

If standard silver–silver chloride (Ag-AgCl) electrodes are being used, it will be necessary to periodically rechloride the electrode surface according to the following steps. Thoroughly clean the surface electrodes using a light scouring powder or silver cleaner to remove the tarnish and dirt. Place the electrodes in a nonmetallic container with 5% salt (NaCl) solution. Attach the electrode that requires chloride to the (+) positive terminal of a 1.5 V battery and the other electrode to the (−) negative terminal. Connect a 100 Ω resistor in series with the electrodes. The chloriding electrode darkens, while the other bubbles. Continue until the darkened surface is evenly coated. Repeat the process for the other electrode.

Bias current also contributes to movement artifact of surface electrodes. Cable motion induces mechanical pressure at the electrodes, which is known to produce time-varying changes in electrode impedance on the order of ± 50 kΩ. The voltage drop across the electrodes is ± 2.5 mV. That is, the magnitude of the motion artifact incorporated within the biological signal will be ± 2.5 mV. A smaller bias current ($1-2$ pA) can help with but not eliminate this problem. Additional methodological controls must be employed to minimize cable motion.

Amplifier Noise

The amplifier electronics generates several different types of noise. The different types of noise are detailed in the next chapter, but we point out here that the amplifier is unable to detect a signal if the signal is smaller than the sum of these noise sources. The root-mean-square amplitude of baseline amplifier noise should be less than 5 μV within the bandwidth of interest. A way to determine the exact value is to short the input terminals and measure the output.

Cabling

Remember that the goal is to have most of the voltage drop occur at the amplifier. The input leads from the electrodes to the amplifier are a major consideration. Current must flow through the input leads to the amplifier. Unfortunately, the input leads have a finite resistance, and there will be some degree of voltage drop between the electrodes and amplifier:

$$V_{drop} = i_{cable} \times R_{cable}$$

Resistance is a function of the conductivity of the material (σ), length (l), and surface area (A):

$$R = \frac{l}{\sigma A}$$

Of these three factors, length is the most critical because it can change to the greatest degree and is under our control. Keeping the length of the input leads and all cables as short as possible will minimize the voltage drop. One can observe a dramatic increase in signal strength by decreasing the length of the input leads to half their original length.

KEY POINT

The amplifier and electrode leads attached to an individual form a circuit. The amplifier draws current into the electrode leads, termed loading down the circuit. Current flowing in the circuit reduces the potential difference between electrodes that is recorded at the amplifier. A high-quality amplifier should therefore be highly resistant at the input terminals to this current flow.

Frequency Response

Amplifiers contain analog circuits that are capable of changing the frequency content of the input signal before it is digitized by the computer. Changing the frequency content is synonymous with filtering the signal. Because the amplifier can filter the signal before it is digitized by the computer, it is called an *anti-aliasing filter.* The significance of anti-aliasing filters is discussed in the next chapter. The exact amplifier implementation of the specific circuits is slightly different from that described in the following sections; however, the basic principles remain the same and are reviewed here.

Bode Plots

The most effective way to communicate how the amplifier alters the frequency content of the signal is through the use of *bode plots.* Bode (pronounced "bo-dee") plots graph the physical characteristics of any system in terms of the input stimulus and the output response. In a process similar to linear regression analysis, data used to construct the bode plots can then be fit with a first- or second-order differential equation whose parameter values quantify the specific physical characteristics.

For example, imagine that you are standing at the base of a thin tree that is least three times your height. The tree is so thin that you are able to fully grasp its circumference with one hand. If you move your arm slowly back and forth in a sinusoidal fashion while grasping the tree, you will notice that the top of the tree moves back and forth with the same amplitude as your arm. Displacement of your arm and displacement of the tree also occur at the same time. If you move your arm back and forth with greater frequency, the top portion of the tree can no longer keep pace. The inertial properties of the top portion of the tree cause it to lag behind the arm. The material properties of the tree also result in a decrease in the amplitude of movement relative to the arm. Thus, the tree behaves as a *low-pass* system. That is, the tree allows the low frequencies to pass through unattenuated. When the arm motion is slow, the tree can maintain the same frequency and amplitude of movement as the arm. With faster arm movement, the movement of the tree lags behind that of the arm, and its amplitude is decreased. Thus, the tree as a physical system attenuates high frequencies.

If movement of the arm is the input stimulus to the system, the frequency (ω) and amplitude (A) of its displacement can be represented by $x(t) = A\sin(\omega t)$. The output response of the tree can then be represented by $y(t) = B\sin(\omega t - \phi)$. Notice that $(\omega t - \phi)$ is used to denote the phase shift (lag) between the stimulus and response. The stimulus–response example is graphically depicted in figure 3.12. In this example, assume that the arm can perform sinusoidal displacements from 0 to 6 Hz in 0.25 Hz increments. At each frequency increment, the amplitude ratio of the output response to the input stimulus is calculated (B_{out} / A_{in}) and plotted. The ratio (B_{out} / A_{in}) is called the gain (G) of the system. This gain is not to be confused with simple multiplication of the input signal to increase its magnitude as discussed in the previous section. Two graphs define the bode plots. The first graph is *system gain* (y-axis) versus stimulus frequency (x-axis). The second is the *phase difference* between the input stimulus and output response (y-axis) versus stimulus frequency (x-axis). The gain and phase relationships together compose the bode plots that describe the physical system.

Decibels

Originally, electrical engineers were concerned with how amplifiers changed the power of a signal. The ratio of the power of the output signal to the power of the input signal was formed to determine the *gain* (G) of the amplifier: $G = P_{out} / P_{in}$, where the unit of power is watts (W). The *power ratio* is by convention expressed in a logarithmic scale of decibels (dB). The decibel scale is $10\log_{10} X$, where X is any number. The system gain in decibels is

$$G = 10\log_{10}\left(\frac{P_{out}}{P_{in}}\right).$$

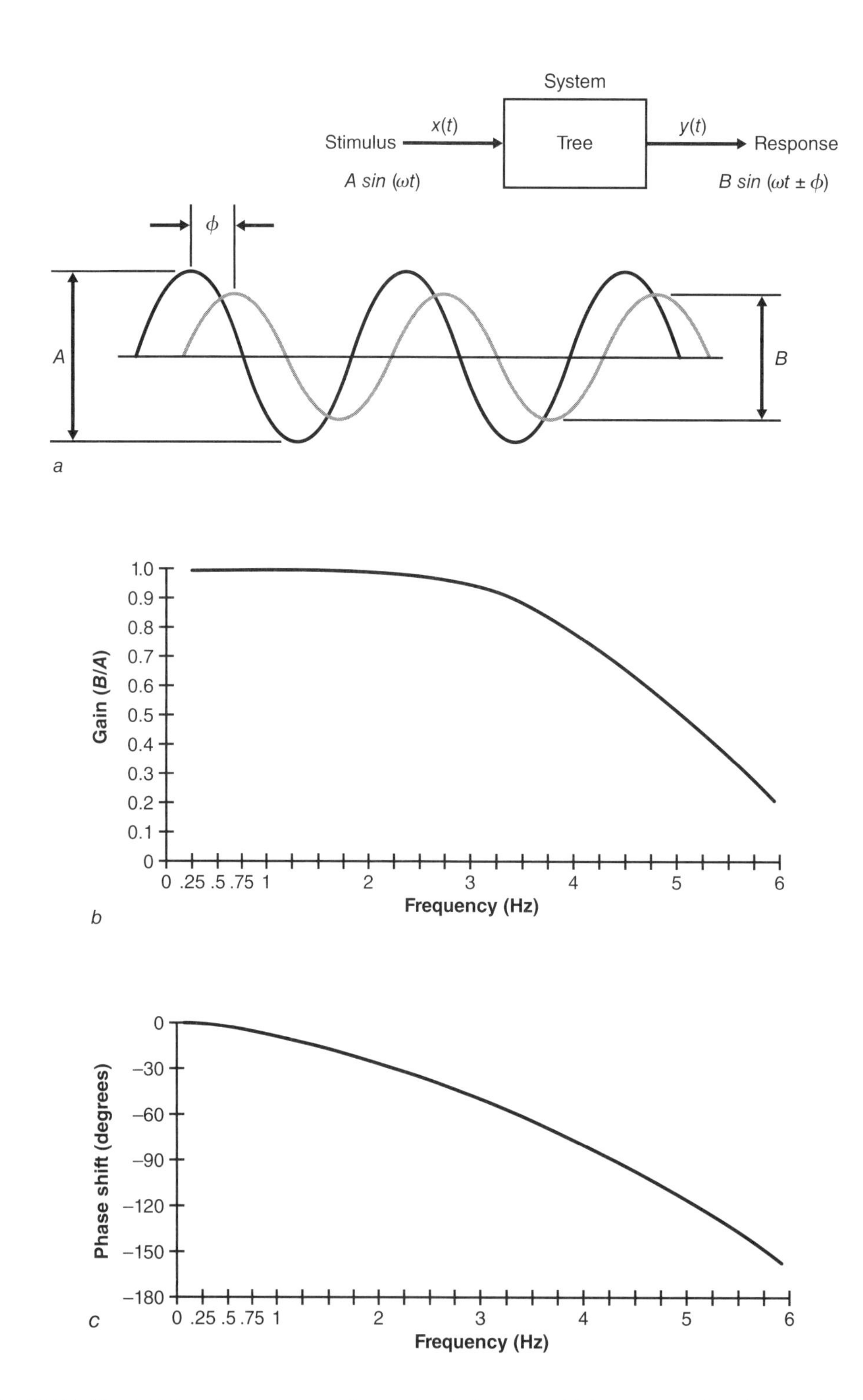

FIGURE 3.12 *(a)* The stimulus input to the system is the arm, which moves the tree back and forth in a sinusoidal fashion (*A*); the system is a tall, thin tree; the response is the amplitude of displacement of the treetop and its frequency of movement (*B*). *(b)* The ratio of the input frequency to the response frequency is the gain (*G*) of the system. *(c)* The phase between the input stimulus and response output is the phase shift. Together, the gain and phase response of the system are used to construct the bode plots of the system.

Recall that in an *alternating current* (AC) signal, the average power of the signal is

$$\overline{P} = \frac{V_{rms}^2}{R}$$

where V_{rms}^2 is the square of the *root-mean-square* (RMS) voltage amplitude. Substituting for average power, the system gain is

$$G = 10\log_{10}\left(\frac{V_{out}^2}{R_{out}} \times \frac{R_{in}}{V_{in}^2}\right).$$

The RMS voltage amplitude is still used. However, after some basic algebra to remove the square of the unit, the system gain in decibels reduces to

$$G = 20\log_{10}\left(\frac{V_{out}}{V_{in}}\right).$$

The ratio is now based only on the RMS amplitude of input and output voltage. As will be detailed in the next section, amplifiers contain analog circuits that consist of capacitors and resistors to filter the signal. Recall that *reactive capacitance* is frequency dependent. As the frequency of the AC signal increases, charges do not have time to build up on the capacitor and resist the flow of current. The characteristic frequency at which the reactive capacitance (X_C) impedes the flow of current to the same degree as the resistor is called the **cutoff frequency** (f_c). It was demonstrated in the previous chapter that the voltage amplitude is 0.707 of its full value at the cutoff frequency.

The ratio 0.707 is an important reference point when one is describing how a filter alters the signal, for the following reason: If the amplitude ratio of the output voltage to the input voltage (V_{out} / V_{in}) has increased or decreased to 0.707, then the notation in decibels is

$$20\log_{10}(0.707) = -3 \text{ dB}.$$

If the amplitude ratio is calculated using the squared unit (V_{rms}^2), the results translate to the power content of signal more easily. For example, an amplitude ratio of 0.50 at the cutoff frequency corresponds to a signal power of 50%. The notation in decibels is

$$10\log_{10}(0.50) = -3 \text{ dB}.$$

Keeping in mind that the gain formula is based on the ratio of two powers, a decrease in the RMS amplitude ratio to 0.707 at the cutoff frequency means that the power of the signal is half of its full value.

Filters

There are electronic circuits within the amplifier that alter the frequency content of the signal. Because these circuits allow certain frequencies to pass through while attenuating others, they are termed *analog filters.* An analogy is often made between analog filters and the simple coffee filter. The coffee filter blocks the coffee grounds (signal frequencies we don't want) from passing through the mesh while allowing the liquid (the signal frequencies of interest) to pass through.

High-Pass Filter The groundwork has been laid for understanding the circuitry underlying the resistor–capacitor **high-pass filter.** In this case, the capacitor is in series with the signal, and the resistor is parallel to the signal and is grounded (figure 3.13). Consider the case in which sinusoidal voltages of the same amplitude but different frequencies from 0 (direct current) to 1 kHz are applied to the input. The ratio of the RMS amplitude of the output voltage to the input voltage is then calculated to determine the gain (G) in decibels.

At low frequencies, the reactive capacitance (X_C) will be high and the signal is directly impeded by the capacitor. The resistor at this point has little influence upon the signal. As the frequency increases, the reactive capacitance decreases (X_C) and more of the signal passes through the capacitor to the output terminal. The resistor can now have a greater impact upon the signal, drawing voltage away from the output terminal to plateau the gain. The magnitudes of the resistance (R) and the capacitance (C) together then determine the slope of the gain curve to the cutoff point (f_c):

$$f_c = \frac{1}{2\pi RC}$$

If we follow the guidelines for high-pass filtering of the sEMG activity set forth by the International Society for Electrophysiology and Kinesiology (ISEK), the cutoff frequency should be set at 10 Hz to remove the low-frequency artifact associated with movement of the wires and electrodes relative to the skin when the muscle contracts. This occurs even during isometric contractions. A 16 $k\Omega$ resistor and a 1μF capacitor placed in the high-pass circuit depicted in figure 3.13 will yield a cutoff of 10 Hz:

$$f_c = \frac{1}{2\pi(1.6\times10^4\ \Omega)(1\times10^{-6}\text{F})} = 10\text{ Hz}$$

The resulting plot shows that the gain increases monotonically until it reaches –3 dB at 10 Hz (figure 3.13). The *transition band* extends from –40 dB to –3 dB. The signal power below –40 dB is so small that the region to the left of this point is called the *stop band.* The filter has functionally blocked the signal below this stop band point. Once the gain has reached the –3 dB point, there is a 50% increase in signal power. The circle in the gain plot in figure 3.13 shows why the cutoff frequency at the –3 dB point is also called the *corner frequency.* Notice that frequency in hertz is plotted on the x-axis. A logarithmic scale is used so that high frequencies can be graphed conveniently. If the frequency range is very large, the x-axis unit may be designated in radians (ω), or in radians normalized to the cutoff frequency, which is also specified in radians (ω / ω_C). Normalization is also useful for comparing different types of filters.

The rate of increase in gain (slope) in the transition band is a measure of how strictly the filter enforces the cutoff frequency and is called the *roll rate.* Ideally, it should be completely vertical so the gain looks like a "brick wall" with no frequency components below the cutoff. However, it is not possible to design such an ideal filter. As shown in the signal-processing literature, the ideal filter is *not physically realizable*. The standard roll rate is 20 dB/decade. As will be detailed later, there are filters with a steeper roll rate, but the result is a trade-off with other filter characteristics. To the right of the cutoff frequency in figure 3.13 is the *pass band.* The ideal filter is *maximally flat* in the pass band. That is, the voltage gain is constant for each frequency

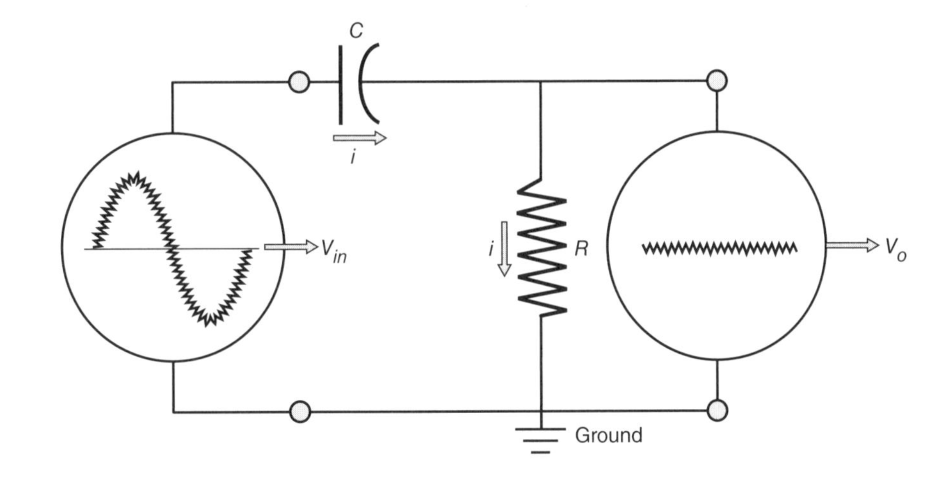

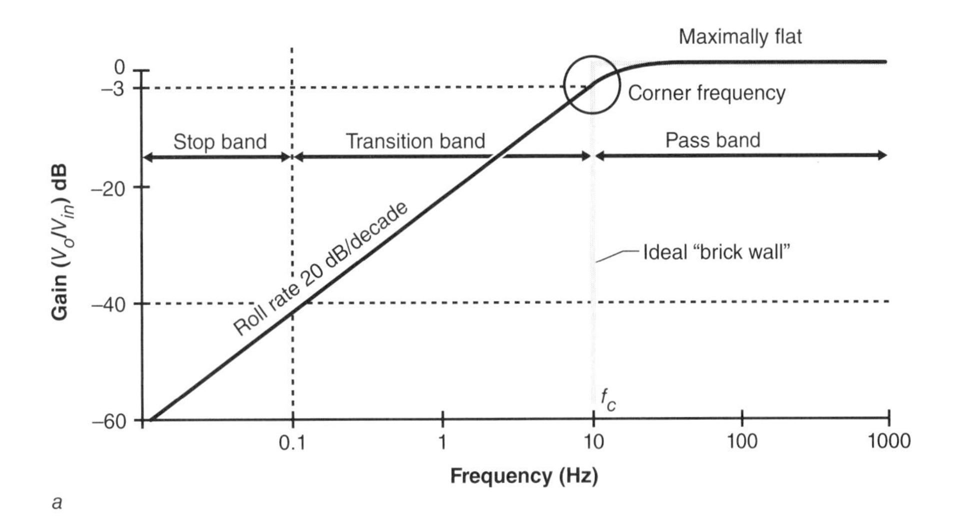

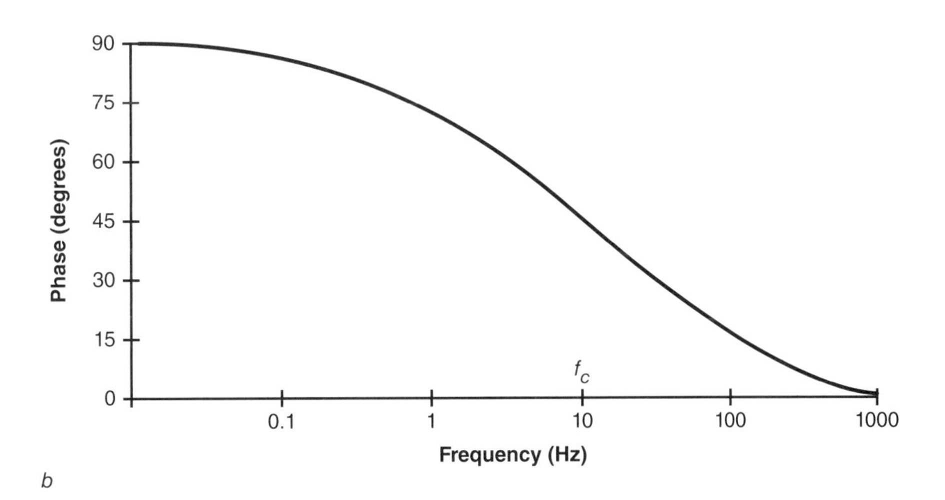

FIGURE 3.13 *(a)* The high-pass analog filter circuit illustrates the relationship between voltage input (V_{in}) and voltage output (V_o) as a function of the specific arrangement between the capacitor (C), resistor (R), and the flow of current (i). *(b)* The bode plots for a high-pass analog filter circuit.

beyond the cutoff. This characteristic is physically realizable for both *Butterworth* and *Bessel* filters. The names of these filter types appear often in the sEMG literature because the voltage gain for each frequency in the pass band is constant, which is essential for correct physiological interpretation of the signal.

The second portion of the bode plot is the *phase delay* between input and output voltages (figure 3.13). The manner in which the filter alters the phase is an important consideration when one is trying to align EMG activity to mechanical events generated by the muscle. Ideally, the filter would result in a zero phase, but this does not occur. For a high-pass filter, the output voltage *leads* the input voltage between 90° and 0°. The magnitude of the phase difference is frequency dependent. The greatest alteration in phase occurs during the transition band, and it is highly nonlinear. The phase difference begins to stabilize within one decade on either side of the cutoff frequency.

Low-Pass Filter The positions of the resistor and capacitor are interchanged to construct a low-pass filter (figure 3.14). At very low frequencies, the reactive capacitance (X_C) is so high that none of the signal will pass through the capacitor to the ground. Most of the low-frequency signals will pass through the resistor to the output. At low frequencies, the circuit behaves as if there is only a resistor in series between the input and output terminals. As the frequencies increase, the capacitive reactance (X_C) decreases, and more of the signal passes through the capacitor to the ground while less is present in the output. In essence, the capacitor is a *shunt,* drawing current away from the output terminals.

It should be apparent now that the arrangement of the resistor and capacitor units within the circuit determines the relationship between input and output voltages. However, their magnitudes still determine the actual cutoff frequency. If their magnitudes remain unchanged from the previous example, the low-pass filter will still have a 10 Hz (−3 dB) point, but the overall shape is dramatically different. The same is true for the frequency-dependent phase delay between input and output voltages. In this case, the output voltage *lags* behind the input voltage between 0° and 90°.

Band-Pass Filter The implementation of a series of high- and low-pass filters in the same circuit is designed to pass signals in the middle-frequency range, termed a **band-pass filter.** Figure 3.15 shows that the bode plot for a band-pass filter is basically a high- and low-pass filter merged with the appropriate cutoff frequencies. Electromyographic signals are most often band passed. However, the selection of appropriate low and high frequencies depends on the nature of the signal. Surface electromyographic activity is typically band passed between 10 and 500 Hz. This means that the high-pass cutoff frequency is set at 10 Hz to remove low-frequency artifact associated with electrode and cable movement. The low-pass cutoff frequency is then set at 500 Hz to minimize the higher-frequency components due to signals in the surrounding environment picked up by the electrode.

These numbers are only a rough guideline. The high-pass cutoff frequency for sEMG ranges from 3 to 20 Hz (−3 dB) in the literature. It is speculated that there is important information about motor unit firing patterns between 3 and 40 Hz in the sEMG signal (Dimitrova and Dimitrov 2003). Careful electrode setup and the use of isometric contractions are required to set the high-pass cutoff frequency as low as 3 Hz. Conversely, a 20 Hz high-pass cutoff frequency is used for more dynamic

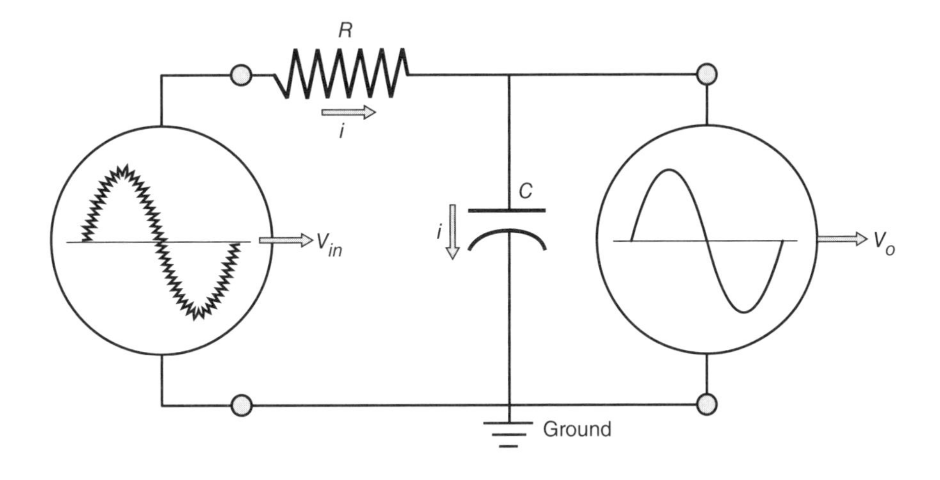

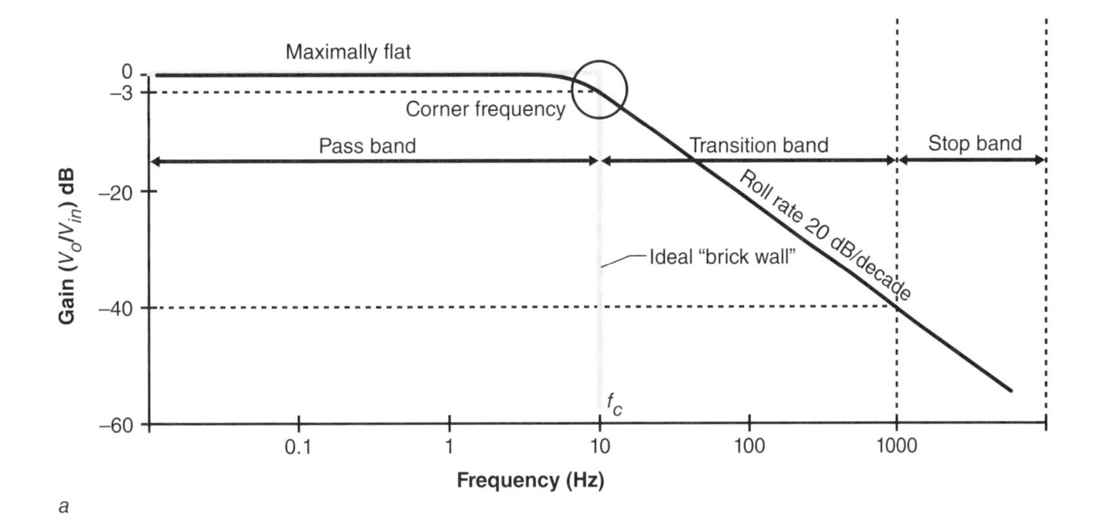

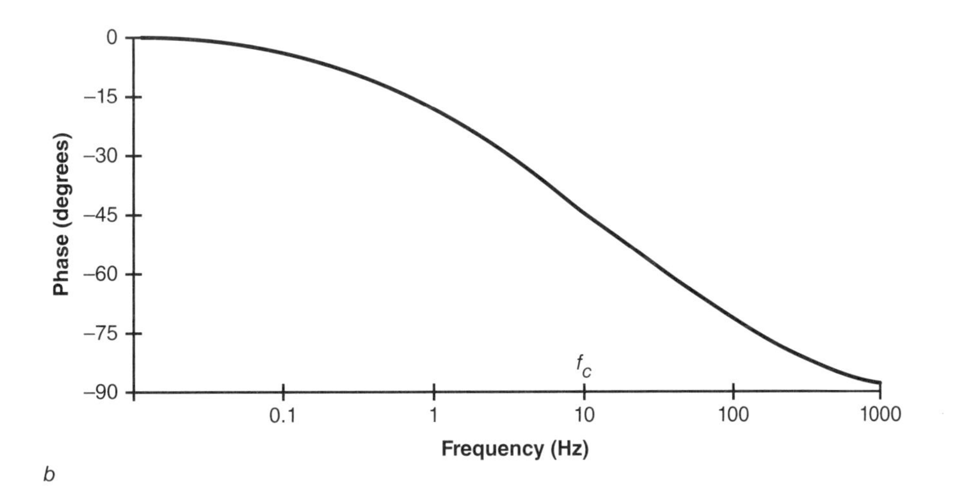

FIGURE 3.14 *(a)* The low-pass analog filter circuit illustrates the relationship between voltage input (V_{in}) and voltage output (V_o) as a function of the specific arrangement between the capacitor (*C*), resistor (*R*), and the flow of current (*i*). *(b)* The bode plots for a low-pass analog filter circuit.

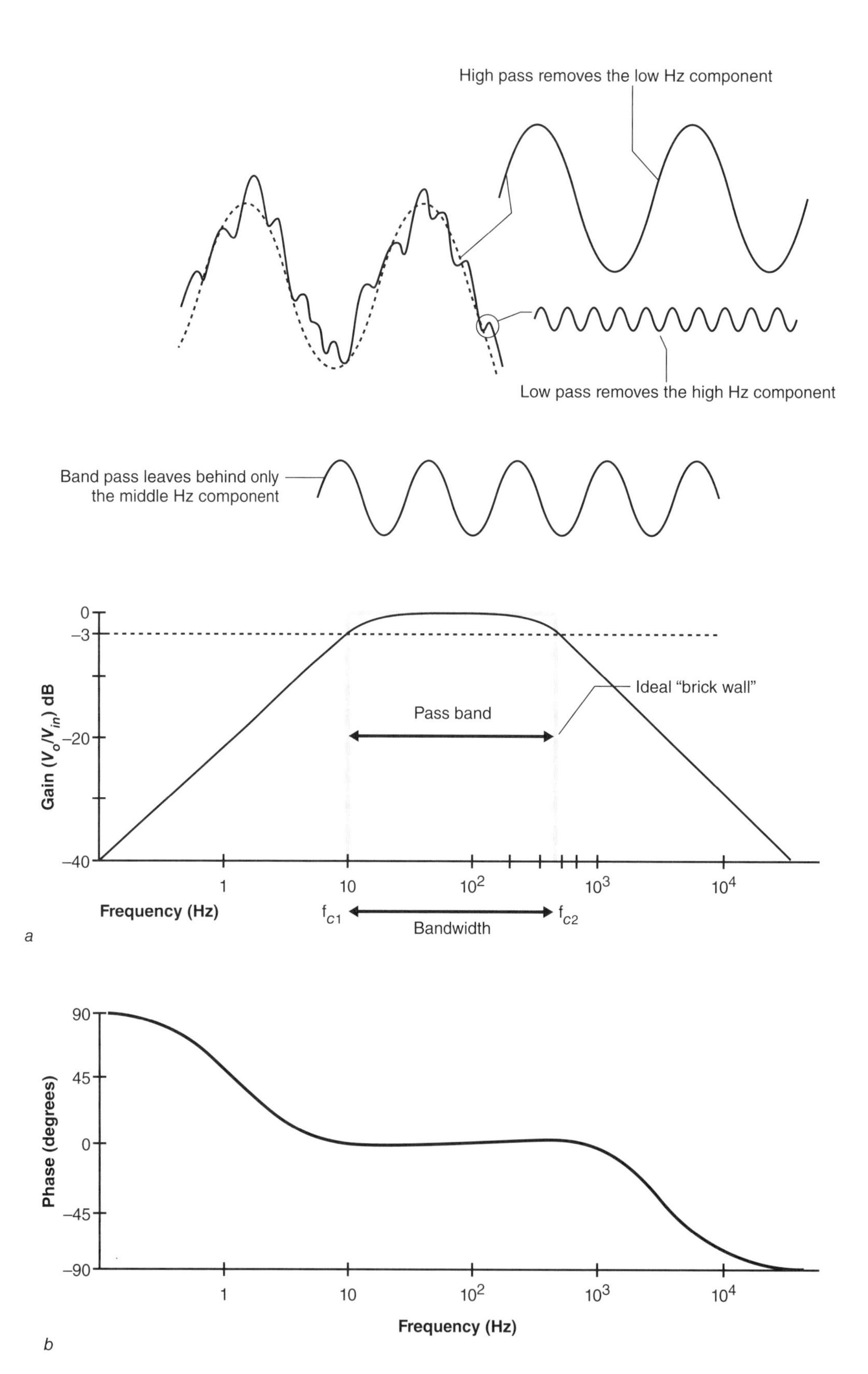

FIGURE 3.15 The band-pass analog filter circuit. *(a)* Electromyographic signals have low-, medium-, and high-frequency components. *(b)* The bode plots for a band-pass analog filter circuit.

contractions. There is very little signal power above 500 Hz in the sEMG signal, so the low-pass cutoff frequency is set at this −3 dB point. Other low-pass cutoff frequencies may be used; the upper frequency limit depends somewhat on the size and function of muscle. Methods to determine the frequency content of the signal will be outlined later. Indwelling wire electrodes directly record muscle activity that is unattenuated by fascia, fat, and skin. Thus, there can be significant power in the interference pattern signal up to 1000 Hz, so the low-pass cutoff frequency is usually set at this −3 dB point. To facilitate motor unit identification from indwelling recordings, the band pass is typically set from 1 kHz to 10 kHz.

Practical Applications

With the introduction of bode plots and the decibel unit, we are in a better position to appreciate filtering effects upon the EMG signal other than those imposed by the amplifier. We mentioned earlier in the chapter that the EMG signal is altered as it travels through muscle tissue. Figure 3.16 illustrates the frequency response of muscle tissue with respect to the EMG signal. As the frequency content of the EMG signal increases, there is a progressive attenuation. The magnitude of the attenuation is also distance dependent, increasing as the source gets farther away from the detection surface, from $h = 0.2$ to 50 mm. Although the lines in the family differ in the magnitude of curvature, they all describe a low-pass filtering function. Interelectrode distance also affects the frequency content of EMG signal. Recall that bipolar surface electrodes function as a comb filter, allowing some frequencies to pass through while attenuating others. Figure 3.17 shows the filtering function for the bipolar surface electrodes with an IED of 2 cm and a muscle fiber conduction velocity of 4 m/s. Notice that the bipolar electrode filtering function resembles a series of progressively narrowing band-pass regions. The location of each lobe depends on both muscle fiber conduction velocity and IED. In contrast, needle electrodes have a very simple high-pass function.

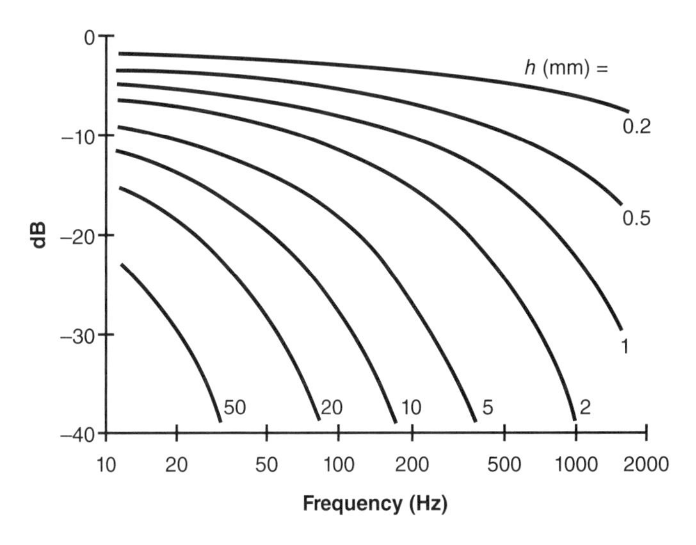

FIGURE 3.16 Low-pass tissue filtering of the electromyographic signal as a function of distance from the active fibers (h).

Reprinted, by permission, from L.H. Lindström and R.I. Magnusson, 1977, "Interpretation of myoelectric power spectra: A model and its applications," *Proceedings of the IEEE* 65: 654. © 1977 IEEE.

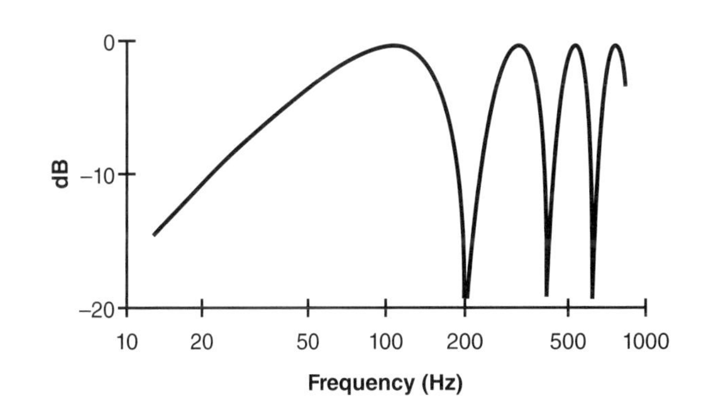

FIGURE 3.17 Bipolar surface electrode filtering function for an interelectrode distance of 2 cm and muscle fiber conduction velocity of 4 m/s.

Reprinted, by permission, from L.H. Lindström and R.I. Magnusson, 1977, "Interpretation of myoelectric power spectra: A model and its applications," *Proceedings of the IEEE* 65: 655. © 1977 IEEE.

Electrode Arrays

There are two other basic electrode configurations: *double differential gain* and *electrode arrays*. Electrode arrays are discussed here because they require a slightly more

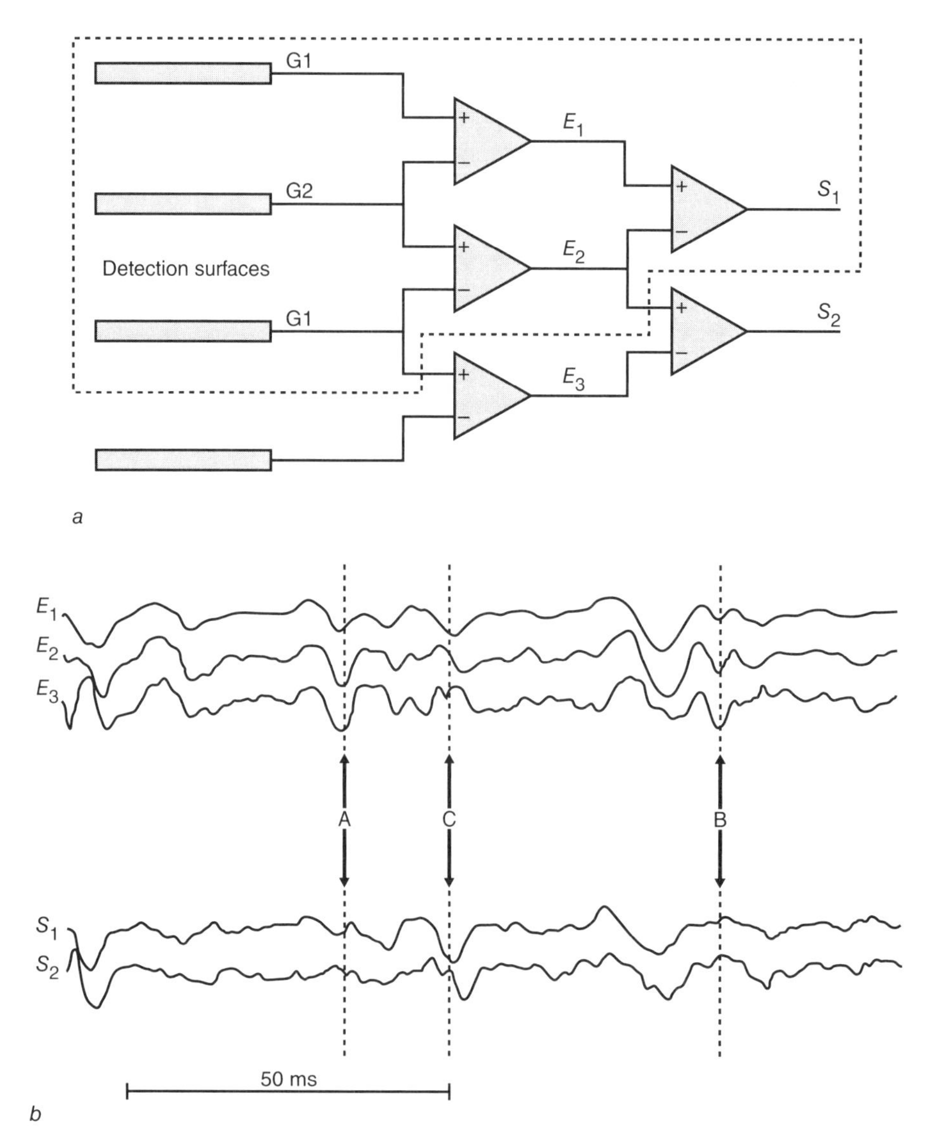

FIGURE 3.18 The double differential electrode. *(a)* Two bipolar signals (E_1 and E_2) are obtained through the first differential gain. The two bipolar signals are then fed to a second differential gain to produce the "double differential" signal (S_1). *(b)* Three possible bipolar signals, E_1, E_2, and E_3, are plotted. The three bipolar signals are then used to create two double differential signals: $S_1 = [E_1 + (-E_2)]$ and $S_2 = [E_2 + (-E_3)]$.

Reprinted, by permission, from H. Broman, G. Bilotto, and C.J. DeLuca, 1985, "A note on the noninvasive estimation of muscle fiber conduction velocity," *IEEE Transactions on Biomedical Engineering* 32: 343. © 1985 IEEE.

complicated arrangement of differential amplifiers. Consider three consecutive electrode surfaces in the dotted box in figure 3.18. Two bipolar signals (E_1 and E_2) may be created if the outer two electrodes function as the active surfaces (G1) and the central electrode is their reference (G2). The two bipolar signals E_1 and E_2 are termed the first differential gain; they are fed to a third amplifier to yield a second level of differential gain:

$$S_1 = E_1 + (-E_2)$$

The double differential gain further decreases any residual common mode signal, resulting in a much "cleaner" signal. The selectivity of these surface recordings is increased because the common mode signal associated with cross-talk from adjacent or distant muscles is minimized (van Vugt and van Dijk 2001). The two bipolar signals can be used separately to evaluate muscle fiber conduction velocity. If the IED is known, the time it takes for the same potential to travel across both bipolar surfaces (E_1 and E_2) can be determined based on the known sampling rate.

Linear electrode arrays contain "stacks" of bipolar electrodes. The length of the linear array is designed to span a significant portion of a specific muscle to track the propagation of action potentials across the membrane (Merletti et al. 2003). The smallest array consists of four electrodes arranged on the skin surface to obtain two sets of double differential gain signals (S_1 and S_2) (Fiorito et al. 1994). Below the schematic, in figure 3.18*b*, are three bipolar signals (E_1, E_2, and E_3) that yield two double differential signals (S_1 and S_2). The baselines of S_1 and S_2 fluctuate much less than E_1, E_2, and E_3 because of additional common mode rejection. The ability to observe the propagation of MFAPs between S_1 and S_2 is greatly enhanced by the further reduction of common mode sources. Notice that S_2 is time shifted relative to S_1.

The detection surfaces in figure 3.18 have a rectangular shape, but other geometries may be used. Separate electrodes can be used to construct an array, but it is much easier to obtain constant IEDs if the electrodes are mounted within a fixed structure. A fixed electrode array decreases the application time considerably. In either case, the electrode configuration for double differential gain is limited to muscle groups large enough to support the application of an array of electrodes.

KEY POINTS

- The amplifier has circuits that alter the frequency content of the signal. Low frequencies can be allowed to pass through the system unattenuated while the higher frequencies are removed (low pass). If higher frequencies are allowed to pass through unattenuated but lower ones are removed, the system is set for a high-pass filter. A band-pass filter removes both the low and high frequencies, allowing only the intermediate ones through.
- The frequency response of the system characterizes how well the amplifier performs in terms of its low-pass, high-pass, and band-pass filtering. That is, it characterizes how well the amplifier removes the unwanted frequencies.
- A bode plot is a graph that describes the frequency response of the system by plotting the relationship between signal input and output for sinusoids over a broad range of frequencies.
- The ratio of the signal amplitude output to the input at a given frequency is the system gain. The system gain is expressed in decibels. The y-axis of the bode plot is the system gain, and the x-axis is the range of frequencies of interest.

Grounding

Electrical measurements are based on a difference in potential (voltage) that requires a reference point. The most convenient reference point from which to make standardized measurements is the earth potential, which we assign the value of zero voltage, though nonzero earth potentials are possible. The earth can actually conduct electricity, so the earth potential can vary between different locations. This fact has important implications that will be discussed in later sections. In any case, the reference potential is literally obtained through a wire connection to a metal stake driven into the earth. The resistance of the conduction path to the earth is usually very low. Because the earth potential serves as the reference value, the reference electrode in EMG is also referred to as the *ground*. There is a link between the actual ground and the three-wire electrical outlet. The first wire (black) is *hot* because it delivers the current to the apparatus. The second wire (white) is considered *neutral* because it carries the return current from the apparatus to the ground. The third wire (green) carries no current and is connected directly to the ground for safety purposes.

Safety Grounding

Laboratory equipment is constructed so that the chassis that houses the circuitry is electrically isolated from the wires that carry power to the circuits. This is accomplished through the use of insulated wires so that the chassis does not become electrically "hot." Wear and tear (or general abuse) of the wires can cause the loss of insulation. A section of the wire can become exposed and actually touch the chassis. The chassis will then have the same voltage as the exposed wire. If a person touches the chassis while in contact with the earth, the body will function as a resistor between the chassis and the earth and will form a complete circuit. Current will flow through the body while it experiences a most unpleasant voltage drop.

To guard against this shock hazard, the chassis includes a very low-resistance connection to the earth through the third (green) wire. This low-resistance path offers less impedance to the flow of current than any other conducting path through the appliance. When the exposed wire touches the chassis, the current will flow to the ground rather than through the person (figure 3.19). There is always a certain amount of **leakage current** flowing through laboratory equipment. This is due to the capacitive coupling between the chassis and the power-line "hot wire," internal circuitry, and

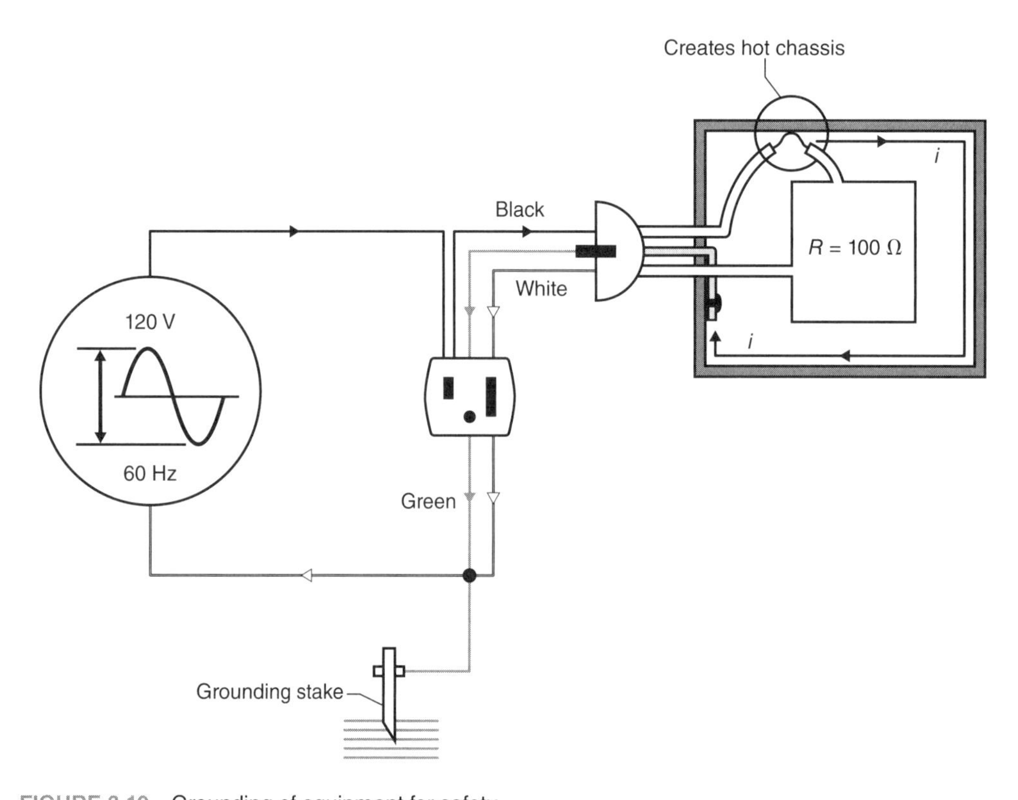

FIGURE 3.19 Grounding of equipment for safety.

WOLF, STANLEY, GUIDE TO ELECTRONIC MEASUREMENTS AND LABORATORY PRACTICE, 2nd Edition, © 1983, pg. 50. Adapted by permission of Pearson Education, Inc., Upper Saddle River, N.J.

TABLE 3.1 **Effects of Current Size**

Current	Effect
1 mA	Just perceptible
Up to 10 mA	Tingling sensation
15 mA	Muscular contraction—inability to "let go"
15-100 mA	Pain, fainting, difficulty breathing
100-500 mA	Ventricular fibrillation
>500 mA	Restarting of heart when current stops
6 A	Sustained muscular contraction of the heart
>6 A	Temporary respiratory paralysis—serious burns

other external cabling. It is important to ensure that leakage currents are always kept to a minimum. The International Electro-technical Commission (IEC) has published written guidelines under the code IEC 60950 (formerly IEC 950). No more than 3.5 mA is allowable for stationary (permanently connected) nonmedical equipment. Medical equipment leakage current limits are much lower. Table 3.1 shows the effects of various leakage currents on the human body.

Recall that the ground potential can vary dramatically between different locations. If the amplifier is plugged into a different outlet than other equipment in use, there is the possibility that the two instruments will be at different potentials. A voltage drop across the cable extending from the amplifier to the computer can occur that results in a constant offset in the digitized signal. Worse still, the amplifier and analog-to-digital (A/D) computer can form a complete current loop that is mediated by the intervening cable and the ground plane between the separate outlets (figure 3.20). Because the loop is associated with the ground, which by design is supposed

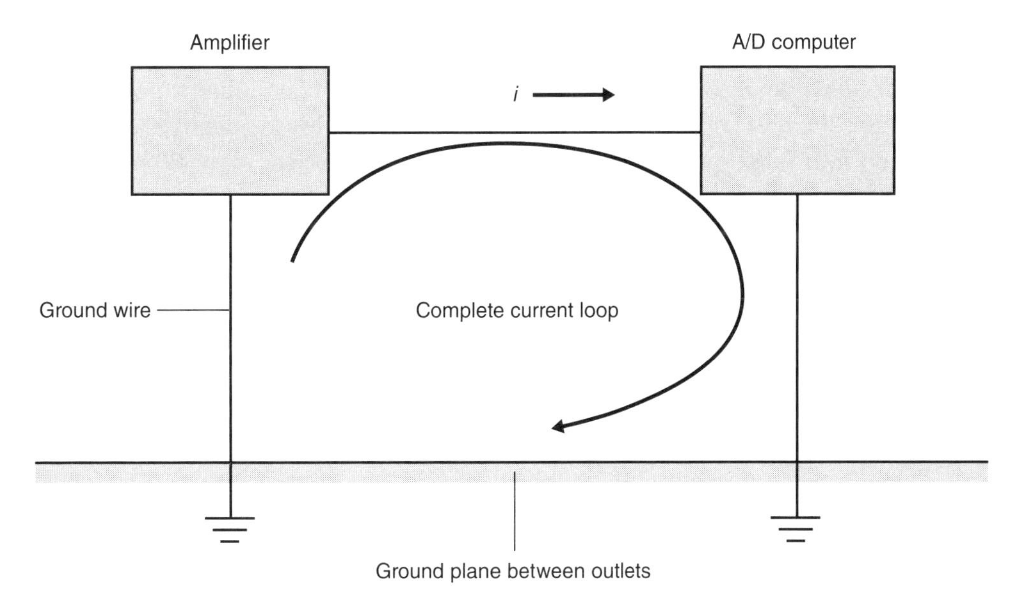

FIGURE 3.20 Ground loop for equipment connected to different electrical outlets.

WOLF, STANLEY, GUIDE TO ELECTRONIC MEASUREMENTS AND LABORATORY PRACTICE, 2nd Edition, © 1983, pg. 399. Adapted by permission of Pearson Education, Inc., Upper Saddle River, N.J.

to be a low-impedance conductive path, even the smallest voltage difference will result in a large flow of current. There is a strong possibility that 60 Hz earth currents associated with other nearby power systems could easily flow into the **ground loop.** Whenever a closed conductive loop encompasses a large area, it is also highly susceptible to inductive pickup of electromagnetic fields. Shielding does not reduce the interference because it is attached to the ground and both the shielding and the ground are part of the loop. The only way to minimize the ground loop is to plug the amplifier and accessory equipment into the same power outlet and place them close together. If there are not enough receptacles to plug all of this equipment in at the same outlet, an isolation transformer may be used to connect a power bar with multiple receptacles.

KEY POINT

The third connector in a three-prong electrical outlet is linked with a metal stake driven into the earth, referred to as a ground. The ground has several purposes. First, it serves as a safety should the chassis of laboratory equipment become electrically hot through an exposed wire or excessive leakage current. The ground provides a highly conductive path for these currents to flow through, rather than through the subject or patient.

Signal Grounding

There are two basic forms of noise: electrical and magnetic. An electric field surrounds a conductor that carries a net charge. The flow of current within the conductor then generates a magnetic field. If the charge or current changes in a cyclical fashion as a function of time (i.e., 60 Hz), the strength of the electric and magnetic fields will follow the same cycle. Two wires in proximity to each other may become *capacitively coupled* even though there is no physical connection between them (figure 3.21). It is important to remember that capacitor plates are separated by a space, so no physical connection is necessary. One wire (the power line) behaves as one plate, and the electrode lead is the other plate. Recall that North American power main voltage is alternating between ±120 V at 60 Hz. When the power line cycles to positive values, as one side of the capacitor plate, it will draw electrons toward it from the opposite plate (electrode input leads). The power-line side of the plate will then repel electrons from the opposite plate (electrode input leads) when it cycles to negative values. The result is that voltage variations in electrode leads will follow the same 60 Hz cycle as the power line. This type of external electric noise is so pervasive that it has its own name, *"hum."* Presumably, this is the sound the noise would make if the cables were also connected to a speaker.

Electric fields have a long range and it is not possible to reduce them simply by moving equipment away from the source. The problem of capacitive coupling associated with electric fields can be reduced only through the use of *shielded cables.* Shielded cables are composed of three layers. A signal-carrying conductor at the center is covered by a flexible insulating layer, which is then surrounded by a braided metal sheath. Static electric fields cannot penetrate the braided metal sheath. If the shield is also grounded through the amplifier, capacitive current will flow through it to the ground rather than the input terminals. A grounded shield may not eliminate the pickup of interference signals by the signal-carrying conductor but can reduce it by a factor of 100 to 1000. *Coaxial cables* are generally used for all other connections in the laboratory, such as that between the amplifier and the A/D conversion board in

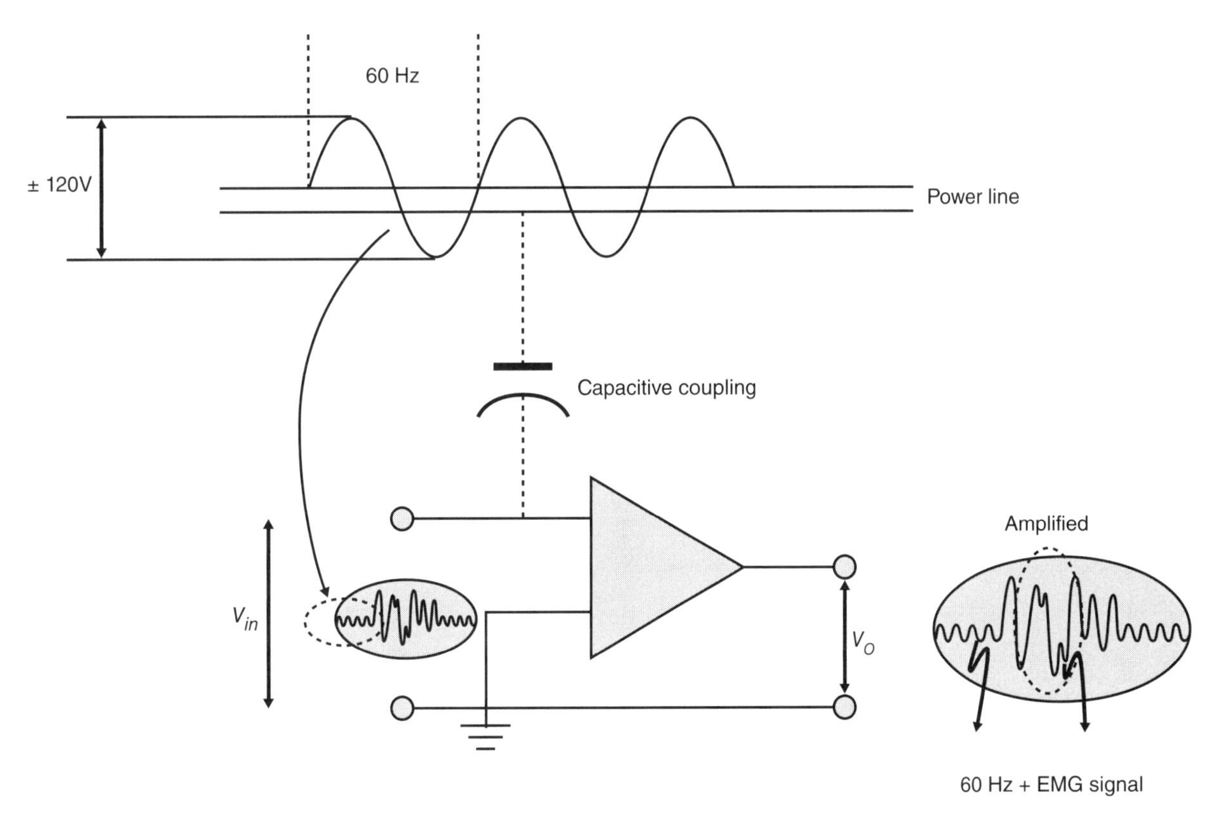

FIGURE 3.21 Power-line interference through capacitive coupling between the power line and equipment wires.

the computer. Coaxial cables are also shielded, but they consist of two conductors. These types of cables have an outer conductor that surrounds an inner conductor. Currents in the two conductors flow in opposite directions so that their respective electromagnetic fields cancel each other. Thus, signals can be transmitted in coaxial cables without causing interference in other laboratory electronics.

Capacitive interference involves the electrostatic field that exists between conductors at different potentials. Inductive interference arises from magnetic fields that are associated with current-carrying conductors. If the current changes with time, so too will the magnetic field. Current flowing in the power cords in laboratory instruments and in the surrounding walls and ceilings is the most prevalent source of noise. Electric motors are the worst offenders for generating magnetic fields. They are actually quite abundant in some buildings (electric pumps, soft drink machines, and treadmills).

Magnetic fields unfortunately cannot be reduced by shielding, but they are greatly affected by distance from the source. The magnetic field generated by a current-conducting wire has very little impact beyond 0.5 to 0.7 m. So, the first line of defense is to keep power cables as far away from other equipment as possible. Wire length and arrangement are the next considerations for noise reduction. *Magnetic induction* of current flow in electrode leads can be considerable; and the larger the leads, the larger the induced current. Decreasing the length of electrode leads is important not only for reducing voltage drops but also for minimizing inductive loop area. Braiding the wires decreases the loop area further and changes their orientation with respect to the magnetic field. The positive and negative currents that are induced by the periodic changes in orientation should cancel each other out.

Suppose now that twisted electrode leads are used and that they are both shielded. A new problem arises. The subject now acts as the other capacitor plate for the power

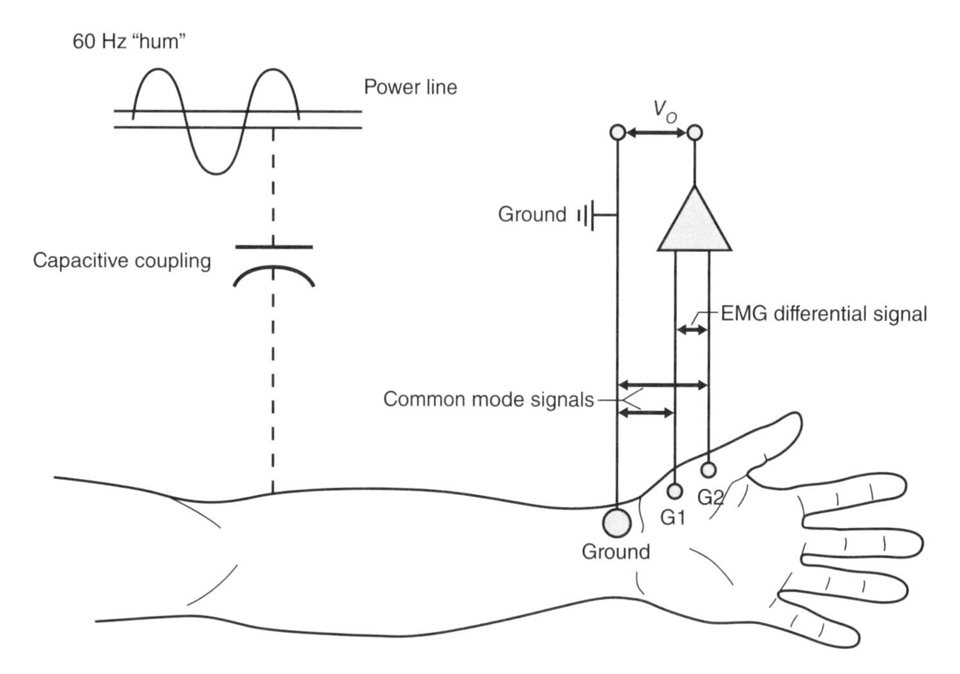

FIGURE 3.22 Power-line interference through capacitive coupling between the power line and the human body.

line (figure 3.22). Fortunately, the differential amplifier assumes that the whole body is one capacitor plate and that any nearby power line constitutes the other plate. The body is an excellent conductor and will propagate a voltage at the power-line frequency; there will be no difference between any two points on the body with respect to the amplitude of the voltage specific to the power-line frequency. Thus, the entire body has a common mode signal. The only difference between two points is due to the generation of muscle action potentials. The signals at G1 and G2 are referenced against the zero potential of the ground. Individually, the situation for each electrode depicted in 3.22 resembles that shown in figure 3.21. The input at both G1 and G2 includes both common mode and biological signals. The CMRR of the amplifier is then used to attenuate the common mode signal. The amplifier obtains the difference between G1 and G2 and increases the signal magnitude which is now predominantly biological.

The use of battery-operated amplifiers is one way to reduce electrostatic coupling (hum). These amplifiers are also beneficial for avoiding grounding loops by reducing electromagnetic induction. However, these forms of interference are so pervasive that battery-operated amplifiers may not result in an appreciable decrease in noise. The cost–benefit analysis is left to the individual consumer.

KEY POINTS

- The earth potential is taken as the zero reference point against which muscle electrical activity is assessed. Thus, there is a direct connection between the ground electrode and the actual ground function of the amplifier.
- Both cables and people can become capacitively coupled with wires and electronic equipment in close proximity. The result is the presence of 60 Hz voltage variations ("hum") in the baseline electromyographic activity. The ground electrode and grounded cables provide a conductive path for these induced currents to minimize their impact on the recorded measurements.

Computer Interfacing

Computer interfacing involves converting the analog EMG signal into a digital waveform. Then a computer program can be written to process the digital signal in a meaningful way that allows for its measurement. Of course, basic amplitude and timing may be extracted directly from an oscilloscope, but more sophisticated measures require a computer program. Computer interfacing can affect EMG measurement and correct interpretation of the signal to the same degree as filtering, probably more so. The reason is that, if the data have been collected properly, mistakes in postprocessing can potentially be fixed. However, if the EMG signal has not been digitally sampled at the correct rate or its vertical resolution has not been set properly, the errors are not so easily fixed and data may have to be collected again.

Sampling

The EMG analog signal is sent from the amplifier to the computer, where it is digitized. The computer uses its internal clock to send a signal that opens an electronic switch to a sample-and-hold circuit (figure 3.23). A capacitor in the circuit is used to hold the analog signal as the computer assigns it a digital value while the switch is open (aperture time). The aperture time is normally very small (i.e. 3 ns). Commercially available software is used to control the computer and specify how frequently to sample the analog signal. Sampling rate is given in hertz (samples per second). A 2 s analog signal sampled at 8 Hz will result in 16 discrete values. The amplitude–time graph of the digital signal follows the general pattern of the analog input. It is not difficult to see that a more representative digital version of the analog signal may be obtained with a higher sampling rate.

Horizontal Resolution

Consider low-, moderate-, and high-frequency waveforms (figure 3.24). The three waveforms are sampled at the same rate. The circles represent the exact sampling times when the electronic switch opens and the computer creates a digital representation of the amplitude. Immediately next to each analog signal is the digital representation of the amplitude at each sample time. It is clear that the sampling rate is most appropriate for the slower-frequency waveform. Both the amplitude and shape are well preserved. The sampling rate is just barely adequate for the moderate frequency. The amplitude and shape are still preserved, but not as well as for the lower-frequency waveform. In contrast, the high-frequency waveform has been completely distorted. First, there is a reduction of the original amplitude, and the frequency of the waveform has been *aliased* from 15 Hz down to 5 Hz. **Aliasing** is the error that occurs when the sampling rate is too low to capture the frequency of the waveform of interest.

Aliasing is a common observation in the movies. For example, the propellers on an airplane increase in speed then appear to slow down, stop, and actually reverse direction. This phenomenon results from a mismatch between the sampling rate (shutter speed) of the camera and the frequency of propeller rotation. The high frequency is aliased down to a low frequency.

Because the sinusoid is a symmetrical waveform, a minimum of two points per cycle is required to define its frequency. If the time interval between these two points is Δt, the sampling rate is $1/\Delta t$, and the highest frequency that can be represented by this sampling rate is $f_c = 1/(2\Delta t)$. The frequency $f_c = 1/(2\Delta t)$ is called the cutoff frequency, the Nyquist frequency, or the **folding frequency.** For a specific

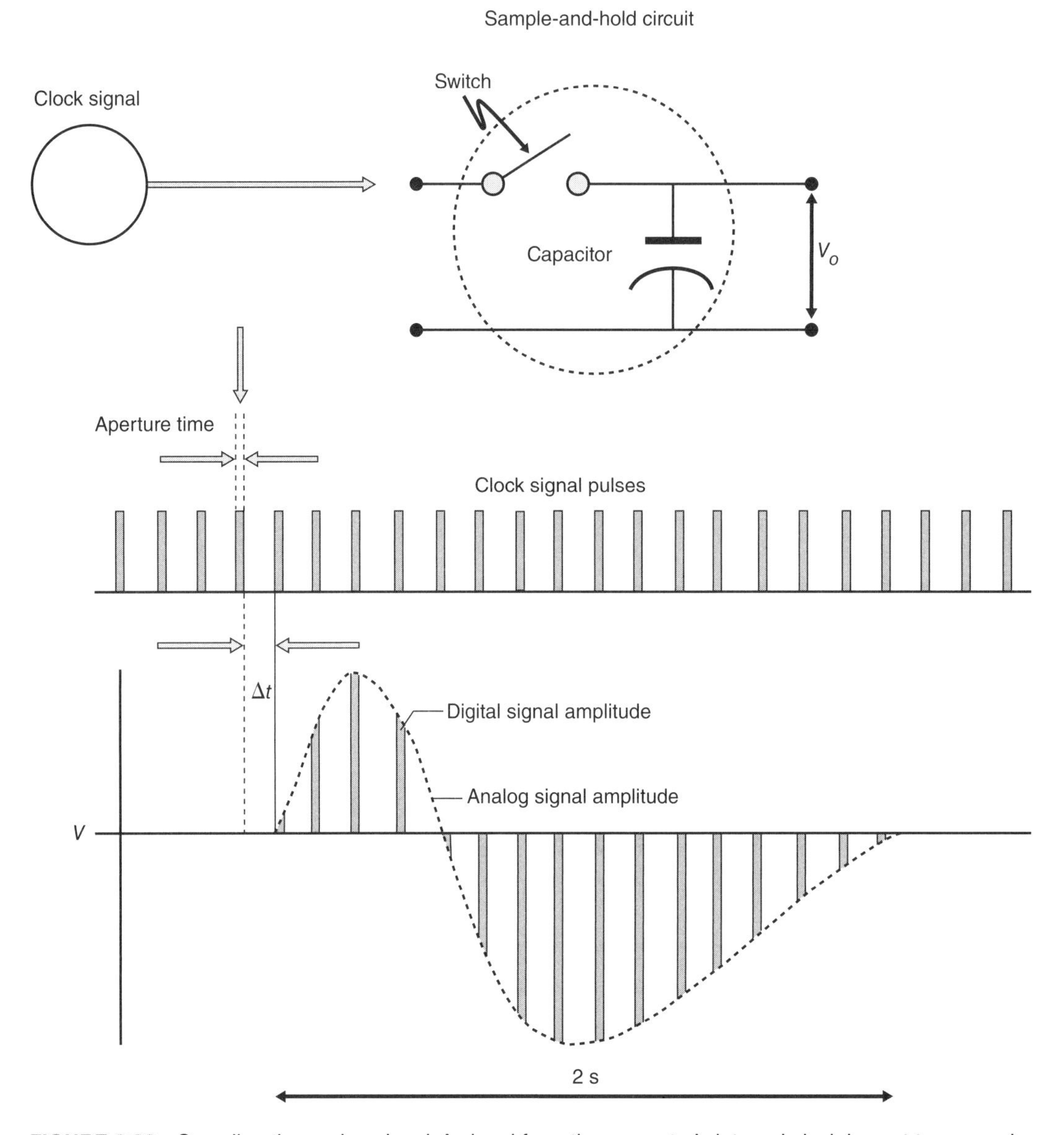

FIGURE 3.23 Sampling the analog signal. A signal from the computer's internal clock is sent to a sample-and-hold circuit. The vertical pulses shown below the circuit represent the rate at which the signal underneath is sampled. The result is an amplitude value of the original analog signal at a discrete point in time.

Adapted, by permission, from S. Wolf, 1988, Computer application in electromyography. In *Practical electromyography,* 2nd ed., edited by E.W. Johnson (Baltimore: Williams & Wilkins), 467.

sampling rate $1/\triangle t$, any frequency greater than $f_c = 1/(2\triangle t)$ will be folded back (or aliased) into the frequency range between 0 and $f_c = 1/(2\triangle t)$. For example, a time interval of $\triangle t = 0.01$ corresponds to a sampling rate of $1/0.01 = 100$ Hz. The highest frequency that can be represented by this sampling rate is $1/(2\times 0.01) = 50$ Hz. $1/(2\times 0.01) = 50$ Hz. Any frequency higher than 50 Hz will be folded back between 0 and 50 Hz, and will be confused with data in this lower range. For any specific frequency (f) in the range $0 \le f \le f_c$, the higher frequencies that will be aliased with f are given by

$$\tilde{f} = \left(2f_c n \pm f\right),\ n = 1,\ 2,\ 3\ldots$$

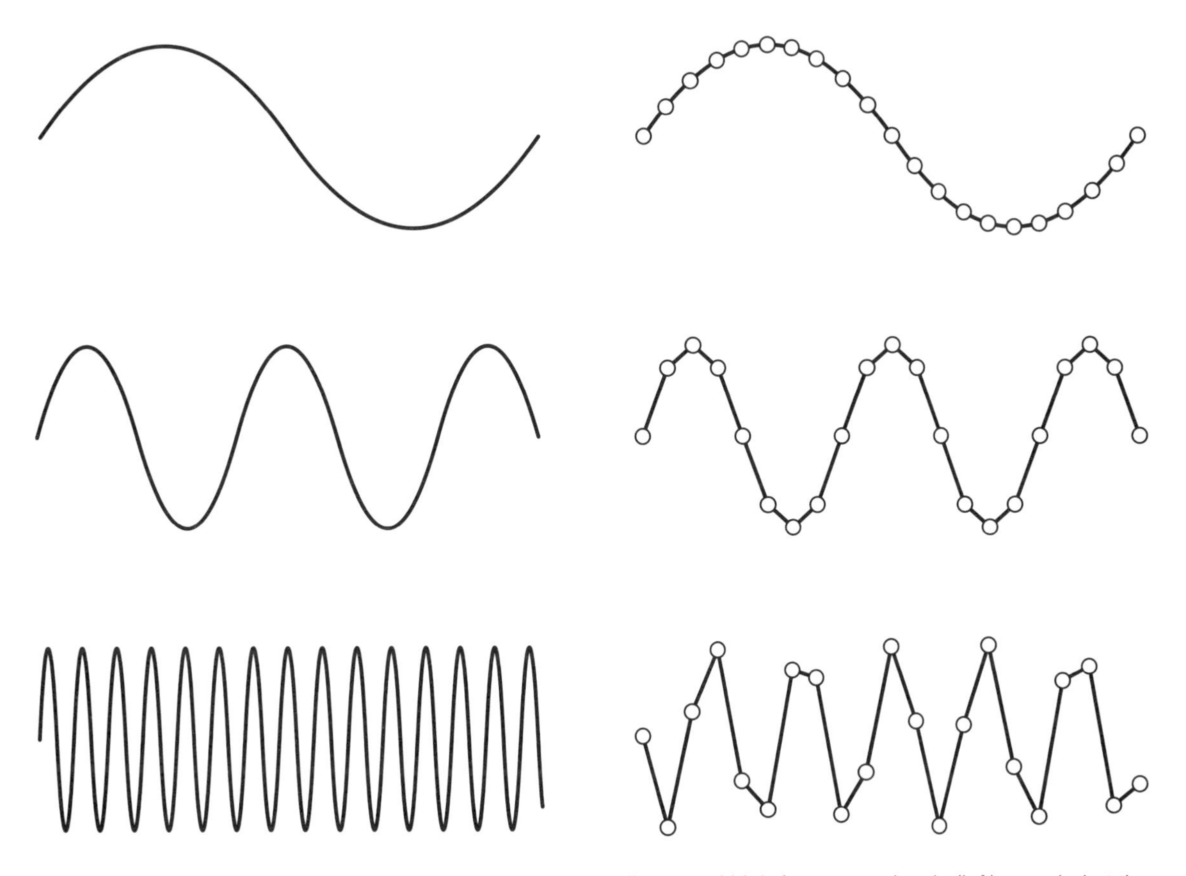

FIGURE 3.24 Horizontal resolution of the signal. Low-, medium-, and high-frequency signals (left) sampled at the same rate (right). Since the signals are sampled at the same rate, there is the exact same number of markers for each waveform, and they all occur at the same points in time.

Using our previous example where $\Delta t = 0.01$ and $f_c = 50$ Hz, the higher frequencies aliased with $f = 20$ Hz are $\tilde{f} = 80, 120, 180, 220, 280$ Hz, and so on. This can be observed as an abnormally high signal power (bulge) in an unexpected frequency region (Marmarelis and Marmarelis 1978; Bendat and Piersol 1971). Aliasing can occur if the filter roll rate is not steep enough to prevent aliasing near the cutoff region (figure 3.25). It is important to account for the transition band when setting the sampling rate.

Named after the engineer Harry Nyquist (1928), the Nyquist sampling theorem states that the sampling rate should be twice the highest-frequency component present in the signal. The sampling rate (frequency) that is twice the highest frequency present in the original analog signal is termed the **Nyquist frequency.** Since this is not always known a priori, it is good practice to remove unwanted high-frequency components (low-pass filter) from the analog signal before it is digitized by the computer. Low-pass filtering at a cutoff frequency (f_c) specified by the sampling rate will prevent higher frequencies from being folded back between 0 and f_c. These filters are usually incorporated into commercial amplifiers and are termed *anti-aliasing filters.* It is important to note that the sampling theorem is a lower limit, analogous to the rate used for the moderate-frequency waveform. In practice, one usually samples a sinusoid between 15 and 20 times in order to faithfully reproduce its shape in both the amplitude and frequency domains.

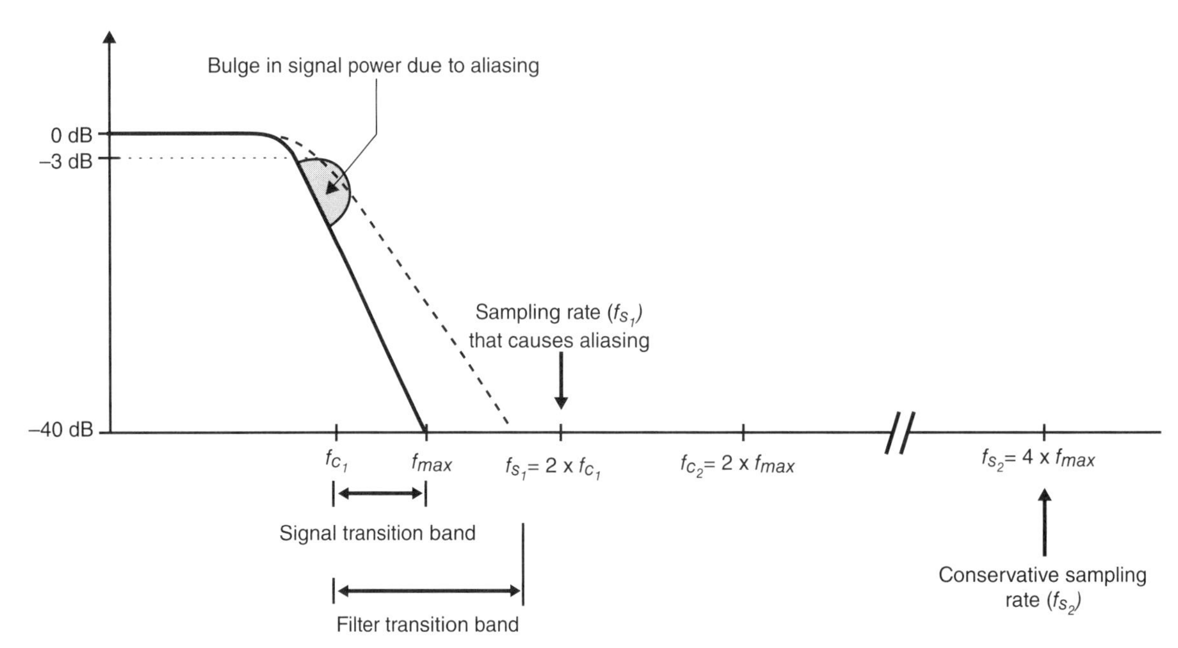

FIGURE 3.25 The increase or "bulge" in signal power due to aliasing near the cutoff frequency within the transition band for the anti-aliasing filter on the amplifier.

The Nyquist sampling rate was originally used because storage space and CPU power were limited, so investigators actually had to estimate the minimum requirements. The following procedures are recommended to determine the cutoff and sampling frequencies from pilot data if there is any uncertainty about the frequency content of the signal. First, oversample the signal (i.e., 100 kHz) to ensure that the folding frequency is well beyond the bandwidth of the highest-frequency signal normally observed in EMG. Second, calculate the power spectrum of the resulting digital signal. The –40 dB point should be used to designate the maximum-frequency component of interest (f_{max}). There is very little usable signal energy beyond the –40 dB point (1%). The low-pass filter cutoff (f_c) should be set 1.5 to 2 times greater than f_{max}, depending on the roll rate of the anti-aliasing filter. Figure 3.25 illustrates that (f_c) is adjusted for the transition band to set a more realistic sampling rate for the aliasing example. The new sampling rate is then $4 \times f_{\text{max}}$. The higher sampling rate also has the advantage of preserving the amplitude structure of the signal.

Multiplexing

In many cases, the EMG signal is obtained simultaneously with other sensors, such as a load cell (force) or potentiometer (displacement). Multiple signals may be sent to the A/D board, which samples each sequentially through an electronic switch called a *multiplexor.* Consider an experiment with two EMG electrodes (channels 1 and 2) and a load cell (channel 3). Channels 1 and 2 are the biceps and triceps brachii, respectively. Channel 3 is used to record maximal isometric elbow flexion force. The A/D board will sample channels 1, 2, and 3 in sequence to obtain the first data point. The second data point will be obtained as the A/D board returns to channel 1 to start the process again, and so on for the duration of the sampling period. The sampling time between channels must be considered.

For example, **electromechanical delay (EMD)** is the time between the onsets of EMG and force. The EMD time is often used to align muscle activity to the mechanical events. Isometric contractions typically have an EMD between 10 and 20 ms. If the sampling rate is 1000 Hz, each data point represents 1 ms. This translates to a 2 ms difference between channels 1 (biceps) and 3 (force). The potential error is 10% to 20% in EMD time. The situation degrades further as the number of channels increases. The desired single-channel sampling rate must be multiplied by the total number of A/D channels in use. In this example the A/D board should therefore be set to sample at 3000 Hz. Commercial venders of data acquisition boards often advertise the maximum sampling rate for a single channel. The maximum single-channel rate must be distributed across the total number of channels being sampled.

Quantization

The process of assigning a digital value to an analog signal is termed *quantization.* The input range of *analog-to-digital* (A/D) conversion boards can vary, but 10 V is quite common. It is more appropriate to measure EMG signals in the bipolar mode with a range of ±5 V. Computers then use the binary system to represent voltage amplitude as a series of 1s and 0s in a base 2 numbering system wherein each 1 or 0 is a single binary digit, or *bit* for short. The resolution of A/D boards is ultimately determined by a factor of 2^n, where *n* is the number of bits. For example, a 4-bit A/D board has $2^4 = 16$ unique combinations of 1s and 0s to yield 16 different numbers. An input range of 10 V is divided into 16 different voltage levels (table 3.2). Because the

TABLE 3.2 The Binary Representation of the Voltage Levels for a Four-Bit A/D Board With an Input Range of 10 V

Input voltage	Voltage level	Binary number
0.000-0.625	0	0000
0.625-1.250	1	0001
1.250-1.875	2	0010
1.875-2.500	3	0011
2.500-3.125	4	0100
3.125-3.750	5	0101
3.750-4.375	6	0110
4.375-5.000	7	0111
5.000-5.625	8	1000
5.625-6.250	9	1001
6.250-6.875	10	1010
6.875-7.500	11	1011
7.500-8.125	12	1100
8.125-8.750	13	1101
8.750-9.375	14	1110
9.375-10.00	15	1111

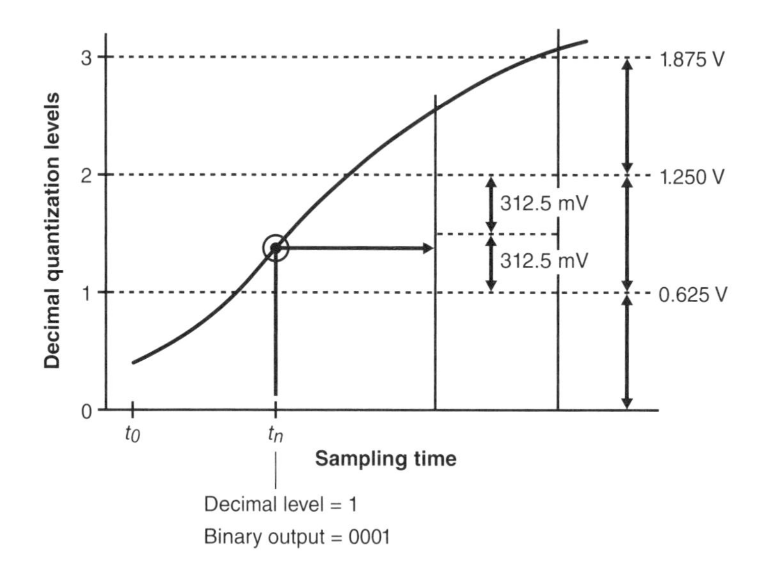

FIGURE 3.26 Quantization error associated with rounding the voltage to the midpoint of each level. The identified decimal level is 1 with a binary output of 0001.

first level is from 0 to 1, the last level is numbered 15. The numbered levels always terminate at $2^n - 1$. The *step size* (resolution) of each voltage level is 625 mV. The *midpoint* of each voltage level is then the voltage level plus 312.5 mV. The A/D board first determines the appropriate voltage level for the observed signal. It then evaluates the amplitude of the signal relative to the midpoint for that level (figure 3.26). The observed signal is assigned the *floor value* if it is below the midpoint, or the *ceiling value* if it is above the midpoint. Note that the ceiling of the preceding level is the floor of the next. The rounding relative to the midpoint is termed **quantization error.** In this case, the quantization error (QE) is 312.5 mV, but a more general formula is

$$QE = \frac{\text{A/D resolution}}{2}.$$

A higher resolution may be obtained with the use of a 12-bit or 16-bit A/D board. For a 12-bit A/D board, the 10 V input range would be divided into 4096 voltage levels with a step size (resolution) of 2.44 mV. The QE is then 1.22 mV. Technically, the limitation to increasing the number of bits for an A/D converter is the additional computing time it takes to evaluate (by successive approximation) the precise quantization level for the observed signal amplitude. However, computing speeds have increased dramatically, and this is no longer a serious consideration. Figure 3.27 illustrates a sine wave sampled by a 4-bit A/D converter. It should be evident that an increase in the number of bits (resolution) would result in a smoother, more accurate waveform.

Vertical Resolution

It is important to have a good estimate of the maximum EMG amplitudes that will be observed in the study. The gain of the amplifier must then be set to match the input

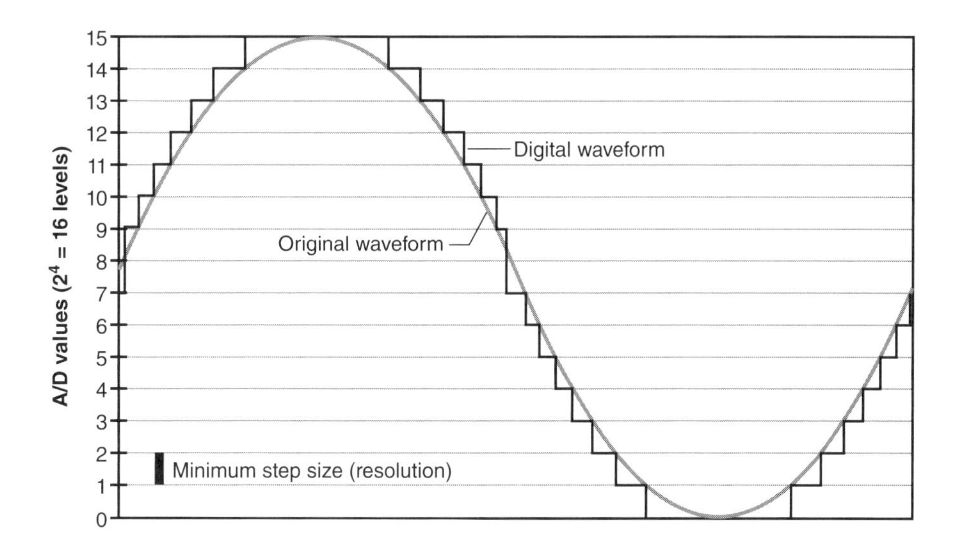

FIGURE 3.27 Analog-to-digital sampling of a sine wave using a 4-bit converter. The original analog signal is superimposed on the 4-bit digital signal.

range of the A/D board. The horizontal lines in figure 3.28 represent the same number of voltage levels for each waveform. The only difference between the graphs is the gain of the input signal.

Consider a 12-bit A/D board with an input range of 10 V (±5 V). There will be 4096 levels with a step size (resolution) of 2.44 mV. An EMG signal with a P-P amplitude of 5 mV will occupy only the middle two levels of the A/D board. Digital representation of the signal will be barely perceptible (figure 3.28*a*). If the amplifier gain is set to 2000, the EMG signal will have a P-P amplitude of 10 V and utilize all 4096 levels of the A/D board. To return to the original voltage magnitude, the output signal must be divided by its gain. However, in practice, it is unlikely that there will be a perfect match between the amplitude of the EMG signal and the vertical resolution of the A/D converter. Consequently, it is advisable to perform a few test contractions and set the amplifier gain so that the EMG signal occupies the middle two-thirds of the input range (figure 3.28*b*). A means of accomplishing this is to monitor the EMG activity on an oscilloscope set to the same input range as the A/D board.

The middle two-thirds rule is used to provide a safety margin. It is not unusual to discover that the test contractions have underestimated the expected maximum P-P amplitudes that will be observed during the experiment. If the gain is set too close to the full input range, the EMG voltages can saturate the A/D board. The digital EMG signals will be *clipped,* and the data from the recording session will be lost (figure 3.28*c*). Calibrating the amplifier gain for dynamic contractions is particularly difficult. The EMG amplitude from a dynamic movement can often exceed that generated by a maximal isometric contraction of the same muscle group.

The vertical resolution actually depends on several factors. The input range of the A/D board and the number of voltage levels (bits) interact to determine the absolute magnitude of the voltage resolution or step size. Increasing the gain to maximize the input range of the A/D board will allow smaller changes in the waveform to be

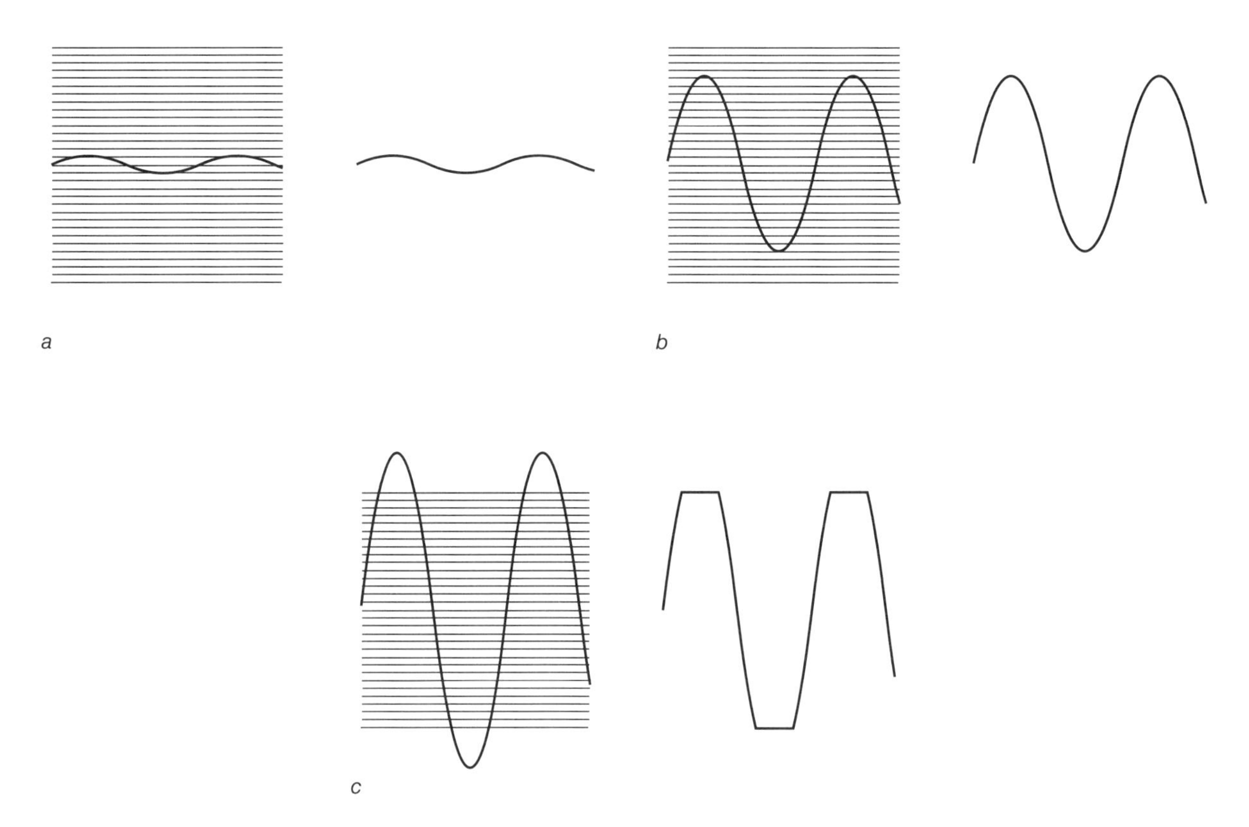

FIGURE 3.28 Interaction between the input range of the analog-to-digital (A/D) conversion board and amplifier gain. The input range of the A/D board is the same for each waveform depicted. The horizontal lines indicate that the number of voltage levels is the same for each condition. The amplifier gain is different for each waveform: *(a)* low, *(b)* moderate, and *(c)* high.

resolved. A failure to adjust the gain properly is the same as voluntarily decreasing the number of bits (resolution). The goal is to increase the gain to the point where the resolution is the best possible for the particular situation.

KEY POINTS

- Horizontal resolution of a waveform refers to the frequency at which it is sampled. The Nyquist sampling theorem states that the sampling frequency should be twice as high as the highest-frequency content of the signal.
- The low-pass frequency filter should be set no higher than half the Nyquist frequency. Undersampling a signal can cause aliasing, in which higher-frequency signals are recorded as lower-frequency signals.
- The process by which the analog signal is assigned a digital voltage value (A/D conversion) is termed quantization. The computer boards (A/D boards) on which this process occurs divide a set voltage range into different levels (resolution). The number of levels is determined by a factor of 2^n, where n is the number of bits.
- Signals should be amplified to maximize the voltage range of the A/D board. If the amplification is too small, the waveform can be represented by only a few voltage levels. With overamplification, the extreme values will not be assigned a voltage value at all, so that the maxima and minima are lost. One may obtain optimal vertical resolution of the waveform by selecting an amplification level such that the waveform occupies the middle two-thirds of the A/D voltage range.

FOR FURTHER READING

Loeb, G.E., and C. Gans. 1986. *Electromyography for experimentalists*. Chicago: University of Chicago Press.

Normann, R.A. 1988. *Principles of bioinstrumentation.* New York: Wiley.

Pease, W.S., E.W. Johnson, and H.L. Lew. 2006. *Johnson's practical electromyography.* 4th ed. Baltimore, MD: Lippincott Williams & Wilkins.

Winter, D.A. 2005. *Biomechanics and motor control of human movement.* 3rd ed. Hoboken, NJ: Wiley.

chapter 4

EMG Signal Processing

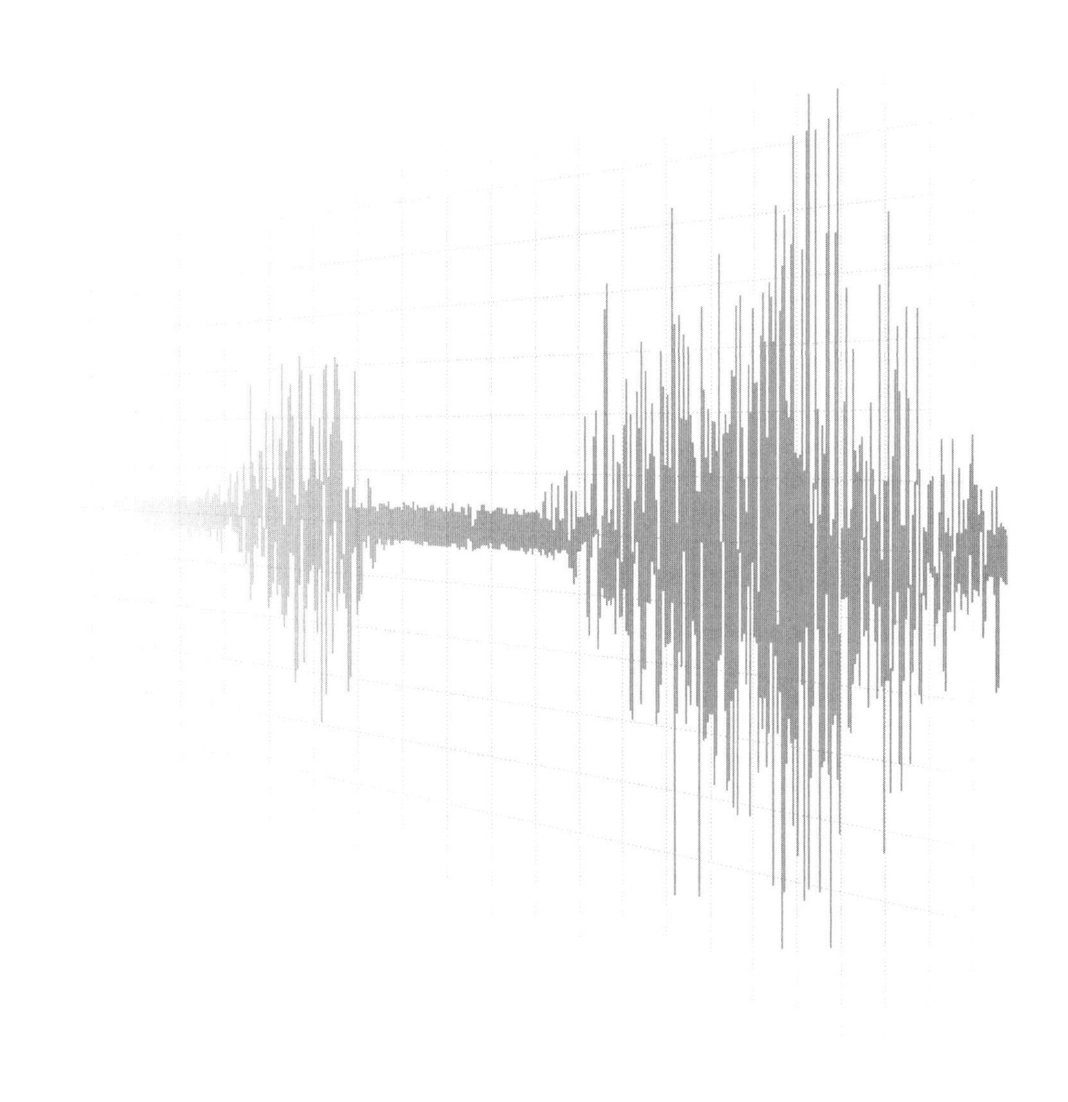

The previous chapter presented information necessary for acquiring high-quality EMG data. This chapter covers basic signal-processing and data reduction techniques used to reduce noise and to extract useful information from the EMG signal. Measures extracted from the EMG signal fall into one of three categories: (1) amplitude, (2) timing, and (3) frequency.

Amplitude

Amplitude of the EMG signal is used as a measure of neural drive to the muscle. In the case of motor-evoked potentials such as the compound muscle action potential, the peak-to-peak amplitude of waveform is proportional to the number of motor units activated by electrical stimulation of the peripheral nerve. The relationship between the number of motor units and the amplitude of the interference pattern generated by voluntary contractions is not straightforward due to wave cancellation (Keenan et al. 2005).

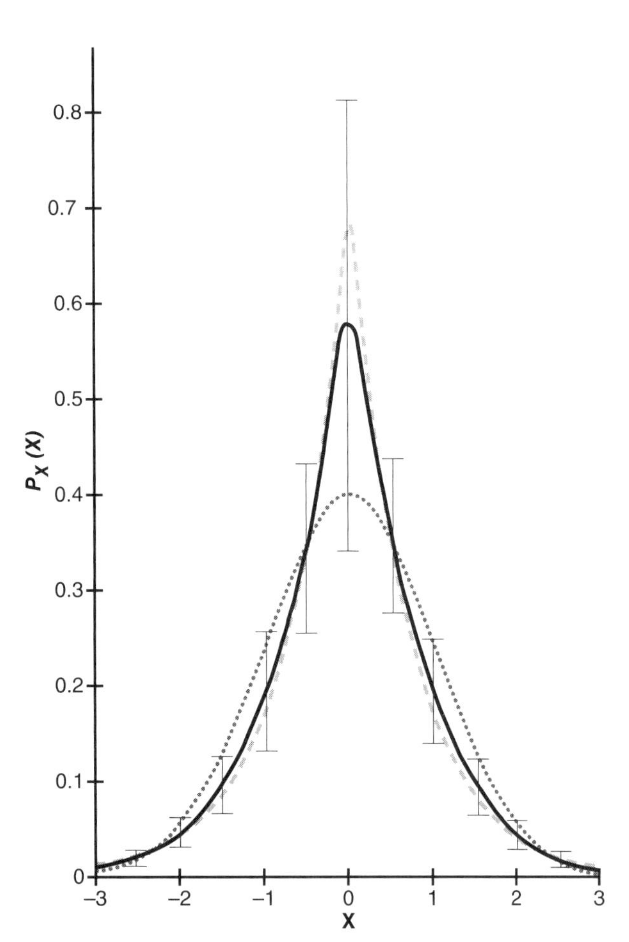

FIGURE 4.1 The frequency distribution curve for surface EMG amplitude. Each EMG amplitude was first converted to a z-score with zero mean and unit variance. A bin was created for a small range of amplitudes. The frequency of occurrence of amplitudes (y-axis) that fit within the range for that bin (x-axis) was counted. The area under the curve is normalized to one unit, so the result is a probability density function (PDF). The smooth PDF is the result of averaging the individual PDFs of EMG activity from 660 trials during which individuals performed constant-force contractions at 50% of maximal voluntary contraction. The vertical error bars each represent one standard deviation, and the dashed line is the Gaussian PDF.

Reprinted, by permission from E.A. Clancy and N. Hogan, 1999, "Probability density of the surface electromyogram and its relation to amplitude detectors," *IEEE Transactions on Biomedical Engineering* 46: 733. © 1999 IEEE.

Nature of the EMG signal

If specific values at future points in time can be predicted, the signal is referred to as **deterministic.** In general practice, the signal from any process that can be described by a mathematical function $y(t)$ that is reasonably accurate is deemed deterministic. If future values cannot be known but do follow a probabilistic pattern, the resulting signal is said to be **stochastic.** Description of a stochastic signal is dependent upon probability statements and statistical measures, such as the mean and standard deviation (Marmarelis and Marmarelis 1978; Bendat and Piersol 1971). Analysis of the EMG signal and its processing is based on the assumption that the EMG signal is a zero-mean random (stochastic) process whose standard deviation is proportional to the number of motor units and their firing rates (Clancy et al. 2002).

Consider a long segment of EMG data generated during a static (isometric) contraction. An amplitude histogram may be created consisting of the total range of amplitude values (positive and negative) divided into equal intervals (bins), and then the proportion of amplitudes in each bin can be determined. If the length of the data segment is increased and the size of the bin sufficiently decreased, the amplitude histogram begins to resemble a continuous function (figure 4.1) (Clancy and Hogan 1999). The continuous function then represents the probability distribution of the signal amplitude. The probability distribution is a mathematical description of the relative likelihood of a possible outcome. It is generally assumed that the probability distribution of the EMG signal amplitude has a *Gaussian* (normal curve) shape:

$$p(x) = \frac{1}{\sigma\sqrt{2\pi}} e^{-\frac{(x-\mu)^2}{2\sigma^2}}$$

where μ is the mean and σ is the standard deviation of x, which is the EMG signal. There is evidence that the probability distribution of EMG signal amplitude is dependent upon strength of the contraction, approaching a more Gaussian shape as contraction intensity nears 70% of maximal voluntary contraction (Kaplanis et al. 2009).

KEY POINT

The amplitude structure of the EMG signal has a Gaussian shape when plotted as a frequency distribution curve.

Linear Envelope Detection

Linear envelope detection of the EMG signal is the most commonly applied *demodulation* technique used to extract information from the observed EMG waveform. The underlying basis for EMG analysis and its processing has its origins in telecommunications, from which much of the terminology is derived. Here we present a brief background to facilitate understanding of the terminology as it appears in the EMG literature.

Radio Signal Demodulation

Consider an *information signal* that is to be transmitted: $S(t) = m\cos\omega_m t$, where m is the amplitude and ω_m is the angular frequency. The signal that contains the information usually has both a low amplitude and low frequency (figure 4.2). Transmission of the information signal then requires a *carrier signal* that has a greater amplitude and frequency: $A(t) = A_c\cos\omega_c t$, where A_c is the unmodulated amplitude and ω_c is the carrier frequency. The information signal and the carrier signal are combined to give a modulated signal:

$$A(t) = A_c\cos\omega_c t\left(1 + m\cos\omega_m t\right)$$

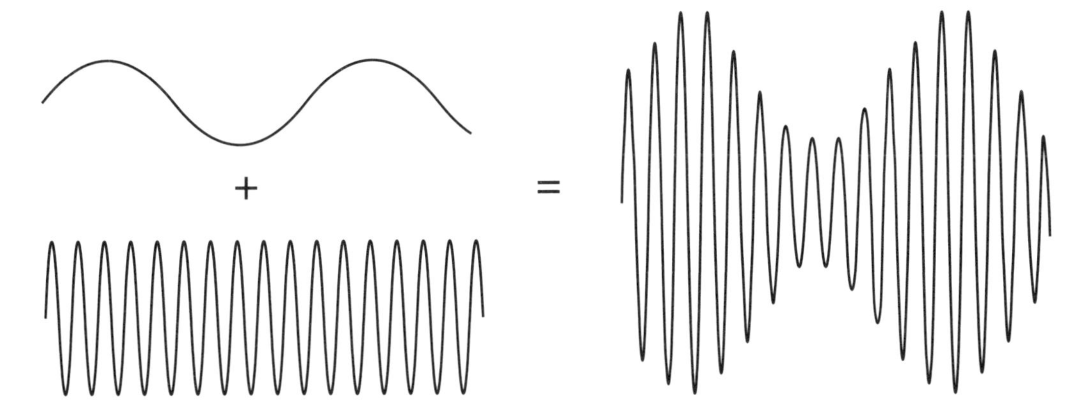

FIGURE 4.2 The creation of an amplitude-modulated signal. The information is contained in a low-amplitude, low-frequency signal superimposed on a high-frequency carrier signal (left). The information signal modulates the amplitude of the carrier signal (right).

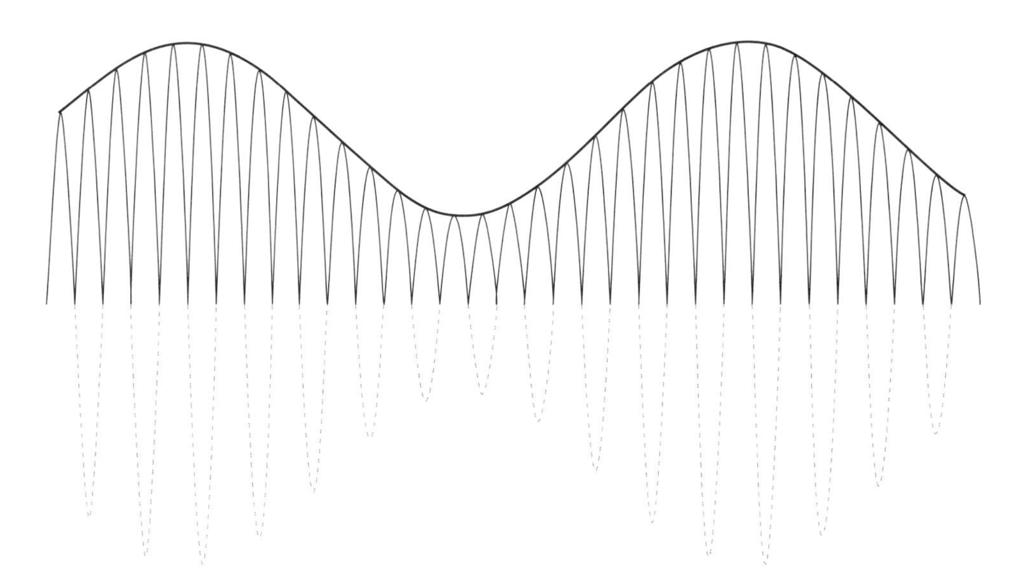

FIGURE 4.3 The envelope of the amplitude-modulated signal. The outer line connecting the peaks of the amplitude-modulated signal is the "envelope." The dotted line indicates the portion of the original waveform where the absolute value has been taken to rectify the signal.

The amplitude of the information signal (m) increases and decreases (i.e., *modulates*) the amplitude of the carrier signal $\left(A_c\right)$ when the two waveforms are combined. The term *envelope* is used to describe the contour of the resulting waveform (figure 4.3). Notice that the envelope mirrors the modulation of the carrier signal. The envelope therefore contains the information signal. If the absolute value of the signal is taken, the envelope still reflects the modulation but with even greater resolution.

Once the signal $A(t)$ is received, the carrier signal must be demodulated so that the information signal can be retrieved from the superimposed signal. There are various ways to *detect* the information in the superimposed signal. Because this is an amplitude-modulated signal, it is the amplitude of the information signal that must be detected. The most prevalent amplitude demodulation technique is *linear envelope detection.* The first step in the process is *full-wave rectification.* Full-wave rectification is the linear detector because it involves simply taking the absolute value of the raw signal (figure 4.3). The dotted lines in figure 4.3 show that the bottom half of the signal is now in the top half of the signal because the absolute value has been taken. The high-frequency component of the carrier signal is then removed by low-pass filtering. This leaves behind the slowly changing waveform associated with the envelope, which is the information content of the signal. The envelope is like a *moving average* because it follows the general trend of the modulated waveform. Demodulation may be implemented as a hardware circuit or through software as part of the offline signal processing.

Moving Average

It is appropriate to delve a little more deeply into the moving average because it utilizes several key concepts related to future topics in this chapter, one of which is the **moving window.** The moving average is applied to the data using a moving window. Consider a three-point moving window as illustrated in figure 4.4*a.* The y-values from t_1 to t_3 are summed, and the result is then divided by 3 to create the first average value $\left(\bar{x}_1\right)$. The calculations continue in the same way for each succeeding point

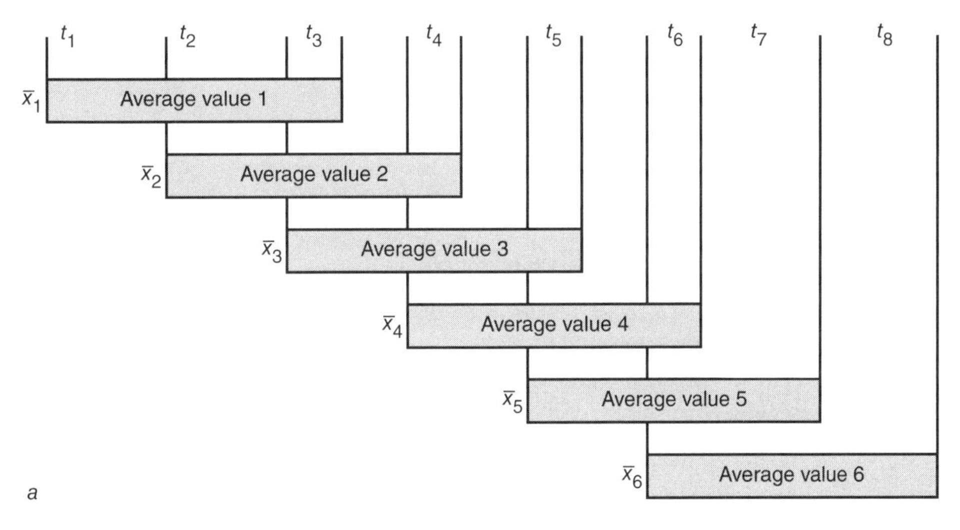

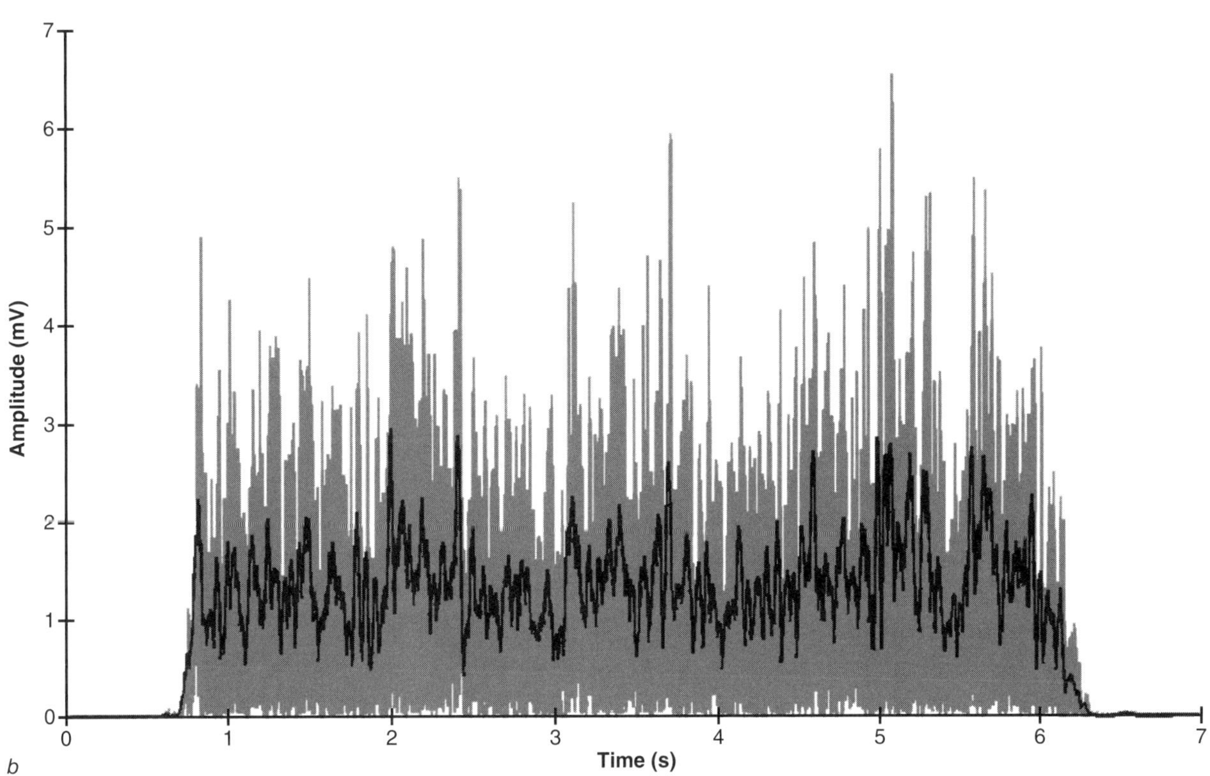

FIGURE 4.4 *(a)* A three-point data window used to calculate a moving average. *(b)* A moving average (black line) was applied to the absolute value of band-passed surface EMG activity (gray line) from the biceps brachii during an isometric maximal voluntary contraction of the flexors.

down the data stream. Because the moving window extends two points forward from the current point, the process must be stopped at two points from the end of the data stream ($N - 2$). Thus, the smoothed (or filtered) signal has two data points fewer than the raw signal, and the time base is different.

The degree of smoothness increases with the length of the moving window, but the length of the filtered signal decreases by an amount equivalent to the length of the moving window. The magnitude of change in the time base depends on both the window length (number of data points) and sampling rate. There are techniques to generate a filtered signal with the same length as the raw signal, but these require a more detailed treatment than is necessary here. The formula for implementing the moving average is as follows:

$$s(\tau) = \frac{1}{n} \sum_{j=t}^{t+(n-1)} x(t)$$

where n is the number of data points used in the window to smooth the data (smoothing factor), $x(t)$ is the raw data value at a specific point in time, and $s(\tau)$ is the smoothed data point in the new time base. A moving average was applied to the biceps brachii surface EMG generated during an isometric maximal voluntary contraction (MVC) of the flexors (figure 4.4*b*). The figure demonstrates that the moving average can be viewed as a basic low-pass digital filter. The low-pass cutoff frequency (f_c) decreases with increases in window length. The moving average is appealing because it is simple and easy to apply, but it has a very poor frequency response and is seldom used. It was presented here only to demonstrate the concept of the moving window.

EMG Signal Demodulation

Electromyographic activity is treated as an amplitude-modulated *noise signal* in which the information is contained in the envelope of the signal (Kadefors 1973). That is, noise is an amplitude-modulated carrier signal. Amplitude modulation is assumed to represent changes in the activation of motor units during the gradation of muscle force. On the basis of the discussion in the previous chapter, the term "noise" might have a negative connotation to the new initiate. In this case, it refers to the fact that the EMG signal is a random process wherein the amplitude distribution is Gaussian.

Linear envelope detection of the EMG follows the steps previously outlined for signal amplitude demodulation (figure 4.5). The first step is relatively straightforward. Full-wave rectification involves taking the absolute value of each EMG

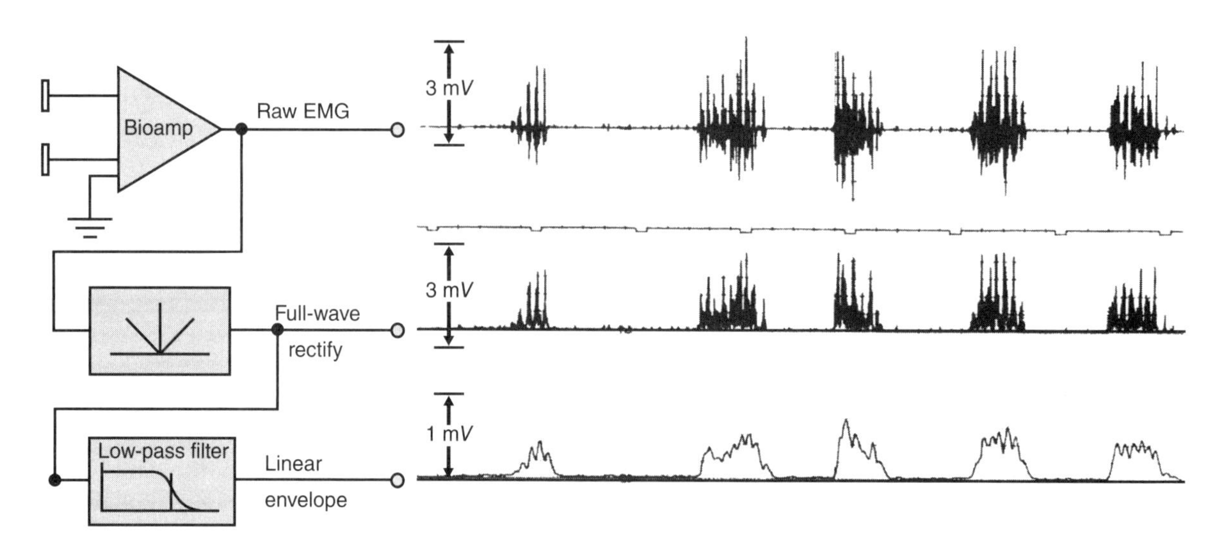

FIGURE 4.5 The steps involved in linear envelope detection of the EMG signal. The top area shows the raw EMG signal. The first step is full-wave rectification (middle). The final step is low-pass filtering (bottom). To the left of the EMG waveforms is the engineering schematic of the steps depicted in the three panels.

data point so that the signal has a positive polarity. The full-wave-rectified signal is then low-pass filtered. A wide range of low-pass cutoff frequencies exists in the literature, from 3 to 60 Hz. However, most EMG researchers and practitioners have used cutoff frequencies below 20 Hz. The exact cutoff frequency depends on the goals of the study. Winter (2005) has argued that the low-pass cutoff frequency should be set so that there is a biophysical basis for interpreting the output signal.

The full-wave-rectified motor unit action potential can be considered as an impulse to constituent fibers belonging to the motor unit that results in a muscle twitch. The motor unit impulse is a high-frequency input, and the muscle twitch is low-frequency output (figure 4.6). This is analogous to a low-pass filter circuit wherein a high-frequency voltage input is changed to a low-frequency voltage output. Linear envelope–detected EMG produced by a low-pass filter with the appropriate cutoff frequency will therefore begin to resemble the force–time curve of the active muscle group. The cutoff frequency is given by the resistor–capacitor time constant for the low-pass analog circuit presented in the previous chapter:

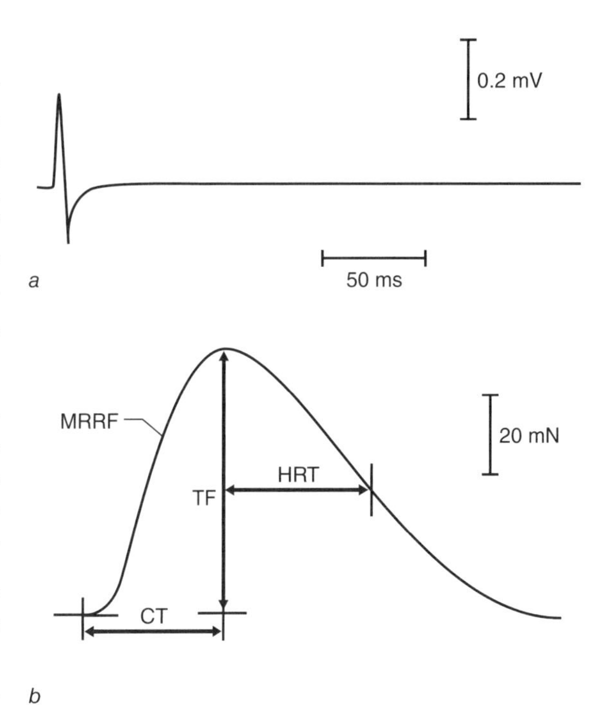

FIGURE 4.6 *(a)* Single motor unit action potential and *(b)* the resulting muscle twitch response produced by intramuscular microelectrical stimulation. Twitch force (TF), maximum rate of rise in force (MRRF), contraction time (CT), and half-relaxation time (HRT) are depicted on the force–time curve in *b*.

Reprinted from *Neuromuscular Disorders* (2)4, J.M. Elek, A. Kossev, R. Dengler, et al., "Parameters of human motor unit twitches obtained by intramuscular microstimulation," p. 262, copyright 1992, with permission from Elsevier.

$$f_c = \frac{1}{2\pi RC}$$

Winter (2005) suggests that the cutoff frequency can be directly linked with the twitch contraction times (CT) of the muscle under investigation. Using the preceding equation, for example, the twitch contraction time for the biceps brachii is on average 52 ms, so the low-pass cutoff frequency should be $f_c = 3.1$ Hz. Twitch times are based on isometric contractions. The low-pass cutoff frequency for dynamic contractions should be based on the frequency content of the movement involved. It is recommended that a low-pass cutoff frequency be set at a level that retains 95% of the total power (defined later) of the movement under consideration. A cutoff frequency specified as 95% of the total power is sufficient to reduce EMG variability while minimizing signal distortion. The logic behind this recommendation is that the frequency of muscle control cannot lie outside the frequency of movement (Shiavi et al. 1998).

KEY POINT

Linear envelope detection of the EMG signal is based on the idea that it is an amplitude-modulated noise signal. Visually, linear envelope detection is analogous to creating a moving average for the EMG signal. The EMG signal is first full-wave rectified and then low-pass filtered. The actual low-pass cutoff frequency depends on the goals of the study.

Linear Envelope EMG Measurement

The slowly changing EMG waveform associated with linear envelope detection is often preferred to facilitate the extraction of area, slope, onset, and shape characteristics of the muscle activity profile. Each measure is thought to reveal specific information about how the muscle is controlled during both static and dynamic contractions.

Area

Because linear envelope–detected EMG is a low-frequency waveform that forms an open area with the baseline, criterion measures obtained from this curve have been mistakenly referred to as integrated EMG (IEMG) activity. The linear envelope–detected EMG curve may be mathematically integrated using a simple algorithm for trapezoidal integration to obtain the area under the curve (see appendix 4.1). This measure is assumed to quantify the amount of muscle activity (or signal energy) within a given period of time, and may be termed IEMG. The units for IEMG are reported as mV·s.

Slope

If the muscle characteristics remain the same, alterations in the slope of the linear envelope–detected EMG may be used to infer changes in neuromuscular control (Ives et al. 1993; 1999). There tend to be a number of "peaks" and "valleys" in the linear envelope–detected EMG during the rising phase of activity, which makes it difficult to fit a line. The stochastic nature of the EMG signal will also make the trial-to-trial estimates quite variable. Linear envelope detection does decrease the variance of the EMG signal, but one must apply caution when decreasing the low-pass cutoff to stabilize the slope measure, because the main effects may be "filtered out" from the study.

A remarkable solution to the problem caused by decreasing the low-pass cutoff was provided by Gottlieb and colleagues (1989). The authors derived a mathematical proof (see appendix 4.1) showing that the slope of the rising phase of the EMG waveform can be approximated by numerical integration of the curve over a very short period of time (T). An integration interval of 30 ms was observed to accommodate a variety of contraction conditions. Thus, the measure is referred to as (Q_{30}). Although the units for the chord slope would be mV/s, it must be remembered that Q_{30} is actually obtained by numerical integration so the units are still mV·s. The problem of having to fit a line to the rising phase of EMG activity is solved because once the onset has been determined, a 30 ms integration interval may be set from that point onward. Furthermore, the low-pass cutoff frequency does not have to be adjusted to facilitate the procedure.

Onset

The correct procedure for the determination of EMG onset has received a great deal of attention and remains the subject of active investigation. Manual inspection of the EMG signal on the computer screen is the only way an investigator can have 100% confidence that every onset was correctly determined. However, this can be a time-consuming and tedious process. There is also the mistaken perception that manual onset determination is subjective and unreliable. Manual determination of EMG onset does require experience, but the agreement between experienced investigators can be quite high (Hodges and Bui 1996). The intraclass reliability coefficient for manual detection of EMG onset is equally high (Ives et al. 1993). Nevertheless, computer-automated determination is advocated as a means to dramatically decrease data reduction time, increase objectivity, and improve the reliability of the data (Micera et al. 2001). Ironically, in the absence of simulated EMG data, manual detection remains the gold standard against which new algorithms are tested (Hodges and Bui 1996).

The *double threshold method* is the most rigorously tested and validated computer algorithm (DiFabio 1987; Hodges and Bui 1996; Micera et al. 2001). It involves calculating the mean and standard deviation of baseline EMG activity in the absence of any muscle contraction. The first threshold criterion is that EMG amplitude must surpass a value that represents the 95% confidence interval ($\mu \pm 1.96\sigma$) for baseline activity. The confidence interval may be extended to 99% ($\mu \pm 2.58\sigma$) or greater if the level of baseline activity is significant. The value of employing statistical threshold criteria is that the Type I and Type II statistical error rates for onset detection are known (Hodges and Bui 1996). As with traditional hypothesis testing, the Type I error is higher for the 95% confidence interval, while the Type II error is higher for the 99% confidence interval (Micera et al. 2001). Detection of EMG onset before its true point is a Type I error, while detection of EMG onset after its true point (delay) is a Type II error. The incorporation of a second threshold has been used to decrease the Type I error. Once the amplitude of EMG activity surpasses the 95% confidence interval for baseline activity, it must stay above the threshold for a critical period of time (t_c). The exact length of time can range from 10 to 50 ms depending on the level of baseline activity. The second threshold minimizes false detection of the erratic departures from baseline that often occur prior to muscle contraction (Walter 1984; figure 4.7). The double threshold method is implemented as a point-by-point *moving window* whose length is determined by the critical time period.

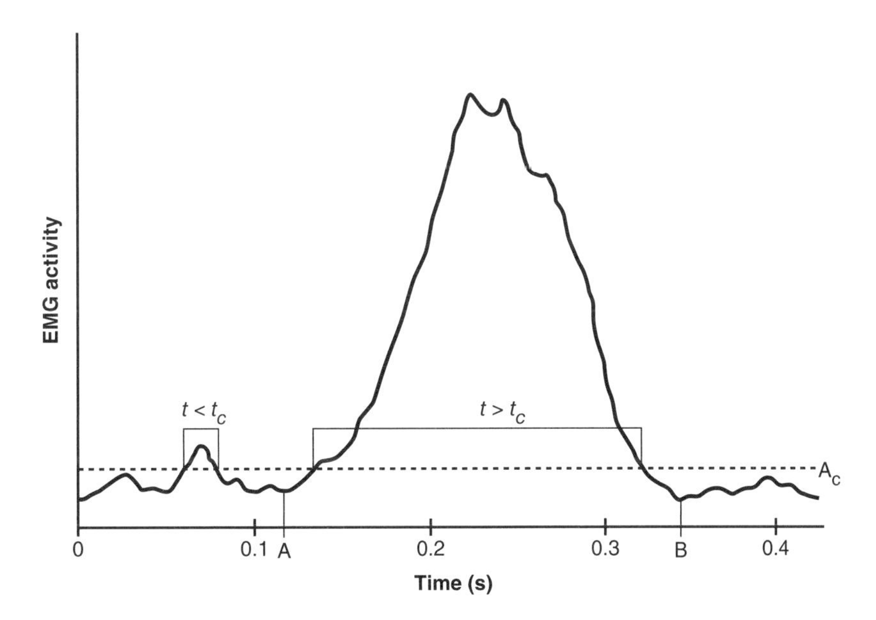

FIGURE 4.7 The onset of a linear envelope–detected EMG burst of muscle activity. The first criterion for the determination of EMG onset is an amplitude threshold (dotted line) that the burst must exceed. The second criterion is a critical time period (t_c) during which the EMG burst must stay above the amplitude threshold for the "true" onset of muscle activity. The smaller burst does not meet the double threshold criteria while the larger one does.

Reprinted from *Human Movement Science* 3(1-2), C.B. Walter, "Temporal quantification of electromyography with reference to motor control research," p. 157, copyright 1984, with permission from Elsevier.

For *computer-automated* determination of EMG onset, the low-pass cutoff frequency should be set at 50 Hz so that the linear envelope–detected EMG *more closely* follows the temporal events of the underlying band-passed signal (Walter 1984; Hodges and Bui 1996). It is very tempting to use a lower cutoff frequency to decrease the variance of the linear envelope, which can reduce the number of erratic departures from baseline. The following example illustrates why linear envelope detection at a very low cutoff frequency is not advisable for automated EMG onset detection. Consider three separate versions of the same surface EMG signal: (1) band passed (10-500 Hz), (2) linear envelope detected at $f_c = 3.1$ Hz, and (3) linear envelope detected at $f_c = 50$ Hz. The EMG activity was generated by the biceps brachii during maximal isometric elbow flexion.

The linear envelope–detected EMG at $f_c = 3.1$ Hz precedes the band-passed signal (figure 4.8*a*). The automated EMG onset would be detected earlier in comparison to the band-passed signal. The situation is not improved with use of the band-passed signal for computer-automated determination of EMG onset. The high-frequency fluctuations prevent the EMG amplitude from staying above threshold long enough to define EMG onset, except well into the progression of the contraction. The linear envelope–detected EMG at $f_c = 50$ Hz removes the high-frequency components so that the double threshold criteria may be used effectively for computer-automated determination of EMG onset (figure 4.8*b*). Moreover, the higher-frequency linear envelope follows the underlying band-passed signal more closely. Ideally, the visual and computer-automated determinations of EMG onsets for the two waveforms should correspond.

An interactive graphics program combines the best attributes of both manual and computer-based detection of EMG onset (Walter 1984). A computer is used to implement the double threshold criteria as a moving window on the linear envelope–detected EMG. The onset point and waveform are both plotted on the computer screen. The band-passed EMG is superimposed for the operator to verify the point obtained by the double threshold criteria. Misidentified onsets can then be corrected using a cursor to digitize the visually inspected data point.

Data reduction time for the interactive method is much faster than for complete manual detection, but it is not nearly as fast as for total automation. Time is indeed sacrificed with use of the interactive method, but the benefit is 100% confidence in correctly identified EMG onset. It is important to acknowledge that the threshold criteria may have to be adjusted to the unique characteristics of the data through trial and error. If the threshold criteria require constant adjustment, then the EMG burst is most likely buried within a large amount of noise around the baseline. Large baseline activity destabilizes the double threshold algorithm, and manual correction will be required more often than not. It is generally acknowledged that double threshold detection is very effective for low levels of baseline activity. The development of new detection algorithms when baseline activity is high is an active area of investigation (Micera et al. 2001).

Shape

The **variance ratio (VR)** was developed to quantify the similarity between linear envelope–detected EMG waveforms generated over multiple trials as part of gait analysis (Hershler and Milner 1978). Although the VR has had its greatest application in gait analysis, it is easily applied to any linear envelope–detected EMG waveform and is an excellent way to document the intraindividual variability (Calder et al. 2005). Unlike the situation with the correlation coefficient, a lower number (i.e., lower variance) is better. The amount of intraindividual variability that can be tolerated depends on the

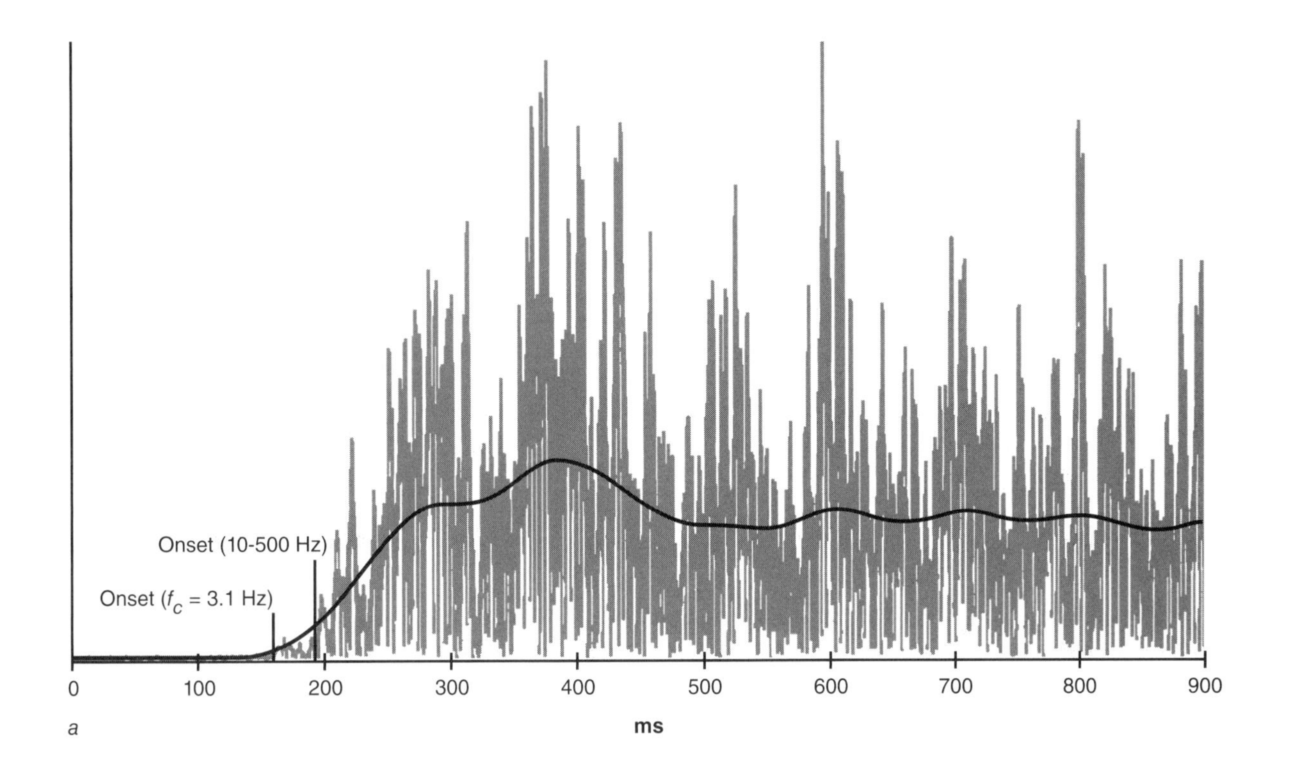

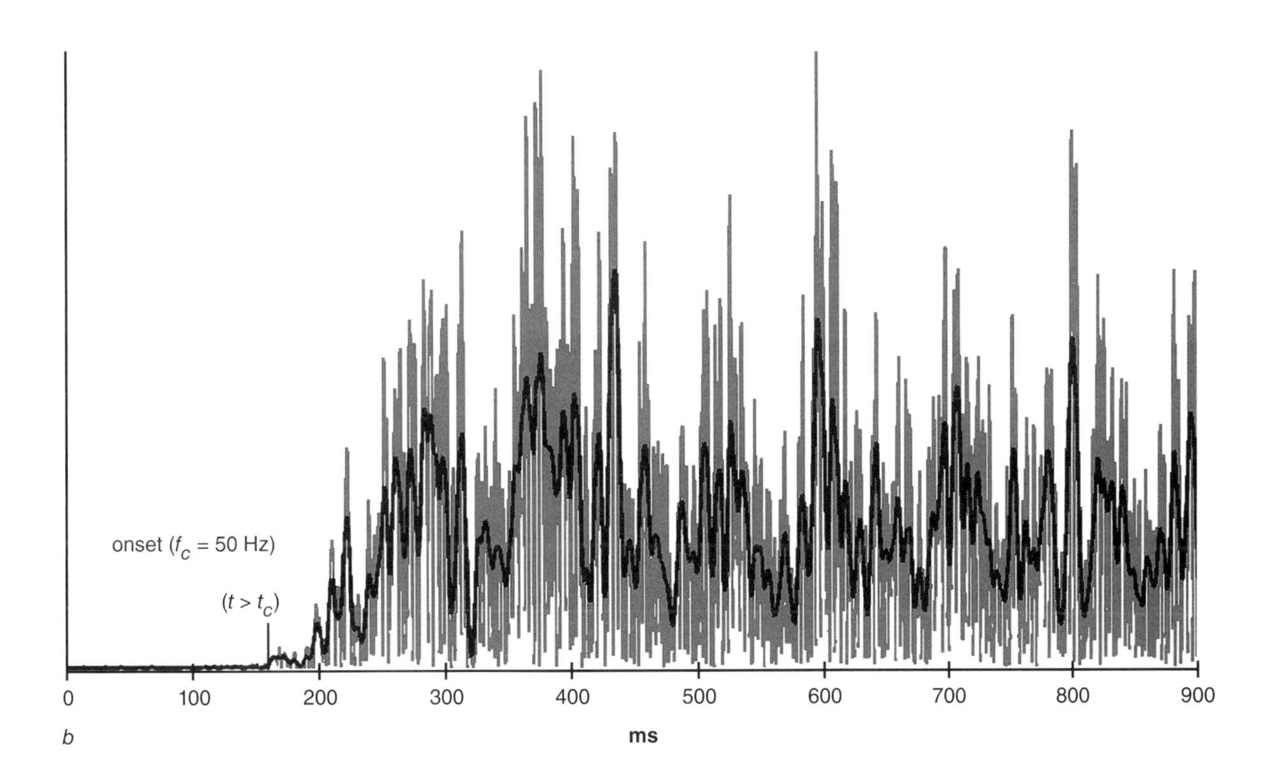

FIGURE 4.8 The interaction between low-pass cutoff frequency and the double threshold criteria. *(a)* The rectified band-passed EMG activity (gray) and the linear envelope–detected EMG at f_c = 3.1 Hz (black). Notice how the linear envelope–detected EMG precedes the band-passed EMG. *(b)* The rectified band-passed EMG (gray) and linear envelope–detected EMG at f_c = 50 Hz (black). The vertical lines denote the EMG onsets for the different signals in both *a* and *b.* Notice that the EMG onset is identified earlier for the linear envelope–detected EMG at f_c = 3.1 Hz. The onset for the band-passed EMG is then delayed with respect to linear envelope–detected EMG at f_c = 50 Hz.

nature of the study. A practical upper limit of VR = 0.40 has been used by Jacobson and colleagues (1995). Calder and colleagues (2005) compared the biceps brachii M-waves from two subjects, corresponding to a high and low VR (figure 4.9). The EMG data defined by the interval between the first point of the EMG waveform (t_1, y_1) and the last point (t_n, y_n) for each individual subject must be interpolated into the same number of data points (T). The formula for calculating the VR is

$$VR = \frac{\dfrac{\sum_{t=1}^{T}\sum_{n=1}^{N}(y_{t,n} - \bar{y}_t)^2}{T(N-1)}}{\dfrac{\sum_{t=1}^{T}\sum_{n=1}^{N}(y_{t,n} - \bar{y})^2}{TN-1}}.$$

To calculate $\bar{y}_t$, the EMG waveforms must be averaged on a point-by-point basis across all trials (N) within each subject to yield a single EMG waveform that is T data points in length. Since an average is calculated using the data points for $\bar{y}_t$, the result is a single value $(\bar{y})$ that represents the entire waveform.

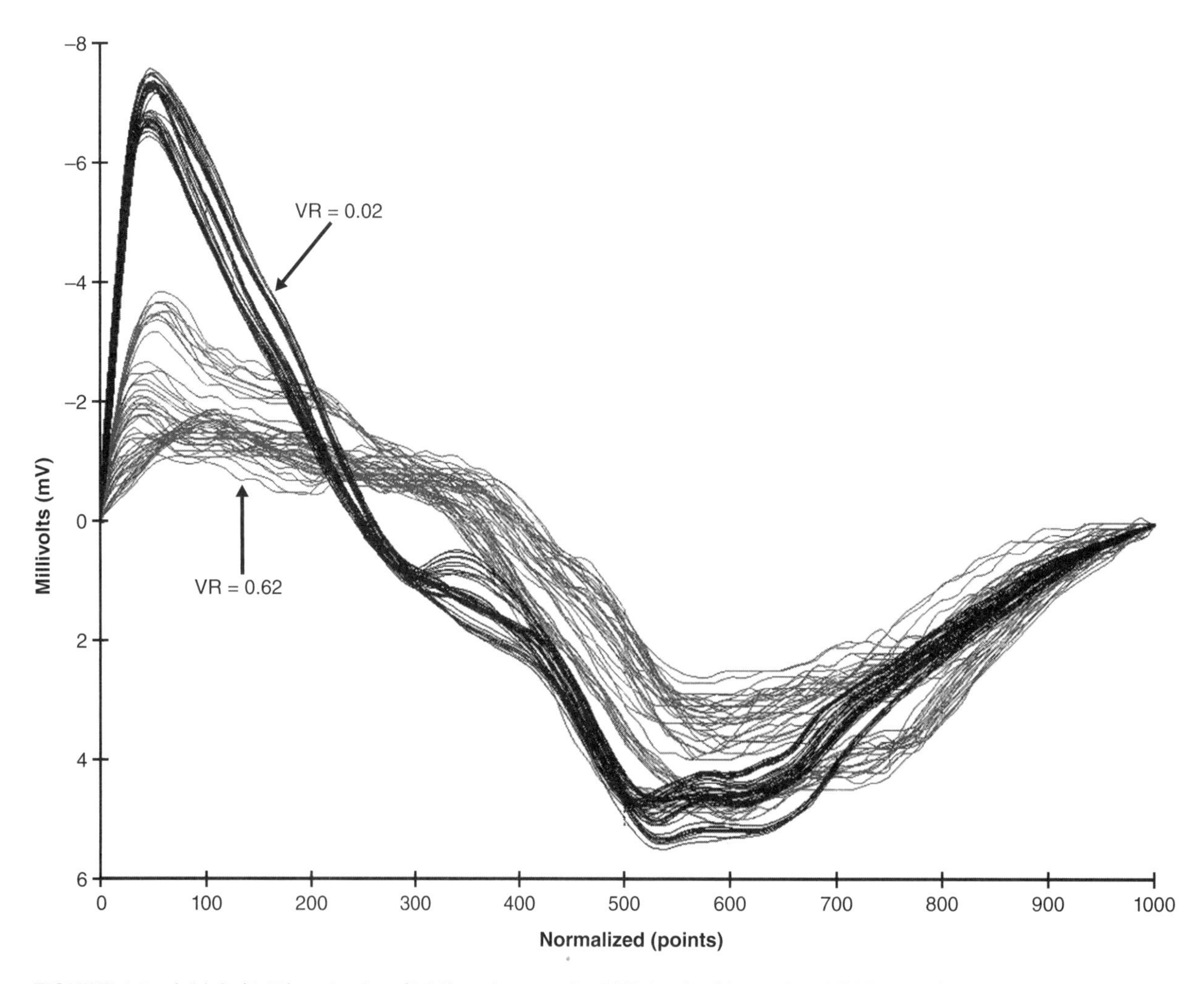

FIGURE 4.9 A high (0.62) and a low (0.02) variance ratio (VR) for the biceps brachii M-wave (compound muscle action potential, or CMAP) as observed for two different subjects. The waveforms were interpolated to fit into 1000 points from the identified onset and end. The waveforms associated with a low VR are tightly grouped, while there is considerably more fluctuation among the waveforms with high VR.

Reprinted from K. Calder, L.A. Hall, S.M. Lester, et al., 2005, "Reliability of the biceps brachii m-wave," *Journal of NeuroEngineering and Rehabilitation* 2:33. © 2005 Calder et al.; licensee BioMed Central Ltd.

KEY POINTS

- Linear envelope detection may be used to quantify the amount of muscle activity as measured by the area of the EMG burst. The rate of muscle activation is reflected in the rising phase of the EMG burst. This may be measured simply by the slope or Q_{30} of EMG activity.
- The timing of muscle activity is given by its onset. The onset of EMG activity may be digitized manually by visual inspection or with use of a computer interactive program that combines algorithmic detection with manual correction.
- The most widely used algorithm is the double threshold, which includes both magnitude and duration criteria for EMG onset. The accuracy of the double threshold method is, however, sensitive to the low-pass cutoff frequency used for linear envelope detection.
- The variability of the linear envelope–detected waveforms may be quantified using the VR.

Band-Passed EMG Measurement

There are two EMG detectors that involve the band-passed EMG signal, necessitating little or no additional processing (De Luca and Van Dyk 1975; Farina and Merletti 2000. The **average rectified value (ARV)** is a linear detector because it is calculated using only the absolute value of each datum of EMG over a specific interval $(0, T)$. Because the specific interval occurs over a given period of time, it is often referred to as either an *epoch* or a *data window.* Where $|EMG(t_i)|$ is the absolute value of a datum of EMG in the data window,

$$ARV = \frac{1}{T}\sum_{t=1}^{T} |EMG(t_i)|.$$

The unit of measurement is mV or µV. The ARV calculation is similar to the numerical formula for integration and is often confused with IEMG, especially if it is calculated on the linear envelope–detected EMG waveform.

The **root-mean-square (RMS)** amplitude is a nonlinear detector based on the square law. There is no need to rectify the EMG signal prior to its calculation. Where $EMG^2(t_i)$ is the squared value of each datum of EMG within the data window,

$$RMS = \sqrt{\frac{1}{T}\sum_{t=1}^{T} EMG^2(t_i)}.$$

The RMS amplitude has both physical and physiological meaning. First, it should be recognized that RMS amplitude is linked with the signal power (V_{rms}^2). Changes in the ARV of the EMG signal due to motor unit recruitment, firing rate, or muscle fiber conduction velocity are highly dependent on amplitude cancellation due to the superimposition of positive and negative phases of motor unit action potentials (Keenan et al. 2005). The RMS amplitude of the EMG signal is considered a better measure than ARV for monitoring changes in muscle activity because it is not affected by cancellation (De Luca and Van Dyk 1975; De Luca 1979). It is important to emphasize that specific motor unit behavior cannot be inferred from amplitude changes alone. In general, the ARV and RMS amplitude of the EMG signal are often reported simultaneously. The ARV is proportional to the RMS of the signal if it has a Gaussian probability distribution (Lowery and O'Malley 2003). However, this is not likely to be the case at low force levels when few motor units are active (Kaplanis et al. 2009).

Mean spike amplitude (MSA) is a linear detector that that has recently been introduced because it has the additional benefit of using noise rejection criteria. As a result, the MSA displays highly stable measurement properties during dynamic contractions (Gabriel 2000), and it is highly correlated with RMS amplitude (Gabriel et al. 2001). The high correlation exists because the MSA is a nonparametric way to calculate the peak-to-peak (P-P) amplitude for a stochastic signal, and there is an inherent relationship between RMS and P-P amplitude:

$$V_{rms} = \frac{V_{pp}}{\sqrt{2}}$$

An EMG spike is defined as a pair of upward and downward deflections that both cross the isoelectric line and are greater than the 95% confidence interval for baseline activity (figure 4.10). Deflections pairs that do not cross the isoelectric line are not a fully developed spike. A peak is a pair of upward and downward deflections within a spike. These peaks do not constitute a discrete spike. Peaks are ignored in the MSA calculation and are denoted by an "x" in figure 4.10. Each spike is identified with a circle at the apex and squares at the base. Where y is the amplitude value of EMG at time (t), single spike amplitude (SA_i) is calculated as follows:

$$SA_i = \frac{\left(B_y - A_y\right) + (B_y - C_y)}{2}$$

The total number of spikes (NS) within the data window is used to calculate the mean:

$$MSA = \frac{1}{NS}\sum_{i=1}^{NS} SA_i$$

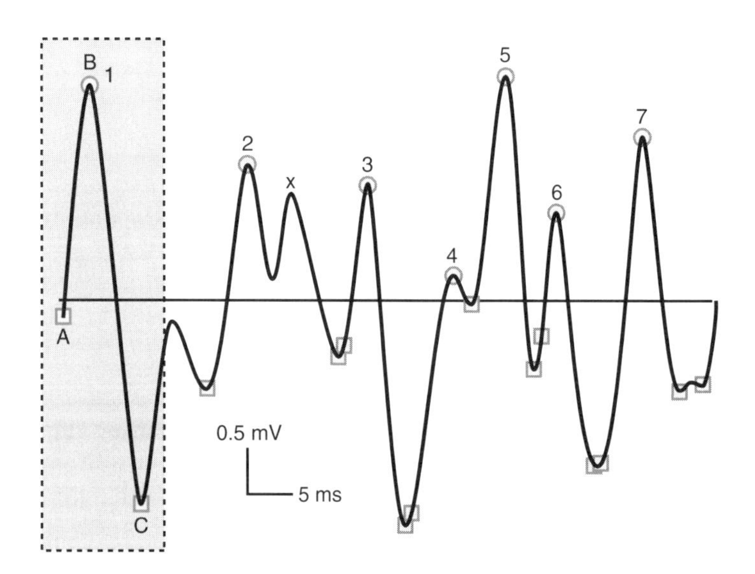

FIGURE 4.10 A short segment of biceps brachii EMG. There are seven complete EMG spikes. Circles denote the apex of each spike, and squares show the base. The "x" is a peak within a spike. The first spike is highlighted to show that the definition includes two vertical deflections that cross the isoelectric baseline. The deflections between spikes 1 and 2 do not constitute a complete or full spike because they do not cross the isoelectric baseline. The first spike is labeled A, B, and C to illustrate calculation of the spike measures (see text).

KEY POINTS

- The ARV, RMS amplitude, and mean spike amplitude of band-passed EMG activity are used to quantify the magnitude of muscle activity.
- It is dangerous to overinterpret what the magnitude of the EMG signals means, as it is affected by many factors.

Cross-Correlation Function

The **cross-correlation function** is most often used to determine the time delay between two EMG signals. The time delay between two pairs of EMG electrodes is used in the calculation of muscle fiber conduction velocity. The EMG may be cross-correlated with either a force or displacement signal to determine the electromechanical delay—the interval between EMG onset and the onset of force or displacement. The second application of cross-correlation is to determine the presence of cross-talk between two electrodes.

Background of the Correlation

Consider the more familiar standard deviation (SD) formula:

$$SD = \sqrt{\frac{\sum_{i=1}^{N}(x_i - \bar{x})^2}{N}}$$

where x_i is a single observation, $\bar{x}$ is the mean of all the observations, and N is the total number of observations. The basis of this formula is the deviation score $(x_i - \bar{x})$. The deviation score expresses how far any one observation lies from the mean. Recall that there are always an equal number of observations above and below the mean. Each deviation score must therefore be squared; otherwise their sum would be zero. Dividing the sum of the squared deviation scores by the total number of observations (N) calculates the *average amount by which any one observation deviates (or varies) from the mean.* The square root then returns the magnitude of the result back to its original scale. In one sense, the SD is actually another type of mean. The denominator ($N - 1$) is actually used to calculate the unbiased estimator, but apart from that it is somewhat difficult to recognize that the SD is really another type of average (Glass and Hopkins 1996).

The next step is to consider a normalized version of the SD that is independent of the original units of measurement: the *z-score.* The z-score is the average amount by which any one observation differs from the mean in *standard deviation units:*

$$z_i = \frac{(x_i - \bar{x})}{\sqrt{\frac{\sum_{i=1}^{N}(x_i - \bar{x})^2}{N-1}}}$$

It should be obvious from the preceding discussion that the sum of the z-scores is zero and therefore the mean of the z-scores is zero, and the SD of z-scores is always 1.

There are two main advantages to the use of z-scores. First, the z-score conveys the magnitude of the score relative to the distribution of scores. For example, there is no real way to know if a 1.2 mV_{pp} score for tibialis anterior EMG from a particular individual is large or small. If the z-score is –0.8, it is almost one full SD unit below the mean. The EMG amplitude is low relative to that of the other participants in the study. This leads to the fundamental basis of the correlation coefficient. *The correlation coefficient is a way to quantify the degree to which the z-scores for two different measures coincide.* The same individual may then have a maximal isometric dorsiflexion strength score of 150 N, which might correspond to a z-score of –1.5 in this sample of subjects. The z-scores for both measures are consistent with the idea that a relationship exists between force and EMG activity. That is, a low-strength ($z = -1.5$) individual might be expected to have low maximum ($z = -0.8$) EMG activity. If the correlation between force and EMG were perfect, the individual would have the exact same z-score for both measures. We are now in a position to step through the formula for the correlation coefficient, which is the basis for the digital implementation of the cross-correlation function:

$$r_{xy} = \frac{\dfrac{\sum_{i=1}^{N}(x_i - \bar{x})(y_i - \bar{y})}{N-1}}{\sqrt{\dfrac{\sum_{i=1}^{N}(x_i - \bar{x})^2}{N-1}}\sqrt{\dfrac{\sum_{i=1}^{N}(y_i - \bar{y})^2}{N-1}}}$$

Notice that $1/(N-1)$ can be factored out of the two terms $(1/\sqrt{N-1})$ in the denominator and cancelled within the numerator. Also recall that since $\sqrt{a}\sqrt{b} = \sqrt{ab}$ the term in the denominator can be combined under one radical:

$$r_{xy} = \frac{\sum_{i=1}^{N}(x_i - \bar{x})(y_i - \bar{y})}{\sqrt{\sum_{i=1}^{N}(x_i - \bar{x})^2 \sum_{i=1}^{N}(y_i - \bar{y})^2}}$$

The numerator is the *sum of cross-products* between two variables, x and y. Each score is located relative to its own distribution, and the two deviates $(x_i - \bar{x})$ and $(y_i - \bar{y})$ are multiplied. The denominator thus measures the *covariance*—the degree to which the two measures vary from their respective means at the same time (Glass and Hopkins 1996).

We will use the force (x) and EMG (y) measures to illustrate how covariance is employed to quantify the relationship between two variables. The following covariance pattern describes a *positive relationship* (figure 4.11). High force and high EMG values result in large positive deviates for both force and EMG, so that the cross-products will be large and positive. At the same time, low force and EMG values result in large negative deviates for both force and EMG. The multiplication of two negative values still results in large positive cross-products. The overall sum of the cross-products is positive. The opposite covariance pattern is associated with a *negative relationship.* The reader is encouraged to graph the negative relationship covariance pattern to reinforce the concept. High force and low EMG result in large positive deviates for force and large negative deviates for EMG. The cross-products will be large and negative. Low force and high EMG have large negative deviates for force and large positive deviates for EMG. The cross-products will be large and negative, so that the overall sum of the cross-products is negative. A random covariance pattern results in little or no relationship. If high force is associated with

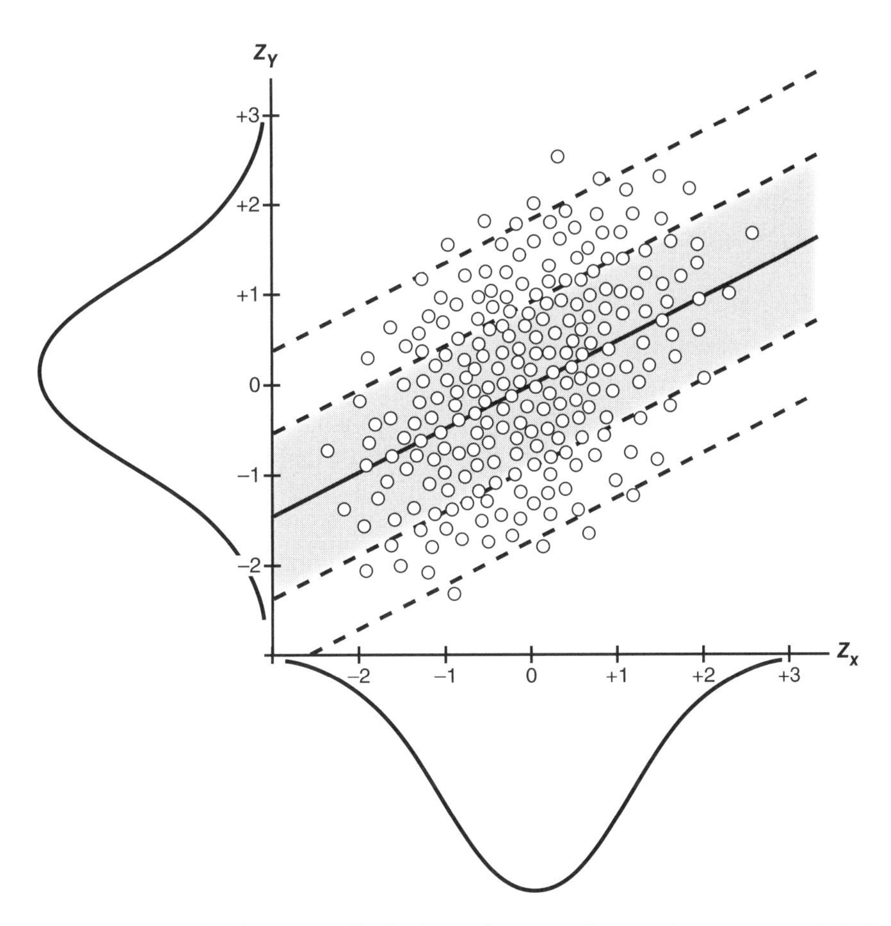

FIGURE 4.11 Hypothetical frequency distributions of z-scores for *x* and *y* measures plotted against each other. Notice that high z-scores on one distribution are associated with high z-scores on the other distribution, and low z-scores on one distribution are associated with low z-scores on the other distribution. The "best fit" line through the data therefore has a positive slope, denoting a positive relationship.

From Glass, S.V., & Hopkins, K.D. *Statistical Methods in Education and Psychology,* 3e. Published by Allyn and Bacon/Merrill Education, Boston, MA.

both high and low EMG, the cross-products will be both positive and negative. Low force linked with both high and low EMG will also have both positive and negative cross-products. The overall sum of the cross-products for both conditions will be close to zero.

The numerator may result in a positive, a negative, or no relationship, but the magnitude is still dependent on the original units of the measures involved. The same principle used for the z-score must be applied in this situation. The denominator contains the SD of both *x* and *y* measures. The sum of the cross-products is then normalized by the SD of both distributions. It is important to recognize that the denominator essentially turns the correlation coefficient into an average z-score for the two variables. Since the SD of z-scores is 1, the correlation coefficient is bounded by those limits. A perfect positive correlation is $r = 1.0$, a perfect negative correlation is $r = -1.0$, and no relationship is $r = 0$.

Calculation of Cross-Correlation Function

It will be easier to understand the cross-correlation function if we develop an intuitive sense by stepping through how it is calculated. Imagine that there are two signals

$x(t)$ and $y(t)$ in separate columns of a spreadsheet, columns 1 and 2, respectively. The signals are 1 s in duration, and each was sampled at 1 kHz. Thus, there are 1000 data points in each column. Calculating the correlation coefficient r_{xy} with the two columns aligned at $x(t_1)$ and $y(t_1)$ is synonymous with calculating the first data point of the cross-correlation function $R_{xy}(\tau)$ at a lag time of $\tau = 0$. This means that the cross-correlation coefficient is indexed starting at $R_{xy}(0)$. The next step is to move $y(t)$ relative to $x(t)$. The signal $y(t)$ is moved down one row so that there is an empty cell right next to $x(t_1)$. The correlation coefficient r_{xy} is then calculated on the remaining overlapping data points. The second data point in the cross-correlation function $R_{xy}(1)$ represents the correlation between $y(t)$ and $x(t)$ at a lag time of $\tau = 1$ ms.

It should be evident that two fewer data points are used to estimate $R_{xy}(1)$. As the calculation of the cross-correlation function continues, shifting $y(t)$ relative to $x(t)$ decreases the number of data points used to estimate $R_{xy}(\tau)$. Thus, as the process continues, there is a progressive increase in the variance of the estimate $R_{xy}(\tau)$ because fewer and fewer data points are used in its calculation. To avoid this problem, we assume that the signals are periodic, with a period equal to the duration of the sample. The signal can then be wrapped around itself so that the same number of data points can be used to calculate each lag time. This minimizes the variance of the estimate and keeps it constant for each lag time. To visualize this, consider a periodic waveform like a sine wave wherein data points from the front are "tacked on" to the back to generate an additional period (figure 4.12).

In the present example, the last data point $y(t_{1000})$ is moved to the first row so that the first row contains $x(t_1)$ and $y(t_{1000})$ and the last row contains $x(t_{1000})$ and $y(t_{999})$. The correlation coefficient is now calculated between $y(t)$ and $x(t)$ at their original lengths (n). The resulting cross-correlation $R_{xy}(1)$ still represents a lag time of $\tau = 1$ m. To continue to the next calculation, the "new" last data point $y(t_{999})$ is moved to the first row so that the first row contains $x(t_1)$ and $y(t_{999})$ and the last row contains $x(t_{1000})$ and $y(t_{998})$. The third data point in the cross-correlation function $R_{xy}(2)$ is the correlation between $y(t)$ and $x(t)$ at a lag time of $\tau = 2$ ms. The

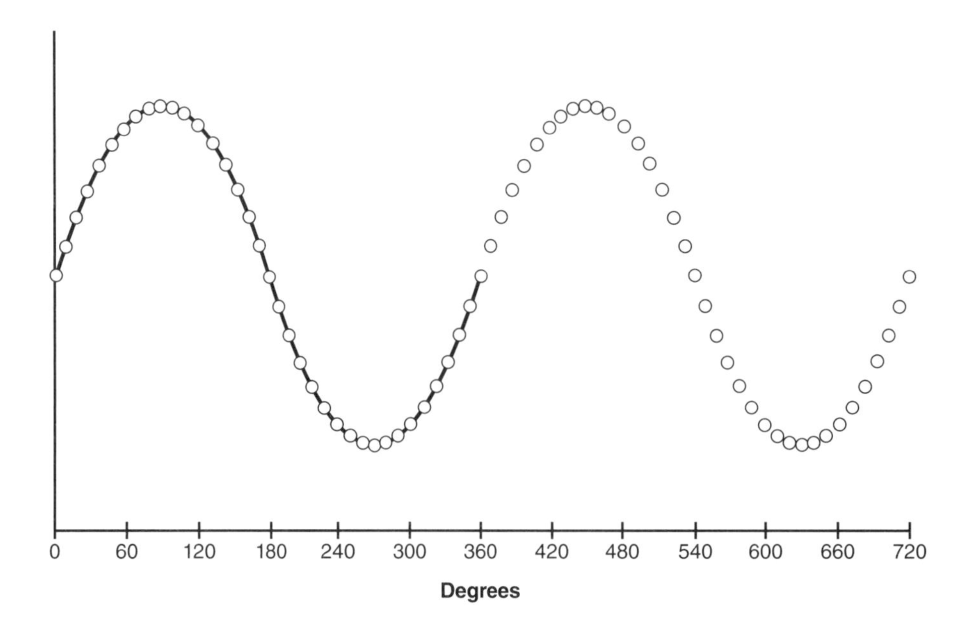

FIGURE 4.12 To extend one period (T) of a sine wave to two periods ($2T$), points at the front of the waveform are copied and added to the back of the waveform.

process of moving $y(t)$ relative to $x(t)$ and calculating $R_{xy}(\tau)$ continues until a full signal period at n data points is reached. The cross-correlation function thus far has a length of n, wherein each point is the correlation between $y(t)$ and $x(t)$ at linearly increasing lag times. By convention, the lag times are considered positive. Calculation of the cross-correlation function is only half complete at this point. The first row should now contain $x(t_1)$ and $y(t_1)$ once again.

The same process must be repeated, except that $x(t)$ is moved relative to $y(t)$. The lag times in this case are designated by negative values. The complete cross-correlation function is actually limited to twice the signal length minus 1 ($N-1$), where N represents the total number of data points used to calculate $R_{xy}(\tau)$ for positive, zero, and negative lag times. The term $N-1$ represents the degrees of freedom. The entire process is called *circular correlation.* The cross-correlation function may be calculated using the convolution integral (appendix 4.2). However, for the sake of simplicity, it may also be calculated using the traditional Pearson correlation coefficient, except that the lag time (τ) is used as an index for each coefficient that is calculated:

$$R_{xy}(\tau) = \frac{\sum_{n=1}^{N}\left[(x_n - \bar{x})(y_{n-\tau} - \bar{y})\right]}{\sqrt{\sum_{i=1}^{N}(x_i - \bar{x})^2 \sum_{i=1}^{N}(y_i - \bar{y})^2}}$$

Muscle Fiber Conduction Velocity

Application of the cross-correlation function to determine muscle fiber conduction velocity (MFCV) assumes that the signal $y(t)$ is the same as $x(t)$, except that it is detected at a slightly different location as the action potentials propagate along the muscle fiber. Conduction time between the two pairs of electrode surfaces is then defined as the lag time (τ) associated with the maximum of the cross-correlation function (Hunter et al. 1987; Fiorito et al. 1994). This should make intuitive sense because the maximum of the cross-correlation function locates the number of data points by which $y(t)$ would have to be shifted back so that the signal peaks of $x(t)$ and $y(t)$ align. When the peaks are aligned, the cross-correlation should be at a maximum (figure 4.13). The MFCV is then simply computed as the interelectrode distance divided by the lag time (τ) associated with the maximum of the cross-correlation function. The maximum cross-correlation is not +1 because the muscle fiber action potentials (MFAPs) change shape as they propagate due to dispersion. There is no standard minimum value for the cross-correlation coefficient that is acceptable for analysis. Literature values range from 0.60 to 0.90 with careful methodological controls (Broman et al. 1985a).

The cross-correlation function is applied to the band-passed EMG signals from two pairs of electrode detection surfaces. If the interelectrode distance is 0.015 m and MFCV is 5 m/s, it takes the MFAP 3 ms to travel from one detection surface to the other. A sampling rate of 2 kHz would result in only six data points to resolve the latency between electrodes. The precision of measurement is clearly not sufficient. Either one of two simple strategies may be employed to increase the precision of measurement. One way is to interpolate the cross-correlation function to a higher sampling rate so that each lag (τ) represents a smaller division of time (Rababy et al. 1989). The desired increase in sampling rate depends partly on the interelectrode distance and the range of expected values. The other method is to actually sample the signals at a much higher rate (for example, 50 kHz per

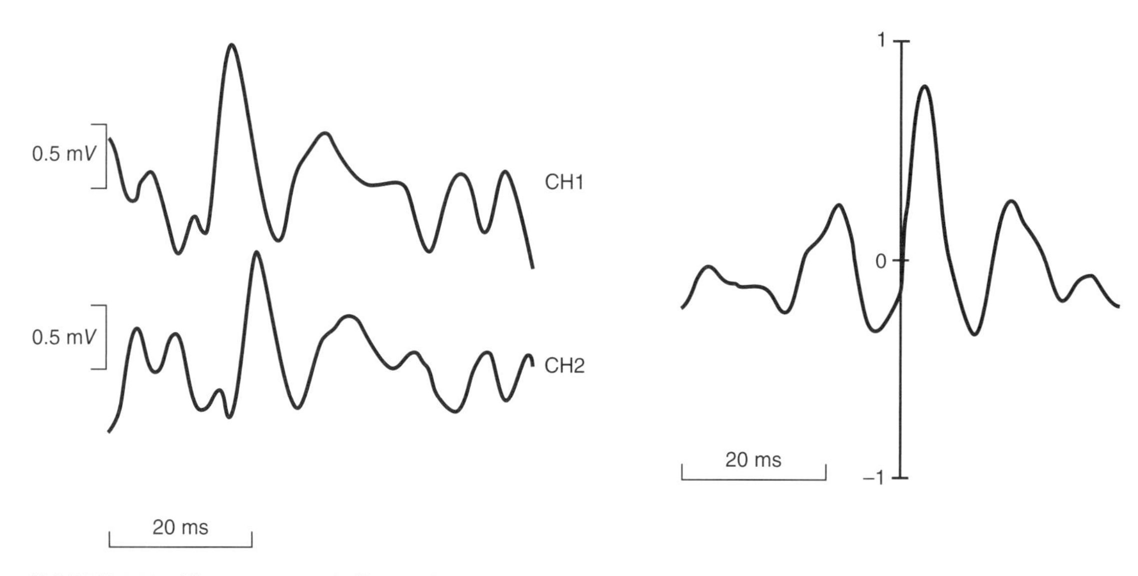

FIGURE 4.13 The same muscle fiber action potential recorded at two different electrodes on channel 1 (CH1) and channel 2 (CH2), respectively (left). Channel 1 is closest to the motor point. The cross-correlation function has a maximum value at a positive lag time (τ), indicating that channel 2 is delayed with respect to channel 1 (right). The lag time (τ) associated with the maximum value corresponds to the amount of time the signal in channel 1 would have to be shifted back so that the peaks between the two channels would be aligned.

channel) to ensure adequate resolution (Fiorito et al. 1994). Technological advances in analog-to-digital (A/D) data acquisition boards and in computing power have made it possible to actually sample the signals at very high rates and perform the postprocessing with relative ease.

Since the goal is to record the same EMG waveform across two different pairs of detection surfaces, the interelectrode distance is kept small (5-15 mm) in order to maximize selectivity and minimize dispersion effects. The use of double differential electrodes is another way to increase selectivity. Nondelayed activity associated with common mode noise degrades the ability of the cross-correlation function to reflect delayed activity associated with MFAP propagation. The result is an overestimation of MFCV values (Broman et al. 1985a). The EMG data from the two pairs of electrodes may also be normalized so as to have zero mean and unit variance, via conversion of each datum of EMG to a z-score. Normalization presumably decreases the impact of dispersion on the magnitude of the cross-correlation coefficient (Hunter et al. 1987). Electrode misalignment with respect to muscle fiber arrangement is a significant source of error in calculating the MFCV. The goal is to have the MFAPs propagate along a straight line between the two detection surfaces. The overestimation error in MFCV is equivalent to the cosine of the misalignment angle multiplied by the interelectrode distance (Sollie et al. 1985b). A way to reveal muscle fiber orientation is to evoke small, local twitch responses such as those that occur when the motor point is electrically identified (Arendt-Nielson and Mills 1988).

Electromechanical Delay

Voss and colleagues (1991) and Vint and colleagues (2001) have advocated the use of the cross-correlation function to detect EMG onset relative to mechanical signals. The main advantage is that determination of electromechanical delay (EMD) is completely objective, unlike what occurs with the interactive graphics method described earlier. The precision of measurement is still important in calculating the cross-correlation function between EMG and mechanical signals. The EMG signal must be linear envelope detected to produce stable estimates of the cross-correlation function. Unfortunately, there is no general rule about the most appropriate low-pass cutoff frequency. Low-pass cutoff frequencies between 2 and 20 Hz have been reported in the literature (Vint et al. 2001; Voss et al. 1991; Li and Caldwell 1999). In the absence of any specific recommendation, it would be prudent to calculate the EMD for a range of low-pass cutoff frequencies (2-25 Hz). Then, one can evaluate the variability in means relative to the effect size in the experiment (Vint et al. 2001).

Cross-Talk

Cross-talk may be defined as electrical activity from adjacent or distant muscles (or both) recorded by the electrodes over the muscle of primary interest. The maximum of the cross-correlation function is used to determine the amount of signal in common between two electrodes. The r^2 value of the cross-correlation is termed the *coefficient of determination,* and its value multiplied by 100% gives the percent *variance* of one measure that is *accounted for* by the other. Likewise, where R_{xy} is the magnitude of the cross-correlation coefficient at lag time (τ), R_{xy}^2 is *common signal* or *common power* between two electrodes (Winter et al. 1994).

It is generally accepted that the cross-correlation function is not the best way to evaluate the presence of cross-talk because it cannot separate the effects of motor unit synchrony or common drive from volume-conducted muscle activity (Lowery et al. 2003). The cross-correlation function has, however, been used effectively to study the magnitude of common signal as the distance between adjacent electrodes increases. It is interesting to note that two adjacent electrodes placed 2.5 cm apart on the knee extensors can contain half the amount of common signal (25%; Winter et al. 1994) found in two adjacent electrodes placed 3 cm apart on wrist extensors (50%; Mogk and Keir 2003). The size of the volume conductor is clearly a moderating factor. The most effective way to evaluate the magnitude of cross-talk is through the use of evoked potentials (De Luca and Merletti 1988; Solomonow et al. 1994). Since the magnitude of antagonist EMG activity can be small, it is important to demonstrate that low-amplitude antagonist EMG activity is not volume-conducted activity from the agonists or synergists. The peripheral nerve of the agonist or synergist, or of both, should be stimulated, and the P-P amplitude of the evoked potential at the electrodes over the antagonist can be measured. The same method can be used to evaluate the magnitude of cross-talk from neighboring muscle.

There are only a limited number of ways to reduce cross-talk. Double differential electrodes have a greater ability to reduce common mode signal than bipolar electrodes. If the muscle is not large enough to support the application of a higher-order electrode configuration, then the bipolar interelectrode distance should be kept small (5-10 mm) to increase selectivity. A detailed knowledge of anatomical kinesiology is extremely important. Electrode location can have a significant impact in decreasing the pickup of common signal from adjacent muscles. The electrodes should be placed in the center of the muscle belly, away from the muscle borders. Placement can then be verified through test contractions of muscles suspected of generating cross-talk and monitoring the EMG activity of the muscle of interest on an oscilloscope (Winter et al. 1994).

Subcutaneous fat is the main risk factor for cross-talk. The reason is that subcutaneous tissue increases the distance between the electrode and the source potential (Dimitrova et al. 2002). The greater distance increases the magnitude of the end effects at the muscle–tendon junction *relative to* the overall size of the motor unit action potential (MUAP) (see appendix 3.1). The potential associated with the end effects at the muscle–tendon junction is called the terminal phase of the MUAP, and it is a high-frequency source. It is also termed the *nonpropagating* component of the MUAP because its location does not depend on the axial position of the electrode along the muscle fiber (Dimitrova et al. 2002; Farina et al. 2002). The high-frequency terminal phases are not reduced by the low-pass filtering effects of tissue. *Low-pass filtering property* of tissue is actually a misleading term. The origin of this tissue filtering property reflects the well-known observation that the main *(propagating)* phases of extracellular potentials produced by the motor unit become wider with increasing electrode distance from the source, the muscle fibers. What is the actual cause of the widening of the extracellular potentials? Extensive modeling demonstrates that the distance-related changes in the MUAP are associated with a volume conductor that is simply resistive, not a low-pass filter (Dimitrova et al. 2002). The main implication is that high-pass filtering cannot be used to reduce cross-talk as has been previously recommended (Winter et al. 1994).

KEY POINTS

- The cross-correlation function determines the time delay at which two signals are maximally correlated. The cross-correlation function may be calculated on the signal passing underneath two different electrodes placed a known distance apart to determine MFCV.
- The same may be done between EMG activity and a mechanical measure of muscle contraction to determine EMD. Although the magnitude of the cross-correlation function has been used to assess the degree of cross-talk from adjacent muscles, this is not advisable, as there are better methods.

Frequency

The frequency content of the EMG signal is used to provide both physiological and nonphysiological information. Physiological aspects of the EMG signal provided by frequency analysis include MFCV and to a lesser extent motor unit firing rates. Nonphysiological information relates to certain types of noise contamination from electrical interference within the EMG signal, which can be easily identified by frequency analysis. Frequency analysis is therefore a powerful tool, and its basic principles are well worth understanding.

Fourier Series

The French mathematician J.B.J. Fourier (1768-1830) showed that a periodic waveform $f(t)$ of period T can be represented as a linear combination of sines and cosines, each having different amplitudes and frequencies:

$$f(t) = \frac{a_0}{2} + a_1 \cos(\omega t) + a_2 \cos(\omega t) + \cdots a_n \cos(\omega t) + b_1 \sin(\omega t) + b_2 \sin(\omega t) + \cdots b_n \sin(\omega t)$$

The sines and cosines are termed the alternating current (AC) components of the waveform. Combining the two trigonometric functions is a convenient way to account for the phase component of the waveform, rather than using $\sin(\omega t+\phi)$. The term $a_0/2$ represents the zero frequency or the direct current (DC) component that may be present in the waveform; it may also be considered as the average magnitude over the period T. For example, if a sinusoid oscillates around a value of 1, the mean value of the sinusoid is 1. Since DC has a constant value, the DC component of the sinusoid is 1. The first term in the Fourier equation can also be called the DC offset because the sinusoid does not oscillate about the zero isoelectric line but is offset by a value of 1.

The *Fourier series* of sines and cosines is usually expressed in summation notation:

$$f(t)=\frac{a_0}{2}+\sum_{n=1}^{\infty}\left[a_n\cos(n\omega t)+b_n\sin(n\omega t)\right]$$

where $\omega=2\pi/T$. The n takes on all positive integer values. However, some terms will actually be missing depending on the waveform. The expansion then includes terms that share a common period *(T)* for the overall waveform. For example, the first *harmonic* terms ($n=1$) are $a_1\cos(\omega t)$ and $b_1\sin(\omega t)$, and they have T as their basic period. The second harmonic terms ($n=2$) have a frequency of 2ω or a basic period of $2/T$. The period T is fundamental for the second harmonic terms because in that time they will undergo two complete cycles. More generally, the nth harmonic terms will have a basic period of T/n and go through n cycles in the common period T. This property ensures that the component waveforms are the same length, and they will be the same length as the original waveform that they are being summated to represent.

The selection of appropriate values for coefficients a_n and b_n and the right combination of sines and cosines is the most mysterious aspect for the beginner. Recall that the formulas for calculating the intercept (b_0) and slope (b_1) for linear regression analysis were derived from a least squares criterion minimizing the error between the best fit line and all the individual observations. The exact same process was followed for Fourier analysis. Minimizing the squared error between the original waveform and the Fourier series waveform determines the formulas for calculating the values of a_n and b_n for the combination of sines and cosines used to reconstruct the original waveform.

$$a_0=\frac{1}{T}\int_{-T/2}^{+T/2}f(t)\,dt$$

$$a_n=\frac{2}{T}\int_{-T/2}^{+T/2}f(t)\cos(n\omega t)\,dt$$

$$b_n=\frac{2}{T}\int_{-T/2}^{+T/2}f(t)\sin(n\omega t)\,dt$$

The integration must take place over any complete period –*T*. The range does not necessarily need to be from 0 to T. Periodic waveforms may be aligned with respect to the y-axis due to symmetry and integrating from $-T/2$ to $+T/2$. The symmetric alignment facilitates the integration of more difficult functions. The term a_0 represents the average value over $f(t)$. Integrating $f(t)$ involves simply finding the area underneath the curve of $f(t)$ and the t-axis for the time interval $t=0$ to $t=T$

seconds. If $f(t)$ is a signal, then the resulting number has the dimensions of volt-seconds. Dividing this number by T seconds gives the average magnitude over time in volts. The a_n terms are obtained by multiplying $f(t)$ by $\cos(n\omega t)$, then taking the area between the product of these two terms and the t-axis over the time interval T, and multiplying the result by $2/T$. The b_n terms are determined in the same way. A worked example is presented in appendix 4.3.

KEY POINT

Fourier analysis is used to determine the amplitude and frequency of the sine and cosine waves that, when added together, will best fit (reproduce) the original EMG signal in a least squares sense.

Frequency Spectrum

Calculation of the Fourier coefficients has a use other than providing the best fit to the original waveform in a least squares sense. The coefficients a_n and b_n reveal exactly what frequencies were used to construct the observed signal, and the magnitude of the coefficients indicates the relative contribution of each frequency to the overall waveform. This is accomplished via plotting of the value of the coefficient at each particular frequency to obtain a **frequency spectrum.** In the example presented in appendix 4.3, the Fourier coefficients for a square wave were calculated, and it was found that there were no a_n coefficients. The magnitude of only the b_n coefficients for each frequency contributing to the square wave is therefore plotted in figure 4.14. There is a way to combine the a_n and b_n coefficients, when they both exist, to obtain a single magnitude value for each frequency; this will be described later. The

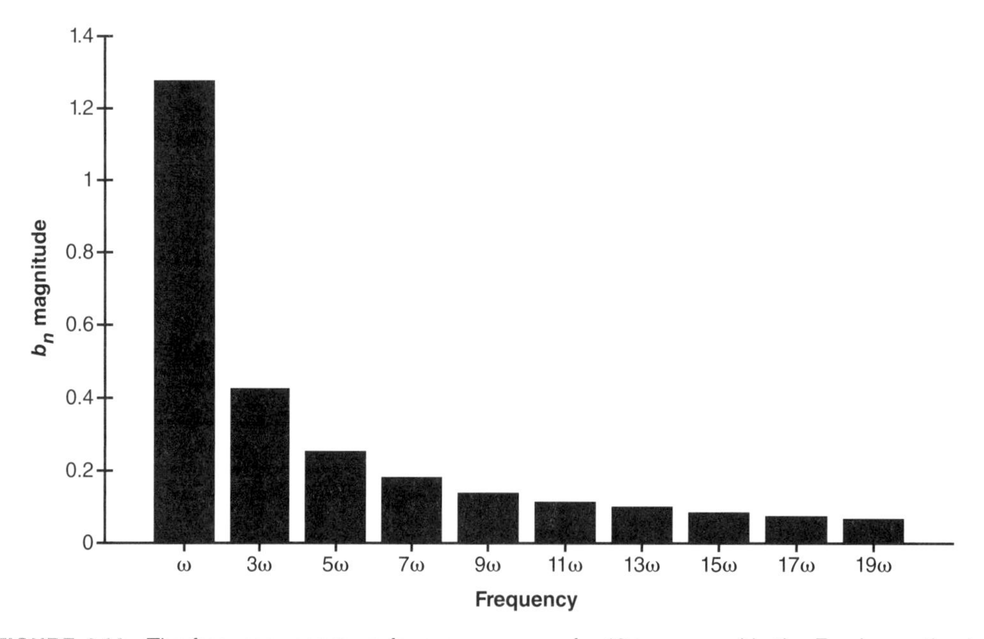

FIGURE 4.14 The frequency spectrum for a square wave for 10 terms used in the Fourier synthesis of a square wave. Each bar corresponds to the magnitude of the b_n coefficient for a given harmonic.

magnitude of the DC component *(mean value)* of the square wave is zero because its amplitude ranges between +1 and −1.

The physical interpretation of the frequency spectra is based on identification of the **signal bandwidth.** The bandwidth of the signal is the range of frequencies over which the signal contains components of appreciable magnitude. In a convergent Fourier series, the coefficients become progressively smaller with increasing n. The bandwidth of the signal will usually extend from the DC component through the lower-order harmonics out to the nth harmonic, which is the highest frequency of appreciable strength. A signal may theoretically contain an infinite series of harmonics, but components of very small magnitude will probably not have any appreciable physical importance. The exact definition of bandwidth depends on the particular application and requires a decision as to what constitutes an "appreciable magnitude." The definition of signal bandwidth in EMG is based on a consideration of signal power.

KEY POINT

A plot of the frequency spectrum involves the following. Frequency is on the x-axis from zero to the Nyquist frequency of the signal. At each frequency, the y-axis shows the relative magnitude of the contribution of that particular frequency (sine and cosine waves combined) to the overall construction of the signal.

Power Spectrum

The concept of signal power is expressed in relation to power dissipated by a resistor for a constant current (i^2R) or constant voltage (V^2/R). This was extended to the instantaneous power of any signal as $x^2(t)$. In the following example, we will show that, if the a_n and b_n coefficients in the Fourier series are squared, the frequency spectrum then becomes the **power spectrum.** The mean squared (MS) value over a given interval of time (T) is then the mean power of the signal:

$$MS = \frac{1}{T}\int_0^T x^2(t)\,dt$$

The mean power for a periodic signal $x(t) = X\sin(\omega t)$ is

$$MS = \frac{1}{T}\int_0^T x^2 \sin^2(\omega t)\,dt;$$

$$MS = \frac{1}{T}\int_0^T \frac{X^2}{2}(1-\cos(2\omega t))\,dt;$$

$$MS = \frac{X^2}{2}.$$

Extending these concepts to the Fourier series for a periodic signal,

$$MS = \frac{1}{T}\int_0^T \left(\frac{a_0}{2} + \sum_{n=1}^{\infty}\left[a_n\cos(n\omega t) + b_n\sin(n\omega t)\right]\right)^2 dt$$

This forbidding equation may be broken up into manageable parts:

$$MS = \frac{1}{T}\int_0^T \left(\frac{a_0}{2}\right)^2 dt + \frac{1}{T}\int_0^T \left(\sum_{n=1}^{\infty}\left[a_n \cos(n\omega t) + b_n \sin(n\omega t)\right]\right)^2 dt$$

The first term reduces to

$$\frac{1}{T}\int_0^T \left(\frac{a_0}{2}\right)^2 dt = \frac{a_0^2}{4}.$$

The second term expands to

$$\frac{1}{T}\int_0^T \left(\sum_{n=1}^{\infty}\left[a_n^2 \cos^2(n\omega t) + 2a_n b_n \cos(n\omega t)\sin(n\omega t) + b_n^2 \sin^2(n\omega t)\right]\right) dt.$$

The cross term integrates to zero when taken over the completed period:

$$\frac{1}{T}\int_0^T \left(2a_n b_n \cos(n\omega t)\sin(n\omega t)\right) dt = 0$$

and because

$$\int_0^T \cos^2(n\omega t)\,dt = \int_0^T \sin^2(n\omega t)\,dt = \frac{T}{2}$$

the expression for mean square of the Fourier series over time, or its mean power, is

$$MS = \frac{a_0^2}{4} + \sum_{n=1}^{\infty}\left(\frac{a_n^2}{2} + \frac{b_n^2}{2}\right).$$

Continuing the square-wave example from appendix 4.3, the following calculations demonstrate that 90% of the total signal power is contained in the fundamental and the third harmonics. The mean square value of the fundamental frequency is

$$MS = \frac{(4V)^2}{\pi^2}\frac{1}{2} = 0.8106V^2.$$

Approximately 81% of the total signal power is located within the fundamental frequency. A high percentage for the fundamental frequency is typical for a convergent Fourier series. The third harmonic contains only 9% of the total power:

$$MS = \frac{(4V)^2}{\pi^2}\frac{1}{2}\frac{1}{9} = 0.0900V^2$$

The fact that 90% of the signal power is contained in the fundamental and third harmonics is evident in figure 4.15. Notice how the basic shape of the square wave is obtained with just the first frequencies of the series.

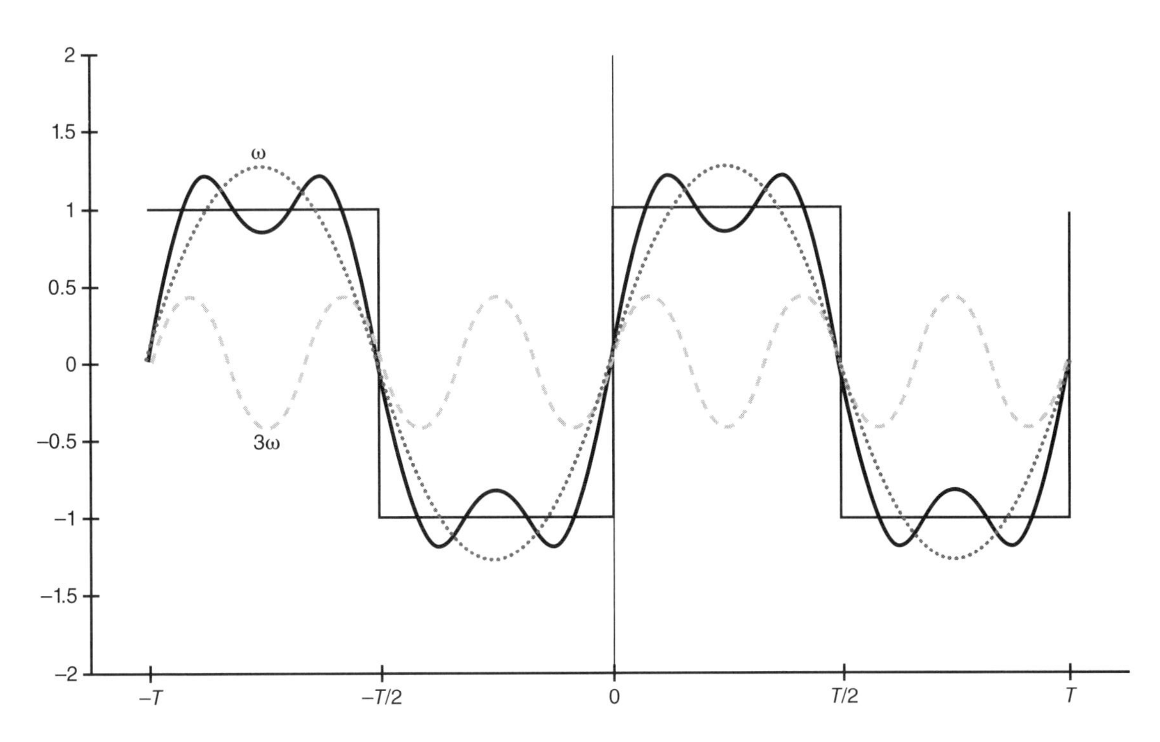

FIGURE 4.15 Comparison of a square wave with a Fourier synthesis that includes only the fundamental frequency and the first harmonic (solid black line). The fundamental (ω) and its first harmonic (3ω) are also presented separately (dotted lines). The basic square-wave shape may be obtained with only these two frequencies. The fact that the basic square-wave shape is obtained with only the first two frequencies is consistent with the observation that they correspond to 90% of the total signal power.

KEY POINT

Squaring the a_n and b_n coefficients in the Fourier series used to construct the frequency spectrum will give the power spectrum.

Fourier Transform

Frequency analysis was introduced using the *ordinary real form* of the Fourier series for well-behaved periodic functions for which an analytic solution is available using nothing more than the tools of a first-year calculus course. The mathematical complexity increases exponentially for real signals like EMG. The time invested in working through the Fourier series example in appendix 4.3 should be returned with an intuitive understanding of the more difficult case. The basis of the problem is that sines and cosines are defined extending from positive to negative infinity, and infinitely long signals cannot be used to synthesize something that is *aperiodic* and finite in length. This can, however, be accomplished through the use of the Fourier transform. The Fourier transform (FT) $X(\omega)$ of any aperiodic signal $x(t)$ is given by the following equation:

$$X(\omega) = \int_{-\infty}^{\infty} x(t) e^{-j\omega t} dt$$

The exponent of e is an imaginary (i.e., complex) number $j = \sqrt{-1}$. The exponential expression invokes Euler's identity wherein the term can be written as

$$e^{-j\omega t} = \cos(\omega t) - j\sin(\omega t).$$

So,

$$X(\omega) = \int_{-\infty}^{\infty} x(t)\cos(\omega t)\,dt - j\int_{-\infty}^{\infty} x(t)\sin(\omega t)\,dt.$$

The Fourier transform therefore has real (R_e) and imaginary (I_m) parts. The most confusing aspect of this formula is the fact that the real (R_e) part corresponds to even functions of ω while the imaginary (I_m) part represents odd functions of ω. No complex numbers are used in the calculations; they are all real numbers:

$$X(\omega) = R_e(\omega) + jI_m(\omega)$$

where

$$R_e(\omega) = \int_{-\infty}^{\infty} x(t)\cos(\omega t)\,dt$$

and

$$I_m(\omega) = -\int_{-\infty}^{\infty} x(t)\sin(\omega t)\,dt.$$

The magnitude of the Fourier transform is therefore

$$X(\omega) = \sqrt{R_e^2(\omega) + I_m^2(\omega)}$$

and the phase angle $\phi(\omega)$ is given by

$$\phi(\omega) = \tan^{-1}\frac{I_m(\omega)}{R_e(\omega)}.$$

The signal $x(t)$ and its Fourier transform are uniquely related by its inverse:

$$x(t) = \frac{1}{2\pi}\int_{-\infty}^{\infty} X(\omega)e^{-j\omega t}\,d\omega$$

Thus, there is a way to reconstruct the original signal from the Fourier transform by taking the inverse of the Fourier transform.

The expressions $X(\omega)$ and $\phi(\omega)$ allow a determination of the magnitude and phase angle, respectively, of the amplitude density function of constituent frequency components for any aperiodic signal. For example, a plot of $X(\omega)$ is the frequency spectrum of the signal $x(t)$. The distribution of frequencies that constitute a given signal $x(t)$ can therefore be assessed by visual inspection. The frequency spectrum is useful not only for identifying the bandwidth of the signal but also for identifying sources of noise. There may, for example, be a peak at line frequency (50 or 60 Hz) indicating electromagnetic interference. In North America, since 60 Hz is the fundamental frequency for this type of noise, smaller peaks at multiples of this frequency (i.e., harmonics at 120, 180, 240 Hz) might be present but be buried in the higher spectra of the signal. It is also possible to see peaks from other interference sources.

KEY POINTS

- The Fourier transform is used to calculate the frequency spectrum of aperiodic signals. The EMG signal is aperiodic.
- The frequency spectrum of the signal is used for multiple purposes. Sources of electrical noise interference have a characteristic frequency that may be identified in the frequency spectrum.

Frequency Spectrum of EMG

Computing the Fourier transform and its inverse can be accomplished in two ways. The most well-known method is the computational method referred to as the *Fast Fourier transform* (FFT). The FFT is "fast" because the number of operations required to calculate the frequency spectra can be greatly reduced if the segment of data (N) used for analysis is limited to the power of 2 (e.g., 256, 512, 1024, or 2048 points). If the data segment is not quite a power of 2, it may be zero padded up to the next power of 2. *Zero padding* literally means adding a string of zeros to the data segment; it doesn't matter where the string is added. The second method is the *Discrete Fourier transform* (DFT), which may be used when the length of the signal is not a power of 2, but at the expense of computational time.

The Fourier transform essentially treats the data segment as one realization of a periodic signal. That is, the period T is the length of the data segment. By restricting the length of the data segment to a power of 2, the FFT is also forcing the fundamental and its harmonics to fit neatly into the period T prescribed by the length of the data segment. The problem is that the chosen data segment will contain partial waveform periods. It is very reasonable to assume that some waveform periods will straddle adjacent data segments. If the FFT is performed on this partial period, the spectra will contain false frequency components. The following analogy may help explain what happens to the frequency spectrum. The Fourier transform attempts to complete the partial period, and the result is a discontinuity, a partially completed waveform. The Fourier series exhibits oscillatory error both near and at discontinuities (see *Gibbs phenomenon* in appendix 4.3). Similarly, the discontinuities due to partial waveform periods generate false frequencies that spread out across legitimate frequencies. The resulting peaks of the frequency spectrum appear spread out, an effect that is termed **frequency leakage.**

Frequency leakage may be minimized by the application of a *window weighting function* to the data segment before the FFT is performed. Consider a Gaussian window that has a bell shape. The magnitude of the weighting coefficients determines its shape. That is, the middle weighting coefficient will have the greatest magnitude. The magnitudes of the coefficients on either side of the center then taper in bell-shaped fashion toward the end. When the data segment is multiplied by the coefficients of the window, the values of the data segment are tapered to zero toward the end. Frequency leakage is therefore minimized as the impact of partial periods is decreased toward the end of the data segment. There are several different weighting functions; the Hamming and Hanning windows are examples. Each window has its own performance characteristics on the frequency spectra. Details about window weighting functions may be found in more specialized texts on signal processing in the references at the end of the chapter.

Because a sine wave is symmetrical, only two points are necessary to define its frequency. Thus, the signal must be sampled at twice the highest-frequency component present in the signal. This has an impact in the calculation of the Fourier transform. This highest frequency that may be determined in the signal is therefore constrained to the Nyquist frequency.

$$f_{max} = \frac{1}{2 \times \Delta t}$$

One of the most difficult concepts to accept is that increasing the sampling rate does not increase the *frequency resolution* (Δf). Frequency resolution (Δf) depends on the length of the data window (milliseconds) used for analysis at any given sampling rate. The length of the data window ultimately determines the number of data points (N) used to represent frequency components up to the Nyquist criterion (f_{max}), wherein two points are still necessary to define a sine wave.

$$\Delta f = \frac{1}{N \times \Delta t}$$

Consider a 1 s data window and a sampling rate of 1024 Hz. A 1 s window means that 1024 points (N) can be used to represent frequency components up to the Nyquist frequency ($f_{max} = 512$ Hz). If the window is decreased to 500 ms in length, 512 points are available to represent frequency components up to the Nyquist frequency ($f_{max} = 512$ Hz). Each frequency bin (Δf) in the spectra represents 2 Hz. If the window is increased to 2 s, there are 2048 points used to represent frequency components up to the Nyquist frequency ($f_{max} = 512$ Hz). Each frequency bin (Δf) in the spectra represents 0.5 Hz. Hopefully, an additional benefit of zero padding can now become apparent. Zero padding increases the number of data points in the time domain. Since the Fourier transform considers Δt constant, zero padding essentially increases the length of the window, which decreases (Δf). The result is an increase in the frequency resolution.

The frequency spectrum for biceps brachii surface EMG is presented in figure 4.16. The frequency spectrum is not normally plotted in the literature as a bar graph. It is done so in this case to facilitate visual inspection of the connection between theory and practice. Each bar denotes a frequency resolution of $\Delta f = 1$ Hz. Also, notice that the unit of the x-axis is hertz. Frequency analysis of signals is usually expressed in terms of angular frequency (ω, $\text{rad} \cdot \text{s}^{-1}$) to accommodate very large numbers. However, magnitudes encountered in EMG are small enough that it is not necessary to reduce their scale.

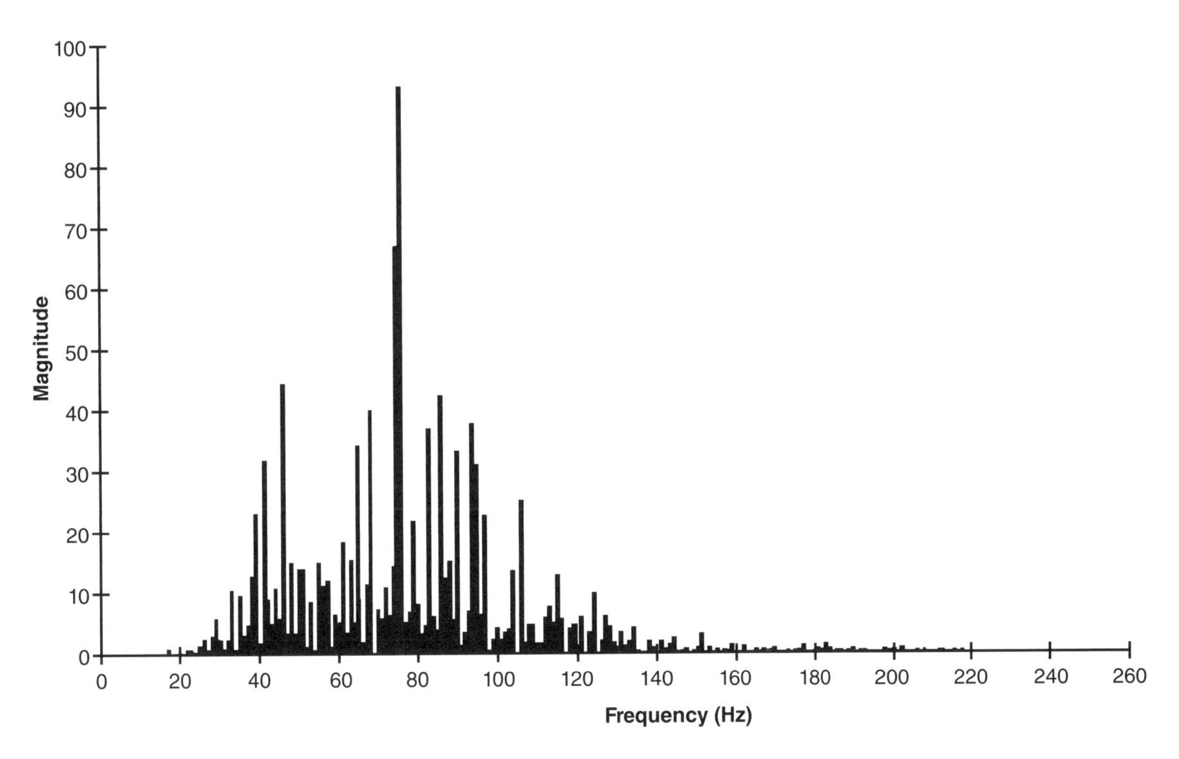

FIGURE 4.16 The frequency spectrum for the biceps brachii surface EMG generated during maximal isometric actions of elbow flexion. Each bar has a frequency resolution of $\Delta f = 1$ Hz and corresponds to the magnitude $\sqrt{a_n^2 + b_n^2}$. The magnitude indicates how much of a particular frequency contributes to the overall surface EMG signal.

KEY POINTS

- Calculation of the Fast Fourier transform requires attention to length *(n)* of the signal, which must be (2^n). If the data do not meet the length requirement, they may be zero padded to the next power *(n)*.
- The Fourier transform of EMG data segments treats the entire data segment as one realization of a periodic signal. It is reasonable to expect that any single data segment will inevitably include partial periods that would be completed across adjacent data segments. The abrupt end to the partial period that occurs within a single data segment can cause frequency leakage across the spectrum.
- A window weighting function must be applied to the data segment to minimize the impact of the data toward the end of the segment. Ultimately, the choice of data segment length (window) and its location during the contraction is a result of several competing criteria.

Power Spectral Density of EMG

The frequency spectrum of EMG is not reported in the literature because it is a stochastic signal. The Fourier transform will therefore depend on the sample selected. Furthermore, the mean value of the spectrum will be zero. The reason is that the amplitude distribution of a stochastic signal has a zero mean and unit variance. The solution to the problem resembles the solution applied to the deviation scores to prevent them from summing to zero. The magnitude of each frequency component of the Fourier transform can be squared. Where $X(\omega)$ is the Fourier transform of $x(t)$ and $X^*(\omega)$ is the complex conjugate of $X(\omega)$, the two can be multiplied to obtain the squared magnitude of the Fourier transform:

$$\Phi(\omega) = X(\omega)X^*(\omega) = |X(\omega)|^2$$

The quantity $\Phi(\omega)$ is the *power spectral density* (PSD) *function* for the signal $x(t)$. This is a natural extension of the squared magnitude $(a_n + b_n)^2$ of each frequency component of the Fourier series, which was demonstrated to yield the power spectrum. Figure 4.16 is the frequency spectrum for biceps brachii surface EMG. The power spectrum for the same signal is presented in figure 4.17. The squared magnitude values are smaller because an averaging process was used to produce a smooth estimate, but the overall shape is identical to the "raw" estimate presented in figure 4.16. The PSD function was calculated on a band-passed (10-500 Hz) surface EMG signal. The 10 Hz high pass is evident by the absence of signal power below this point.

The problem is that small amounts of noise in the time domain representation of the signal result in large errors in the spectral estimates. There are several ways to reduce the variability of the spectral estimates. One method is to divide the data into a number of segments, compute the power spectrum for each data segment, and then take the average of the estimates. This procedure is no different than signal averaging, and it is referred to as the *average periodogram* technique, or the *Bartlett method.* The main limitation is that EMG recordings are not usually of sufficient duration to allow the data window to be divided into a number of segments without compromising frequency resolution.

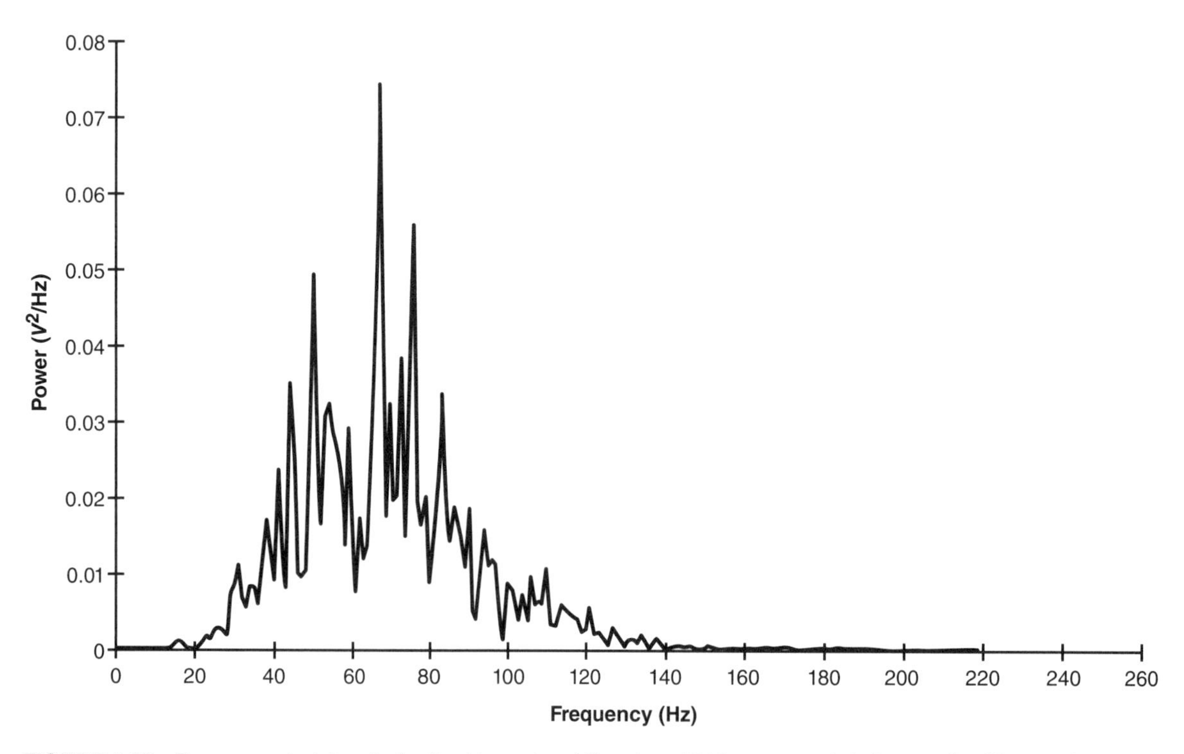

FIGURE 4.17 Power spectral density for the biceps brachii surface EMG generated during maximal isometric actions of elbow flexion. The data analysis window was divided into three smaller segments, and the power spectrum was calculated using a 50% overlap across data segments. Each of the smaller data segments was zero padded prior to calculation of the power spectrum to retain the same frequency resolution ($\Delta f = 1$ Hz). The five separate spectral estimates were then averaged. Each data point of the power spectrum corresponds to the squared coefficients $(a_n^2 + b_n^2)$, which is analogous to (V^2).

The basic issue here is the onset of muscle fatigue and signal stationarity. The most widely used method in EMG practice is the *Welch periodogram* technique. The data window is divided into separate segments, and the spectral estimate is calculated on data that overlap adjacent segments by a specified percentage, typically 50%. The additional power spectral estimates gained by the overlap are then averaged. For example, if the window used for analysis is divided into three data segments and the power spectrum is calculated with a 50% overlap, five power spectrum estimates are available for averaging. Dividing the window into smaller data segments can result in a decrease in frequency resolution, but the Welch method allows for some degree of averaging given the short windows typically used in EMG practice. The original frequency resolution can be retained, however, if the smaller data segments are zero padded up to the appropriate length. The *Daniell method* involves simply computing the spectrum once, then averaging the spectral estimates from adjacent frequencies in the same way as with the moving average procedure. As might be expected, there is a loss of frequency resolution with this method as well. The loss depends on how many adjacent frequencies are used in the average.

KEY POINT

The squared frequency spectrum of the EMG signal yields its power spectrum. The spectral estimate is disproportionally affected by small amounts of noise in the signal. Thus, there are different methods of generating a smooth power spectrum.

Discrete Measures Obtained From the Power Spectral Density Function

The PSD functions from different muscle groups all share the same basic shape, which is generally related to the shape of the underlying MUAPs (Lindström and Magnusson 1977). The main difference between muscle groups is the bandwidth of the signal, which is related to the physiology of the muscle. These properties include the percent MVC over which motor units are recruited, maximum firing rates, motor unit size, and muscle fiber lengths, to name a few (Weytjens and van Steenberghe 1984; Dimitrov and Dimitrova 1998). Surface recordings have a narrower bandwidth than indwelling EMG due to low-pass tissue filtering of the signal (Basmajian et al. 1975).

By convention, the bandwidth of the EMG signal is designated with respect to the half-power points on the PSD graph (Bendat and Piersol 1971). There are some applications for which the PSD function is sometimes represented as decibels versus log frequency (Sinderby et al. 1996), in which case the bandwidth is referenced with respect to the –3 dB point. Signal bandwidth is ultimately connected with the band-pass filter settings on the amplifiers presented in the previous chapter. The low- and high-pass filters are supposed to be set to match the signal bandwidth so that frequencies outside that range do not contaminate the signals of interest.

Stulen and De Luca (1981) showed that the PSD function of the EMG signal may be characterized by two parameters: **mean power frequency (MNF)** and **median power frequency (MDF).** The mean and median frequencies are located on the PSD function in the same way as a positively skewed distribution, with the mean greater than the median (Farina and Merletti 2000). The median power frequency is sometimes referred to as the center or centroid frequency. To compute median power frequency, it is first necessary to calculate total power (TP). Where N is the number of data points in the window, the TP is considered only over that part of the PSD function that describes the power in the positive frequencies and is computed by

$$TP = \sum_{k=0}^{\frac{N}{2}-1} PSD(k)$$

where k is the index for the frequencies $\omega[k] = 2\pi k / N : k = 0, 1, \ldots N-1$. Notice the conversion of the frequency units from rad/s to hertz. The frequency at which the mean power occurs is then computed by

$$MNF = \frac{1}{TP} \sum_{k=0}^{\frac{N}{2}-1} \left(f[k] \cdot PSD[k] \right).$$

The frequency at which the median power occurs is defined as the frequency that divides the power spectrum into two parts of equal power and can be obtained with the following formula:

$$0.5 = \frac{1}{TP} \sum_{k=0}^{MDF} PSD[k]$$

These measures are usually reported together because they are believed to provide both overlapping and different information with respect to motor unit behavior (Farina and Merletti 2000). The MNF and MDF have been shown to be sensitive to muscle fiber characteristics such as conduction velocity. It is also possible that these measures reflect motor unit firing behavior such as motor unit recruitment and rate coding (Stulen and De Luca 1981; Solomonow et al. 1990), though additional research is needed to relate these variables to motor unit discharge statistics.

The number of times the signal crosses the zero isoelectric baseline is related to the frequency content of the signal. A measure that is related to the number of zero crossings is **mean spike frequency (MSF)** (figure 4.10). Mean spike frequency is a nonparametric way to describe a stochastic signal in periodic terms, by simply counting the number of complete EMG spikes that occur per second. The MSF is highly reliable during dynamic contractions (Gabriel 2000), and it is strongly correlated with MNF (Gabriel et al. 2001). One calculates the MSF by taking the number of spikes (NS) and dividing by the total duration (TD) of the data analysis window. If the data window is less than 1 s, then the ratio must be expressed in relation to that time length in hertz:

$$MSF = \frac{NS}{TD}$$

KEY POINTS

- The median and mean frequencies are traditional measures of the frequency content of the signal that require signal stationarity for their use. Mean spike frequency quantifies the number of EMG spikes that occur within a second. It is related to the frequency content of the signal.
- The frequencies at which the mean and median power occur are used to quantify changes in motor unit recruitment, firing statistics, and MFCV. Mean spike frequency is a nonparametric way to calculate the frequency content of the EMG signal. The reader is cautioned about overinterpreting the meaning of the measures, as they are still subject to debate.

Data Window Length

The minimum data window (epoch) required to obtain stable estimates of EMG activity during isometric contractions has been evaluated using the coefficient of variation and the RMS error. Regardless of the specific EMG measure studied, there is a characteristic exponential decrease in the variability of the estimate as the length of the data window increases (Vint and Hinrichs 1999; Farina and Merletti 2000). If the variability is plotted for window lengths from 10 to 1000 ms, the "elbow" of the curve generally occurs before 200 ms. The minimum data window necessary for obtaining stable estimates is between 250 and 500 ms. Longer data windows (>1 s) are of course possible. The potential for fatigue onset during the window increases with the intensity of the muscle contraction.

The location of the data window is as important as its length. If only a single value is being extracted to represent the EMG characteristics of the contraction, then a stable portion of the force–time curve should be used (Vint and Hinrichs 1999). The appropriate length and location to use in order to obtain representative EMG scores during maximal-effort contractions are subject to debate. A short data window (250 ms) may be centered at the location of the peak force value (figure 4.18).

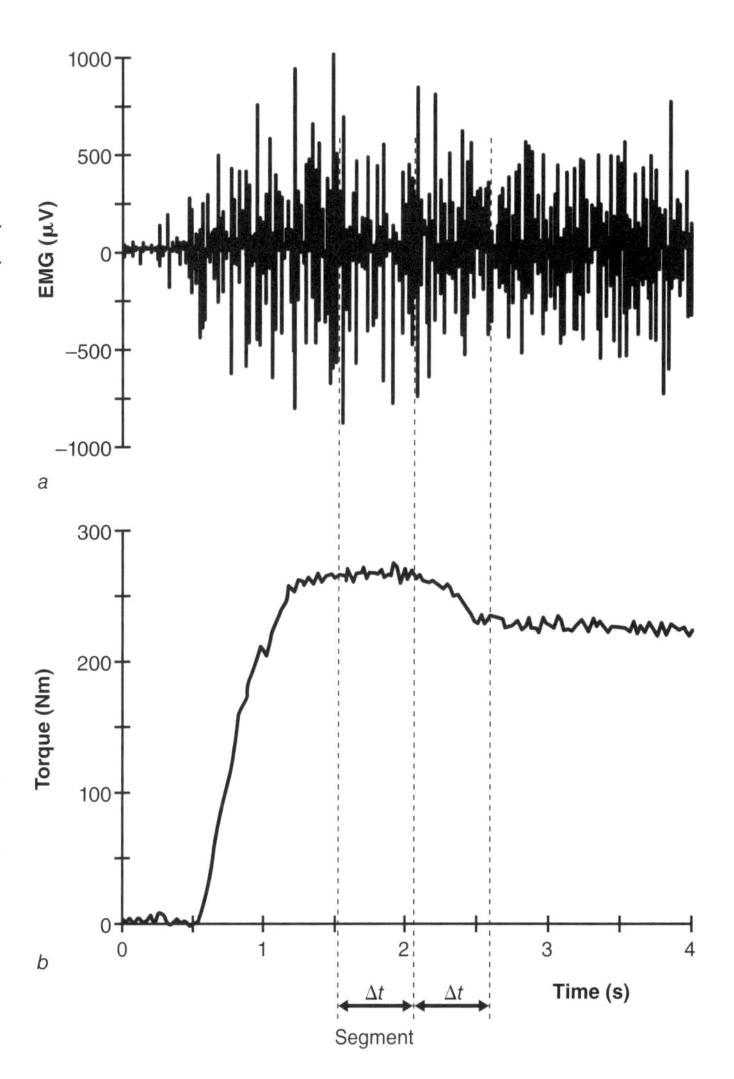

FIGURE 4.18 Surface EMG activity and torque of the knee extensors during maximal-effort contractions. *(a)* Band-passed surface EMG activity of the vastus lateralis and *(b)* knee extension torque. The vertical dotted lines show two data windows (Δt) that include maximal torque values.

Reprinted from *Journal of Electromyography and Kinesiology* 4(3), S. Karlsson, B.E. Erlandson, and B. Gerdle, "A personal computer-based system for real-time analysis of surface EMG signal during static and dynamic contractions," p. 174, copyright 1994, with permission from Elsevier.

This approach is appealing because it yields the greatest values, but it may also be influenced by dynamic overshoot in the measurement system (Karlsson et al. 1994). The alternative is to obtain the average maximal values for both force and EMG for a data window obtained in the middle portion of the contraction (Vint and Hinrichs 1999; figure 4.19). For maximal-effort contractions, however, this procedure bears the risk of obtaining EMG values during an appreciable portion of the contraction when force is not maximal. Both the length and location of the window must be reported in the study.

The choice of appropriate window length becomes a function of several competing factors when the force level is changing, particularly for frequency domain measures. The force level should change slowly during the period specified by the window so that the statistical properties of the EMG signal remain stable. If the data window is divided into smaller segments *(bins)* and the mean and SD do not change significantly from one segment to the next, the EMG signal within the window is **stationary.** Stationarity is a requirement for frequency analysis and may be statistically determined

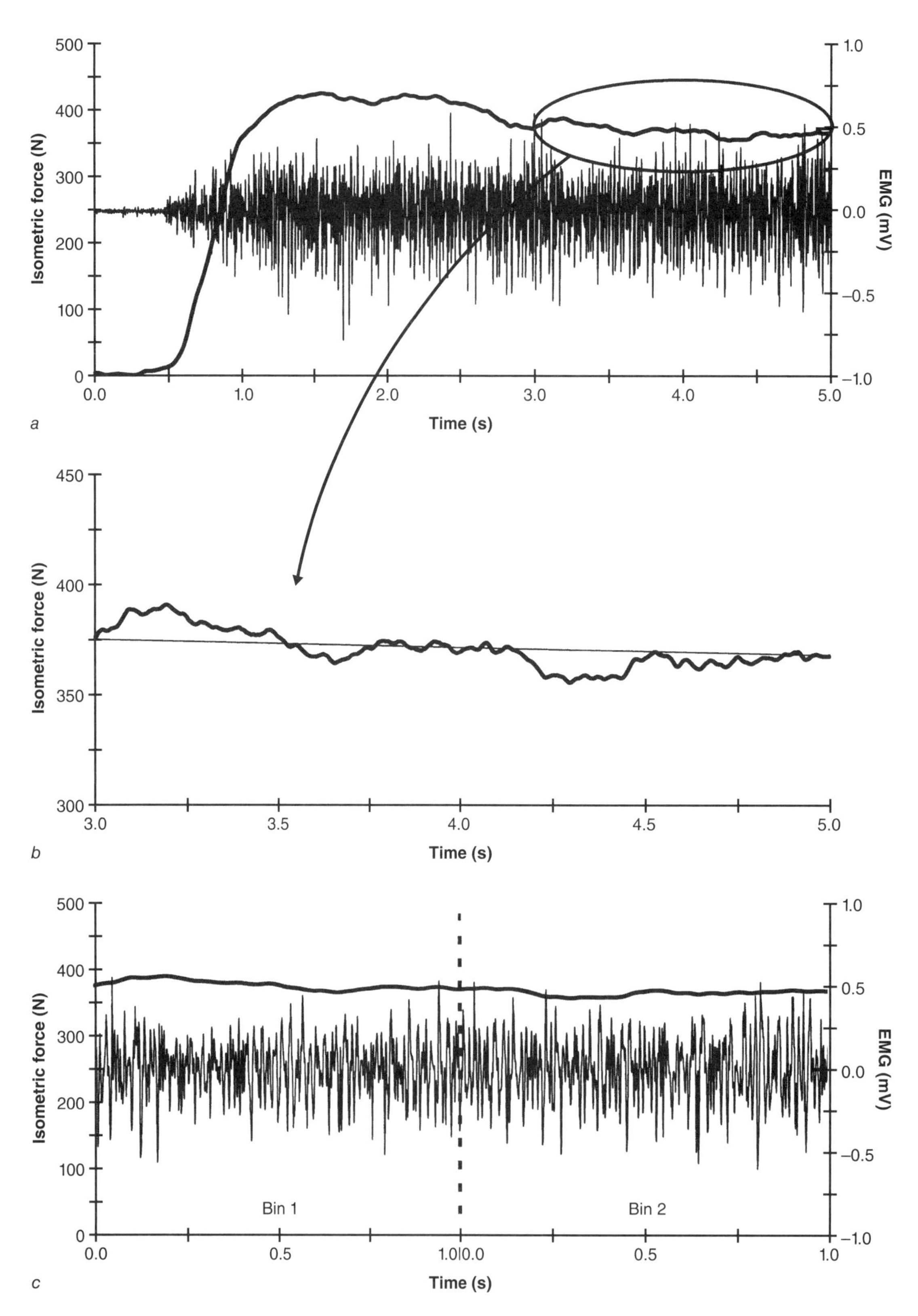

FIGURE 4.19 Maximum isometric actions of the knee extensors. *(a)* Band-passed surface EMG activity of the vastus lateralis and knee extension torque. The ellipse shows that the data window used for analysis occurs after the maximal torque values. *(b)* The maximal torque values within this window are relatively stable. *(c)* One may then segment the data window into smaller bins of data to evaluate the mean and standard deviation of each individual segment within the window.

Reprinted, by permission, from P.F. Vint and R.N. Hinrichs, 1999, "Longer integration intervals reduce variability of EMG derived from maximal isometric contractions," *Journal of Applied Biomechanics* 15(2): 214.

through use of the runs test or reverse arrangements tests (Bendat and Piersol 1971). The EMG signal can remain stationary if force within the window changes by no more than 10% of MVC (Bilodeau et al. 1997). Consider a ramp isometric contraction that increases at a rate of 20% MVC per second; it takes 5 s to achieve maximum. Ten consecutive data windows can be constructed to track changes in EMG activity throughout the contraction if the length of each data window is 500 ms. The number of 500 ms data windows will double for a slower ramp contraction at 10% MVC per second. The risk of fatigue increases for longer contraction durations (Sbriccoli et al. 2003).

Electromyographic activity for anisometric (i.e., dynamic) contractions is not a simple function of force level. Muscle length, the velocity of shortening, and the moment arm length are contributing factors. The data window is usually tailored to the type of contraction, and constraints imposed by multiple criteria may limit data analysis to a single window. For example, during isokinetic contractions, the window may be standardized to a specific location within the range of motion wherein the velocity is constant and it encompasses the apex of the force–time curve (Karlsson et al. 2003).

There are more advanced signal-processing methods for analyzing the time course of EMG activity throughout a dynamic contraction. Short-term time and frequency analysis involves creating very short data windows (i.e., 250 ms) throughout the duration of the muscle contraction (Potvin 1997). However, if the contraction duration is too short to allow obtaining a sufficient frequency resolution (Δf), then instantaneous time–frequency analysis should be used. Time–frequency analysis takes advantage of cyclostationarity. The velocity of contraction, muscle length, and moment arms change throughout the muscle contraction and the same pattern of change may be observed over multiple trials. As a result, EMG activity remains stable across multiple contractions. Cyclostationarity means that the EMG values are consistent over multiple trials. For greater detail on time–frequency analyses, see Knaflitz and Bonato 1999.

KEY POINTS

- The minimum length of time for a data window that is required to obtain stable estimates of the mean and SD of the EMG signal is 250 ms. If the mean and SD are constant within the data window, then the signal is said to be stationary.
- The location of the data window is an important consideration for obtaining muscle activity that is representative of the contraction. For isometric contractions, the data window should encompass the first stable portion of the contraction wherein force is constant. The data window for anisometric contractions entails multiple considerations that involve locating the window at the same point in the range of motion.

Noise Contamination

Always include a minimum data window of approximately 500 ms before and after the muscle contraction to document the magnitude of baseline activity. Noise contamination is most obvious in the baseline activity when the muscle is resting. Acceptable baseline activity can have a P-P amplitude range up to 20 μV in the absence of any contamination (deVries et al. 1976). Noise contamination is then superimposed on baseline activity, and it comes from two basic sources: (1) inherent noise and

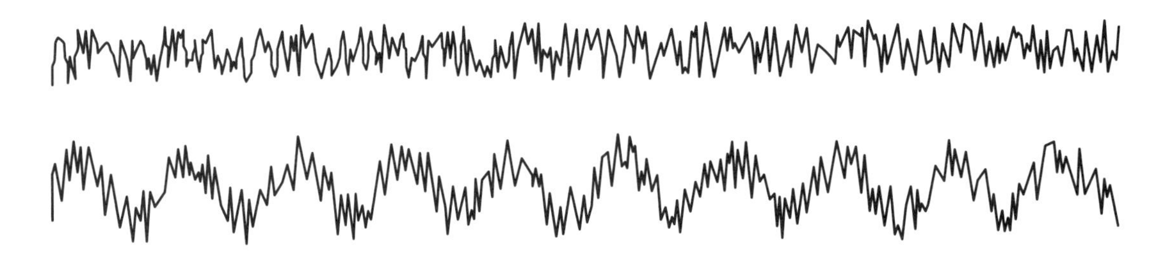

FIGURE 4.20 Two basic types of noise contamination. Inherent white noise is associated with instrumentation (top), and this may be superimposed upon 60 Hz interference noise (bottom).

Reprinted with permission of John Wiley & Sons, Inc. From R.A. Normann, 1998, *Principles of bioinstrumentation* (Hoboken, NJ: Wiley), 182.

(2) interference. The two basic sources of noise in the baseline are depicted in figure 4.20. The following paragraphs review the characteristics of each type of noise and basic signal-processing methods used to minimize their impact on EMG analysis and interpretation.

Signal-to-Noise Ratio

Baseline activity may be conceptualized as background *noise* against which the *signal* must be detected. In this case, the EMG burst is the signal that must be detected. The actual quality of the EMG data is often referred to in terms of the *signal-to-noise ratio* (SNR), which is given in decibels, so it is the ratio of signal power (P_S) to noise power (P_N).

The first practical application of the SNR is the resolution of measurement. Consider a 12-bit A/D board with an input range of 10 V (±5 V). Recall that there are $2^{12} = 4096$ discrete levels, so in this case each level is 10 V / 4096 = 2.44 mV. In this example, the maximum SNR that can be tolerated without deterioration of the resolution for the P-P magnitude of EMG is

$$SNR = 10\log\left(\frac{P_S}{P_N}\right);$$

$$SNR = 10\log\left(\frac{10}{0.00244}\right) = 36\text{ dB}.$$

When test contractions are performed to determine the appropriate amplifier gain, the oscilloscope should be checked to ensure that the P-P amplitude of baseline activity is no greater than 2.44 mV, which is the voltage level corresponding to the resolution of the A/D board. Keep in mind that this value represents an amplified signal. If the amplifier gain is set at 1000, the P-P amplitude of the original noise is 2.44 μV. Notice that the P-P amplitude was used rather than RMS. The input range of the A/D board is expressed in terms of P-P amplitude, and the P-P amplitude of baseline activity can be measured directly from an oscilloscope.

Inherent Noise

The electrodes and amplifier are both sources of inherent noise in the measurement system. Fortunately, the magnitude of inherent noise is generally small and may be further minimized by a combination of methodological controls and simple signal-processing methods as described in the following sections.

Electrode Noise

Electrode noise occurs due to the electrolyte–skin and electrolyte–metal interfaces. Once the electrolyte–metal electrochemical reaction stabilizes, this source of noise is negligible (0.3 μV P-P). The amplitude of the electrolyte–metal noise for Ag-AgCl electrodes decreases dramatically within the first minute of application and stabilizes to a level at or below the amplifier noise. The time to stabilization is much longer for other metals such as stainless steel (180 min; Huigen et al. 2002). The electrolyte–skin interface is more problematic. The noise voltage can range from 5 to 60 μV P-P. The lower limit can be achieved with good skin preparation, but it is also subject dependent (Gondran et al. 1996).

Amplifier Noise Sources

The first type of amplifier noise is referred to as *thermal* or *Johnson* noise and is associated with resistors. The source is the flow of current as electrons randomly collide with the resistive material. The average noise level is

$$V_{TH} = \sqrt{4kTRB}$$

where k is Boltzmann's constant (1.38×10^{-23} J / °K), T is absolute temperature in degrees kelvin, R is the resistance in ohms (Ω), and B is the bandwidth in hertz. The main point to note about the formula is that the magnitude of the thermal noise voltage (V_{TH}) is independent of frequency. That is, the voltage magnitude of this type of amplifier noise is equal across the frequency distribution (i.e., white noise in which there is equal magnitude in each bin of the power spectral density distribution). A key feature of white noise is that it is random and uncorrelated with the magnitude of signal voltage.

The second type of noise is actually present in many electronic devices and physical, chemical, and even physiological systems. In fact, it is also present in electrodes (Huigen et al. 2002) (figure 4.21). When performing noise measurements over long periods of time, one often observes a characteristic frequency distribution. There is a rapid and dramatic increase in the amplitude of the frequency distribution below ≈10 Hz. The amplitude of the frequency distribution is greatest in the low-frequency range. It then decreases according to the function ($1/f$). Thus, the name for this noise is $1/f$ *noise.* In most analog electronics, the transition between Gaussian, white noise, and $1/f$ noise occurs between 1 and 100 Hz. Despite its ubiquitous presence, the actual source of $1/f$ noise remains unknown. In a high-quality amplifier, the two sources of noise within the EMG bandwidth (0.1-400 Hz) can range from 1 to 5 μV P-P. The total sources of amplifier noise represent the smallest level of EMG that can be detected.

Interference Noise

Considerable discussion has been devoted to interference noise (i.e., 60 Hz hum), which can be solved only through shielding, the use of differential recordings, the elimination of obvious interference noise sources (e.g., fluorescent lights), and careful grounding. This type of interference is so prevalent that many amplifiers include a line frequency (for example, 60 Hz in North America) **notch filter** as an optional setting to eliminate signal power at 60 Hz. The bode plot for the notch filter literally has a narrow notch at 60 Hz. The problem is that a 60 Hz notch filter also has a roll rate that results in the reduction in signal power associated with adjacent frequencies

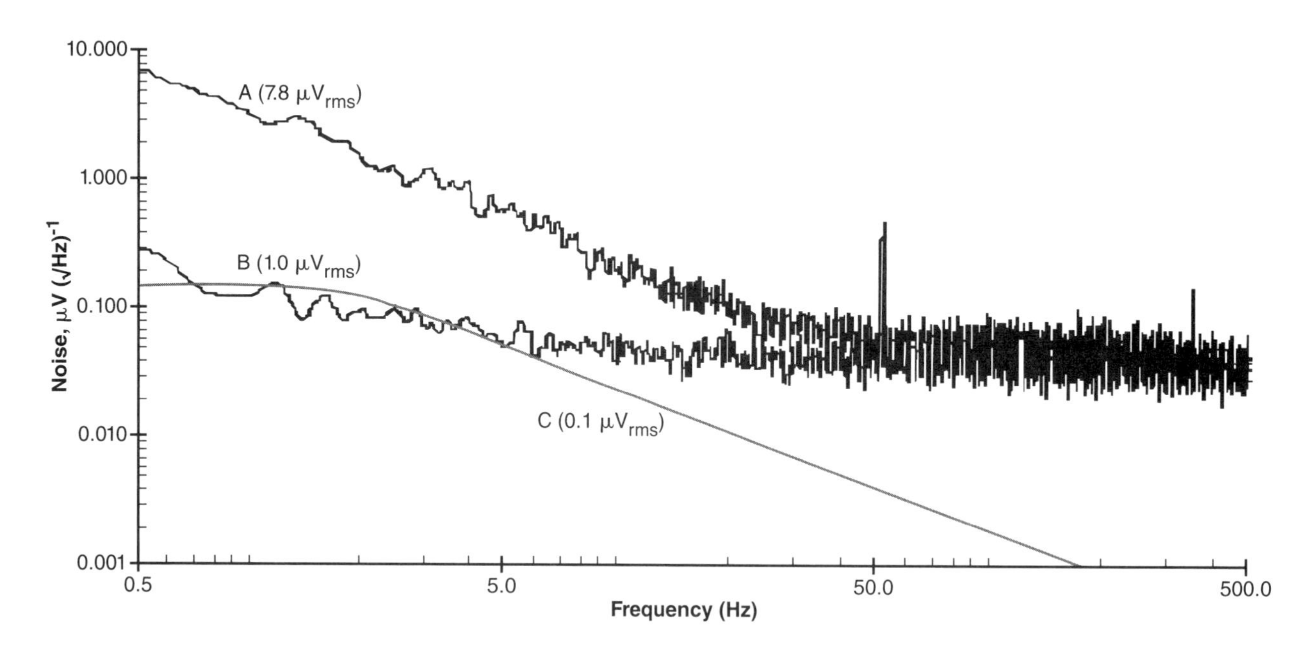

FIGURE 4.21 Noise measurement from Ag-AgCl electrodes (A) and a high-resolution amplifier (B). Both contain 1/f noise, which is an increase in the amplitude of the power spectral density below ≈10 Hz. The thermal noise of the skin–electrode impedance was also calculated (C). The root-mean-square amplitude value for each source of noise is included beside each trace.

Reprinted from E. Huigen, A. Peper, and C.A. Grimbergen, 2002, "Investigation into the origin of the noise of surface electrodes," *Medical and Biological Engineering and Computing* 40(3): 333. With kind permission from Springer Science and Business Media.

in the transition band, not just at 60 Hz. Moreover, a significant amount of signal power generated by muscle activity is found at 60 Hz and should not be eliminated. We recommend disabling the 60 Hz notch filter switch to prevent the well-intentioned person from using it.

Signal Averaging

Inherent noise cannot be reduced by shielding or careful grounding as is done with interference noise. Signal averaging is the easiest way to improve the quality of the signal if it is contaminated by Gaussian noise (Marmarelis and Marmarelis 1978). Consider the *true score* of an evoked response at a specific time point $T(t_i)$. The observed score $x(t_i)$ at a specific time point across multiple trials will be the true score $T(t_i)$ plus or minus an error of magnitude (e_i):

$$x(t_i) = T(t_i) \pm e_i$$

Trial after trial, the observed score at a specific point in time may randomly be above (+) or below (–) the true score. Signal averaging sums the values above and below the true score, which is analogous to adding positive and negative values. Since the error is normally distributed, there will be an equal number of errors of the same magnitude above and below the true score. The noise will then sum to zero after an infinite number of averaged trials, leaving only the true score. Averaging an infinite number of trials is, of course, not feasible. There is a way to calculate the SNR as a function of the number of trials that are averaged:

$$SNR = \frac{S \times \sqrt{n}}{A}$$

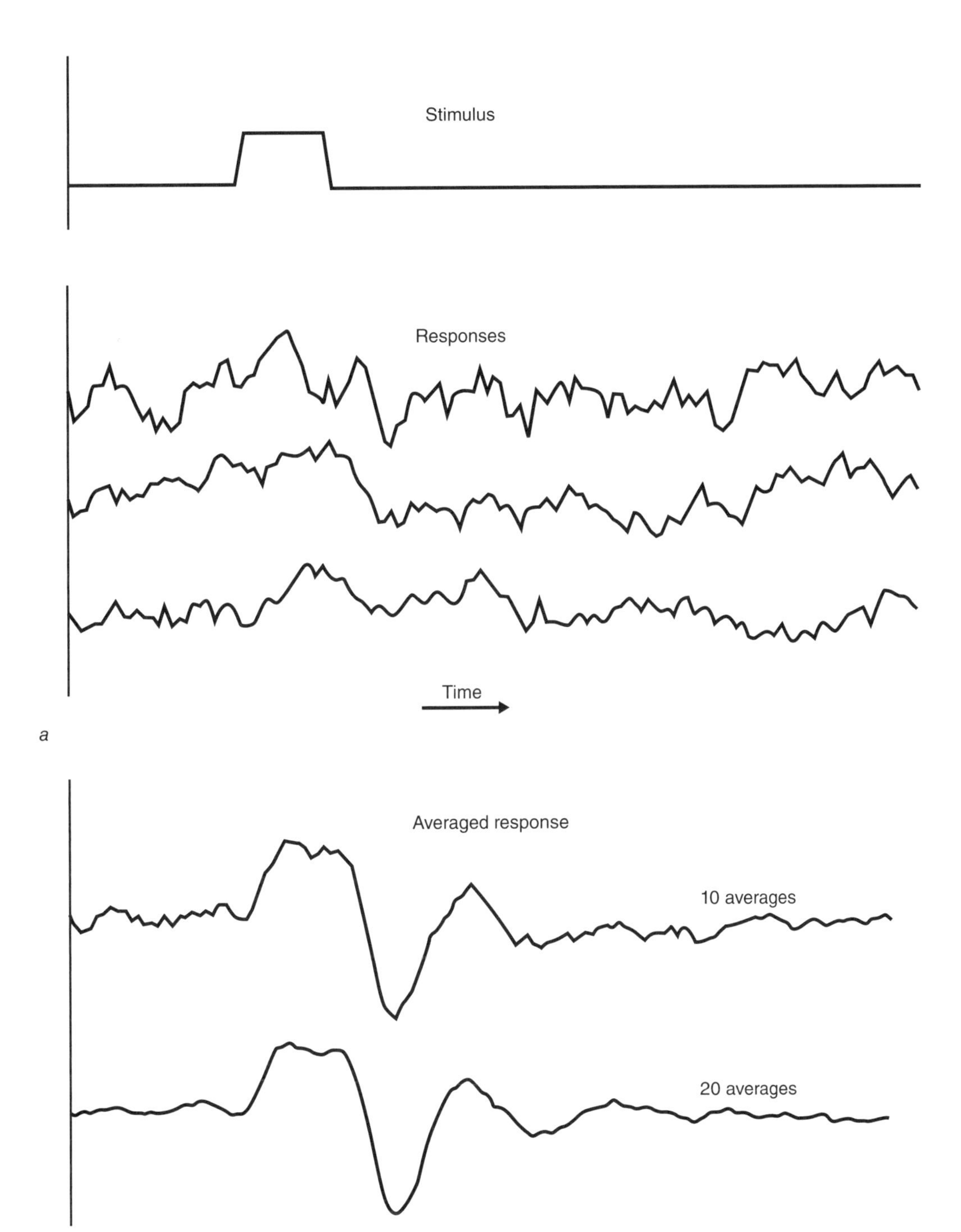

FIGURE 4.22 *(a)* A stimulus and successive responses. *(b)* The mean response becomes smoother as the number of responses averaged is increased from 10 to 20.

Reprinted from P.Z. Marmarelis and V.Z. Marmarelis, 1978, *Analysis of physiological systems: The white-noise approach* (New York: Plenum Press), figure 2.19, page 53, with kind permission from Springer Science and Business Media.

where S is the amplitude of the signal, A is the amplitude of the noise, and n is the number of averages performed. If four signals are averaged, the SNR has doubled because $\sqrt{4} = 2$.

A critical factor in the application of signal averaging is the ability to time-lock the response to an external event (figure 4.22). In the case of evoked potentials, this takes the form of an electrical stimulus applied to the nerve or muscle. Posture and balance studies sometimes apply an external force to perturb the system. The EMG amplitude associated with the reflex response is frequently small in comparison to that of a voluntary contraction. Since baseline activity is the same regardless of the type of contraction, the smaller EMG amplitude for reflex responses decreases the SNR for posture and balance and evoked potentials studies. For example, Newcomer and colleagues (2002) found that a minimum of 16 trials were necessary to identify muscle activation patterns during various footplate perturbations. In this case, signal averaging produced approximately a fourfold increase in the SNR. Signal averaging a greater number of trials may be better but is not always feasible in terms of protocol length.

To take advantage of the benefits of signal averaging for kinesiological studies, there must be a way to time-lock the EMG waveforms. This may be accomplished through use of the A/D board or by software during postprocessing. The EMG waveforms are then averaged based on a fixed number of data points before and after the event of interest. If the responses result in EMG waveforms of different lengths, an interpolator may be used to set each waveform to the same number of data points (Shiavi and Green 1983). It is more difficult to align EMG waveforms using software methods, but this is still possible as long as there is a mechanical or other trigger signal that is recorded concurrently. It is necessary to write a computer algorithm in order to detect an event in the trigger signal, which is much easier to do than for a noisy EMG signal. A data window is then created based on a fixed number of points before and after the detected event to encompass the EMG waveform. The EMG waveforms are then averaged (Darling et al. 1989; Gabriel 2002).

Baseline Noise Spectrum Subtraction

Baratta and colleagues (1998) proposed a simple yet very effective way to minimize the impact of 60 Hz power-line noise after the data have been collected. First, the amplitude and phase of the power-line noise in baseline activity are estimated while the muscle is relaxed. A sinusoidal waveform with the same amplitude and phase as the power-line noise is then subtracted from the entire recorded EMG signal. Two separate linear regressions are performed on the baseline activity, one using the sine function and the other using the cosine function to account for the phase angle of power-line noise:

$$\hat{Y}_1 = a + b\sin(\omega t) + e_1$$

$$\hat{Y}_2 = c + d\cos(\omega t) + e_2$$

where $\hat{Y}_i$ is the predicted baseline activity based on the coefficients *a, b, c,* and *d* computed by regression analysis. The error term e_i is the baseline noise component unaccounted for by the regression model (i.e., the residuals). Subtraction of the power-line noise is then accomplished on the entire signal:

$$EMG_R(t) = EMG(t) - b\sin(\omega t) - d\cos(\omega t)$$

The signal before and after (EMG_R) baseline noise spectrum subtraction is illustrated in figure 4.23. The main advantage of this technique is that it does not affect legitimate frequency components generated by muscle that are at the power-line frequency. Notice that determination of EMG onset can be enhanced by decreasing baseline noise.

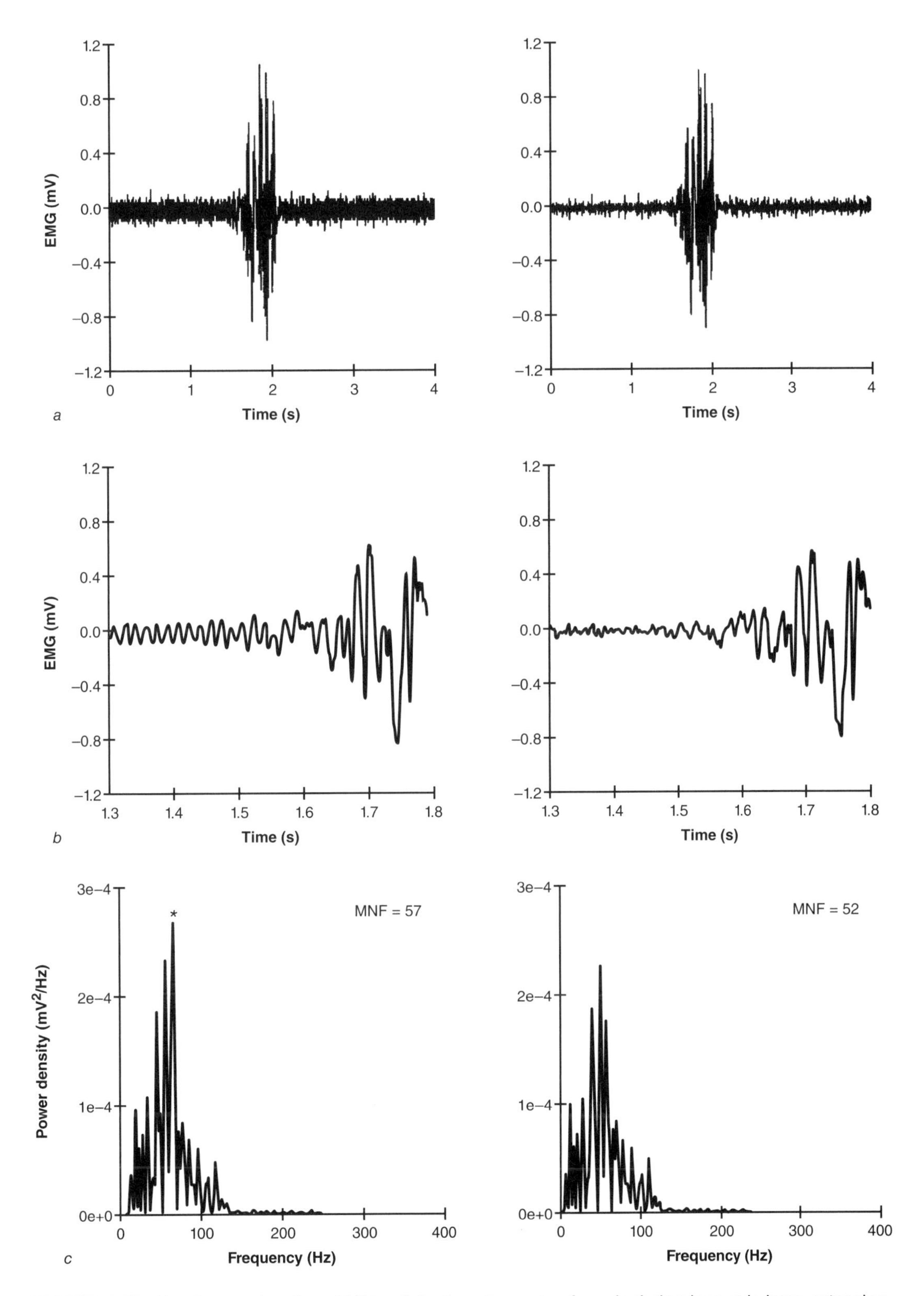

FIGURE 4.23 Band-passed surface EMG activity from the rectus femoris during isometric knee extension. (*a*, left) The signal-to-noise ratio is low because of large baseline activity associated with power-line (60 Hz) contamination. (*b*, left) Power-line contamination is further revealed by an increase in the resolution used to view the baseline. (*c*, left) A plot of the power spectrum of the surface EMG shows a peak at 60 Hz (*), further confirming the presence of power-line noise. (*a*, right) Subtraction of the 60 Hz noise greatly reduces the magnitude of baseline activity. (*b*, right) The baseline appears more random when viewed at an increased resolution. (*c*, right) The peak in the power spectrum at 60 Hz is gone after noise removal.

Reprinted from *Journal of Electromyography and Kinesiology* 8(5), R.V. Baratta, M. Solomonow, B.-H. Zhou, and M. Zhu, "Methods to reduce the variability of the EMG power spectrum estimates," p. 283, copyright 1998, with permission from Elsevier.

ECG Contamination

Perhaps the most troublesome form of contamination for surface EMG recordings obtained from muscles around the torso is electrical activity from the heart, electrocardiographic (ECG) activity (figure 4.24). Fortunately, the impact of ECG contamination decreases as the intensity of the contraction increases. The ECG constitutes one-tenth the total signal power during a 20% MVC. There is then a dramatic decrease to one-hundredth of the total signal power for a 100% MVC (Redfern et al. 1993).

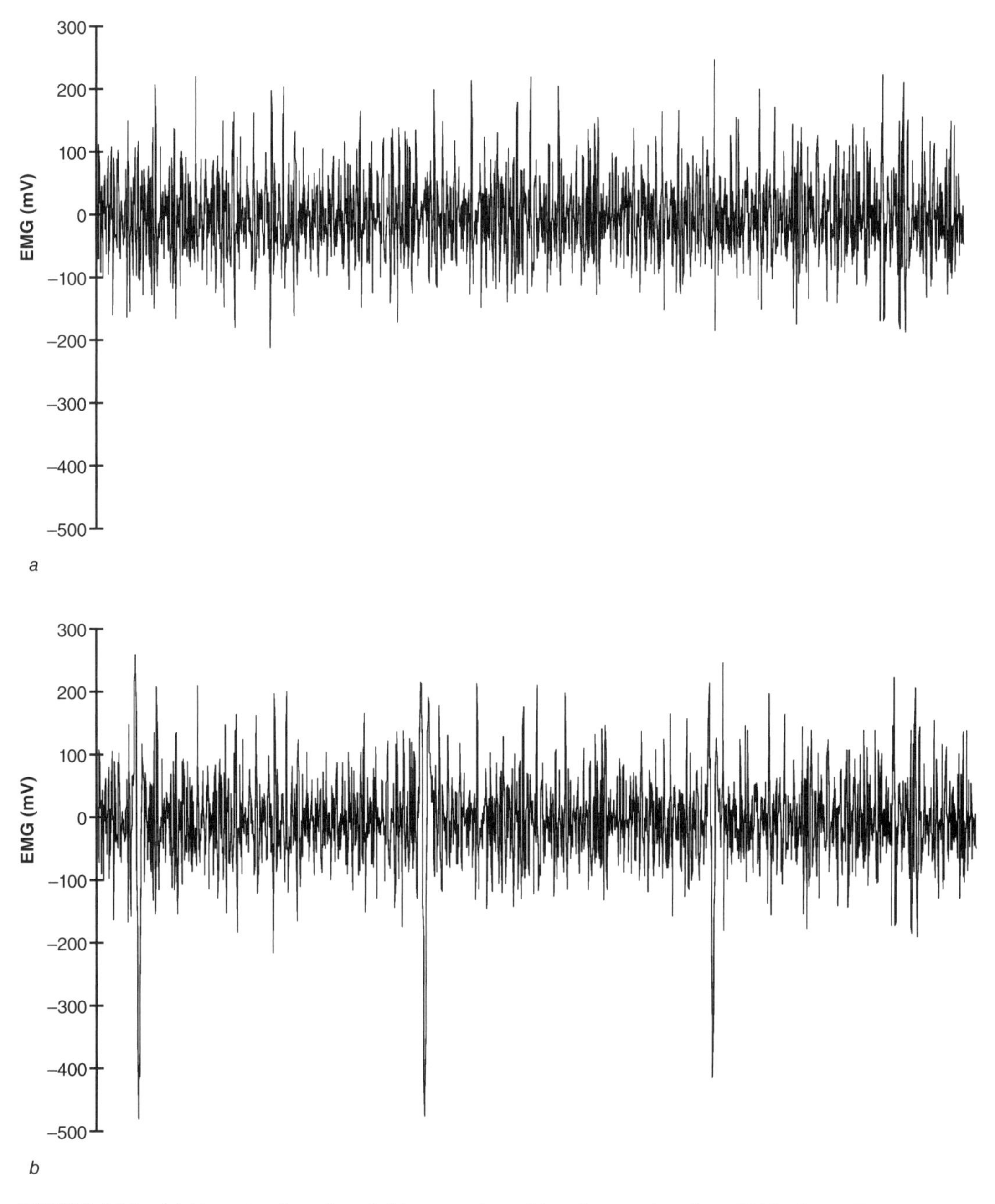

FIGURE 4.24 *(a)* Uncontaminated and *(b)* contaminated band-passed surface EMG signal. The source of contamination is the electrocardiographic (ECG) signal embedded within the surface EMG signal. The contaminated surface EMG shows three ECG waveforms.

There are three basic methods to minimize the effects of ECG contamination. The first technique creates an individualized template of the QRS waveform from baseline muscle activity. The cross-correlation function is then used to locate the QRS waveforms embedded within the EMG generated during a contraction. The template is aligned with the located QRS waveforms, and these are subtracted from the EMG signal. The subtracted waveforms are replaced by zeros. Frequency subtraction using the inverse FFT is the second method, and it is well established in the general signal-processing literature. The frequency components associated with the QRS complex are set to zero before the signal is reconstructed using the inverse FFT (IFFT). This method is appealing because it mimics the "brick wall" effect for the idealized filter described in chapter 3. It is analogous to a notch filter used to eliminate only the QRS waveform from the EMG. The main limitation with this technique is that it produces a *ringing effect* in the time domain associated with the Gibbs phenomenon at discontinuities in the data. This ringing effect causes a distortion in the amplitude structure of the EMG data, and it arises from the abrupt transition between frequencies that are set to zero using the IFFT. Weighted coefficients may be applied to a transition band around the zeroed frequencies to minimize the artifact. The efficacy of this method ultimately depends on the coefficients used and the available width of the transition band. The third method is based on the observation that most of the ECG signal power lies below 30 Hz, so EMG activity is high-pass filtered at that cutoff point. Drake and Callaghan (2006) compared all three methods and found that a high-pass filter at 30 Hz produced the best results.

KEY POINTS

- Inherent noise is primarily due to the physical measurement system, the electrodes and amplifiers. Interference noise may be due to electrical sources (power mains) or may be biological in origin (ECG). Both types of noise may be superimposed upon the EMG signal.
- If the noise has a Gaussian structure as occurs with most inherent sources, then signal averaging can be used to increase the SNR. Noise with a characteristic frequency such as that of power mains (60 Hz) or ECG may be subtracted directly using regression analysis or the IFFT.

Basic Concepts of Digital Filtering

Analog filters were reviewed in the previous chapter in association with settings on the traditional amplifier. The emphasis should be placed on collecting "clean" data upfront without having to treat the signals afterward for noise contamination and artifacts. There are several applications that do require additional software filtering for linear envelope detection.

Residuals Analysis

In the absence of any a priori criteria for setting the low-pass cutoff frequency (f_c), the preferred method is based on an analysis of the residuals between the unfiltered EMG and filtered EMG at low-pass cutoff frequencies (f_c) ranging from 2 Hz up to the Nyquist frequency (Winter 2005). The residuals are evaluated via calculation of the mean square error in the following way:

$$MSE(f_c) = \sqrt{\frac{1}{N}\sum_{i=1}^{N}(x_i - X_i)^2}$$

where x_i and X_i are the raw and filtered EMG at the ith sampled data point, respectively. The main advantage of this method is that it is data driven based on the interaction between the EMG signal and the characteristics of the filter. This technique assumes that the residuals will fluctuate around a fixed value for the noise component (Yu et al. 1999). This is evident in the plot of the residuals as an asymptote (figure 4.25). A regression line (dotted line) is then fit to the asymptote of the residuals to identify the intercept with the y-axis. A straight line is then extended from the intercept to the Nyquist frequency (500 Hz). The area between the horizontal line and the residuals line (shaded) is the noise that is passed through the filter. The low-pass cutoff frequency (f_c) is then the intersection between the horizontal line and the residuals plot. Cutoff frequencies below this point will result in signal distortion. The cutoff frequency identified by this method is f_c = 261 Hz, which is consistent with the power spectrum for biceps brachii surface EMG. Most of the signal power lies below this point. If the low-pass cutoff frequency is set below f_c = 261 Hz, the result is an increase in distortion of the filtered signal.

Digital Filtering

The same characteristics described in the bode plots for analog filters apply to software **digital filters:** (1) the cutoff frequency, (2) flatness in the pass band, (3) roll rate during the transition band, and (4) flatness in the stop band. Since these filter characteristics are all interdependent, the optimization of one is usually performed at

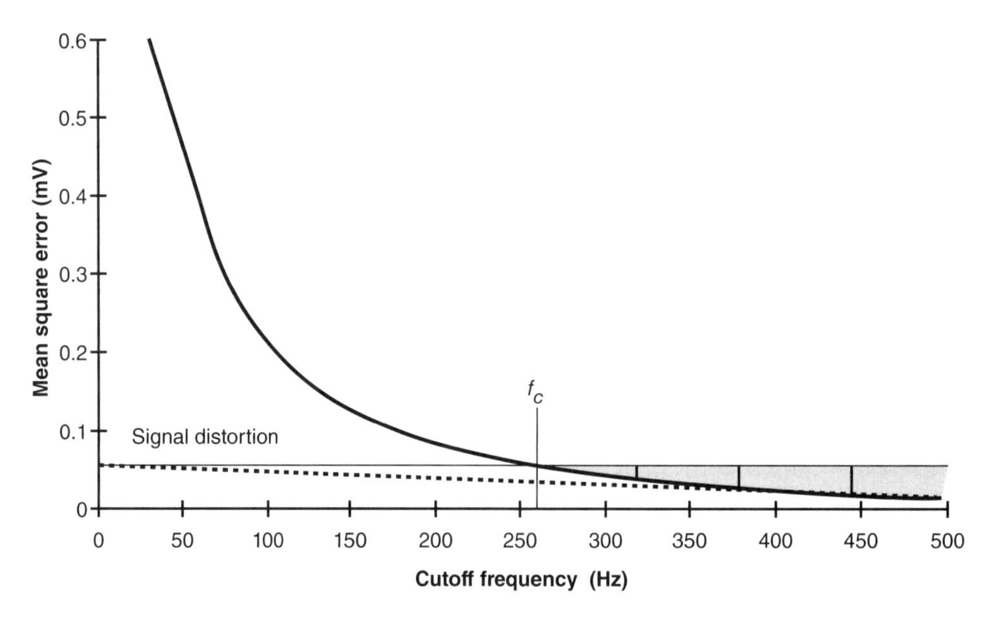

FIGURE 4.25 Residuals analysis of biceps brachii surface electromyographic activity to determine the optimal low-pass cutoff frequency.

the expense of one or more of the others. Thus, there really is no ideal filter. Another critical property of both analog and software digital filters that has yet to be discussed is the phase response. All filters delay the signal to some extent, which is not a problem as long as the delay is linear throughout the range of frequencies within the signal. If the filter delays some frequencies more than others, it will distort the signal. This is why the phase response of the filter is an important part of the bode plot. The *Butterworth* and *Bessel* filters are the two most common filters used in EMG practice because they have the best combination of two characteristics—roll rate and phase lag. The roll rate for Butterworth filters is poor in comparison to other types of filters but is maximally flat in the pass band, and the phase delay is linear in the pass band (see chapter 3). The Bessel filter is also maximally flat in the pass band. Although the roll rate is worse than that for the Butterworth filter, the phase delay is linear throughout the full range.

The delay, or *phase lag,* of the filter is of critical importance in EMG when one is examining the timing of activity with respect to mechanical events generated by the muscle. A way to achieve a *zero phase lag* is to pass the signal through the digital filter twice, in both the forward and backward directions of the data *(dual pass)*. The dual-pass method cannot be applied blindly. Rather, the desired cutoff frequency must be adjusted to account for the dual pass using a formula provided by Robertson and Dowling (2003).

Analog filters stack resistor–capacitor (RC) elements. Each RC element is considered to be a stage in the filtering process; the output of one stage is fed to another until the phase delay has been eliminated. A one-stage RC circuit has two poles and is called a *second-order filter.* A two-stage analog filter has four poles and is called a *fourth-order filter.* A signal that has been passed through a second-order filter, both forward and backward, to eliminate the phase lag is said to be digitally filtered with a fourth-order filter.

A good phase response means that the filter will respond in a predictable way to an abrupt increase in the level of the signal as represented by a step change or a spike. If the output oscillates before settling down to the input voltage value, the response is termed **underdamped.** A digital filter is **overdamped** if the output voltage takes a long time to achieve the input voltage value. If the output voltage follows the step increase in voltage without any overshoot, the filter is **critically damped** (figure 4.26). A critically damped filter is the standard in EMG analysis. These terms originate from a mechanical analysis of how a mass–spring damper system responds to an abrupt perturbation such as a step-input signal. The response to a step increase in voltage for a zero phase lag digital filter is shown in figure 4.27. The zero phase lag filter was dual passed and "anticipates" the step increase. The anticipation effect increases as the low-pass cutoff frequency decreases, which is evident in the linear envelope–detected EMG activity in figure 4.8*a* (p. 115).

We will consider a simple symmetric low-pass digital filter in order to illustrate the basics of how digital filtering is accomplished. Consider a three-point weighted average. The weighted average y_n at time $t_0 + nT$ of the original signal x_n is generated using a weighted average of the data points x_n and its two neighboring samples x_{n-1} and x_{n+1}:

$$y_n = \frac{1}{a+2}\left(x_{n-1} + ax_n + x_{n+1}\right),$$

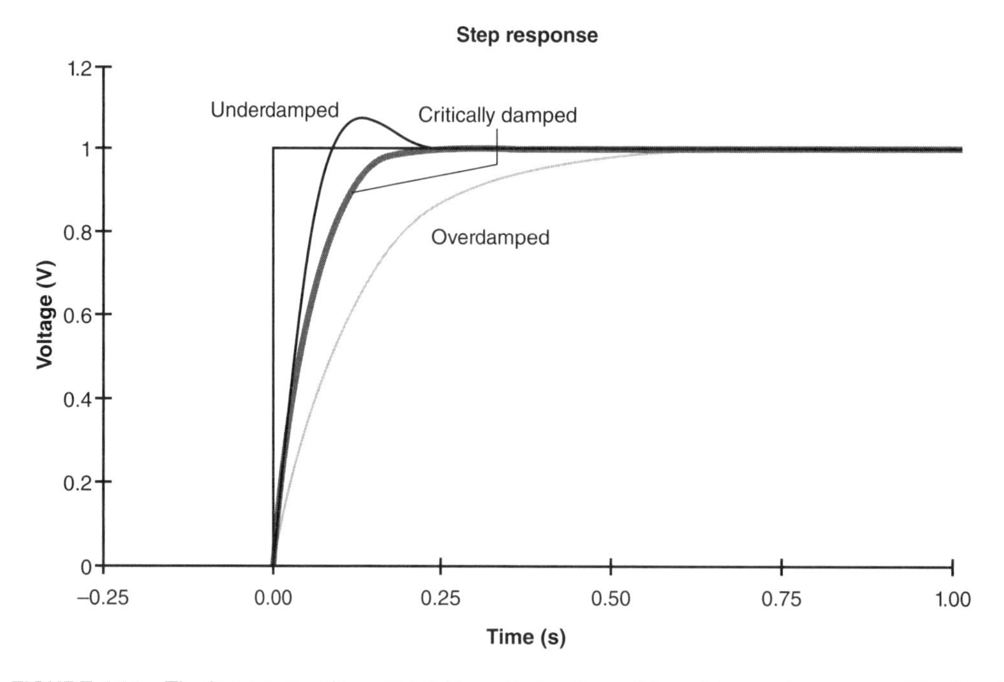

FIGURE 4.26 The input-output for a digital filter with the three different types of responses. The input is a step increase in signal voltage. The output for a critically damped filter is limited to the input voltage level. The underdamped filter slightly overshoots the input voltage while the overdamped filter takes a relatively long time to achieve the input voltage.

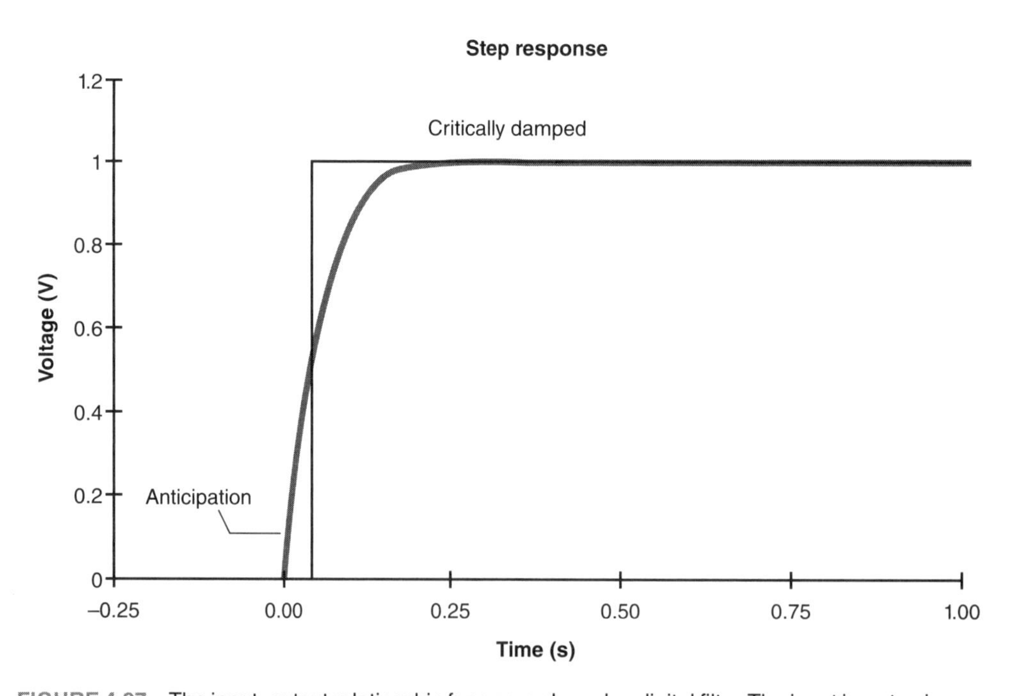

FIGURE 4.27 The input–output relationship for a zero phase lag digital filter. The input is a step increase in signal voltage. The zero phase lag filter anticipates the step increase in voltage. The anticipation ensures that the filtered signal retains the same phase as the original unfiltered signal.

where t_0 is the beginning of the signal and T is the sampling time (Milsum et al. 1973). The weighting coefficient in this case is the digital filter coefficient $a = 2$. If we express the digital filter as an array, the coefficients are (1, 2, 1). The array of coefficients is lined up alongside the data in the signal in the following order (x_{n-1}, x_n, x_{n+1}). Each filter coefficient is multiplied by the data point alongside it, and the result is summed. In this case, because it is a weighted average, the sum is divided by $a+2$. The sample in the center (x_n) is ultimately replaced by the calculated value (y_n). The new value (y_n) is stored until the digital filter has moved down the signal a number of data points equal to the length of the array of coefficients. This ensures that the digital filter is executed only on the original signal. The three-point weighted average moves along the signal, from the beginning toward the end, one point at a time, repeating the same operations. Thus, the operations require a *moving window* consisting of three digital filter coefficients. The same basic procedures are also involved in the application of a window weighting function as outlined earlier for calculating the FFT.

It is important to note that the digital filter can start only at the second index (x_2) of the original signal because there is no x_{n-1} at the start of the signal. There also is no x_{n+1} at the end of the signal. More data points will be necessary as the digital filter becomes more complex, requiring a greater number of coefficients. This problem is exacerbated as the filter "anticipates" the signal for very low-pass cutoff frequencies. Increasing the data collection period to include more points before and after the signal of interest is one way to accommodate both the length of the digital filter array of coefficients and "anticipation" of the signal. When it is difficult to obtain additional data points, the *inverse inflection method* is particularly useful (Smith 1989). The inverse inflection method copies data points from the end of the signal and appends them to the beginning. Consider the case in which 20 data points are padded in front of a signal that starts at (t_i). The last data point (N) is then appended to the front at (t_{i-20}). Each of the end-point values takes on the opposite sign at the front the signal:

$$x(t_{i-20}) = -1 \times x_N$$

$$x(t_{i-19}) = -1 \times x_{N-1}$$

$$x(t_{i-18}) = -1 \times x_{N-2}$$

$$\vdots$$

$$x(t_{i-1}) = -1 \times x_{N-19}$$

The Butterworth and Bessel digital filters are natural extensions of the three-point weighted average. The three-point weighted average is a *nonrecursive filter* wherein the output is a weighted combination of a finite number of past and present samples of the data. The Butterworth and Bessel digital filters are *recursive* or *autoregressive* filters because the output depends not only on past and present samples of the raw data (x_n) but also on past values of the filtered signal (y_n). The following equation is for a second-order Butterworth filter:

$$y_n = a_0 x_n + a_1 x_{n-1} + a_2 x_{n-2} + b_1 y_{n-1} + b_2 y_{n-2}$$

The recursive terms (b_1 and b_2) will require additional data points at the beginning of the signal before the digital filter can be initiated. This is an additional reason to either include more data points before and after the signal of interest or use the inverse inflection method. *The exact values of the digital filter coefficients determine the filter characteristics outlined here.* Formulas for calculating the coefficients for a Butterworth digital filter according to the desired characteristics may be found in Winter 2005, with an additional treatment by Robertson and Dowling (2003).

KEY POINTS

- There are very few options in terms of using digital filters to remove noise. A digital filter is a series of weighted coefficients that, when multiplied with the EMG signal through a moving window, alters the frequency content of the EMG signal, leaving certain frequencies unaltered while minimizing others.
- The emphasis should be on obtaining a "clean" signal through good methodological controls. The Butterworth digital filter is used heavily in EMG processing because it is maximally flat in the pass band, while the Bessel digital filter is favored because it has a linear phase lag in the pass band.

FOR FURTHER READING

Kumar, S., and A. Mital. 1996. *Electromyography in ergonomics.* Bristol, PA: Taylor & Francis.

Shiavi, R. 1999. *Introduction to applied statistical signal analysis.* 2nd ed. New York: Academic Press.

Smith, S.W. 1997. *The scientist and engineer's guide to digital signal processing.* San Diego, CA: California Technical Publishing.

Sörnmo, L., and P. Laguna. 2005. *Bioelectrical signal processing in cardiac and neurological applications.* London: Elsevier Academic Press.

Winter, D.A., and A.E. Ptala. 1997. *Signal process and linear systems for the movement sciences.* Waterloo, ON: Waterloo Biomechanics.

chapter 5

EMG–Force and EMG–Fatigue Relationships

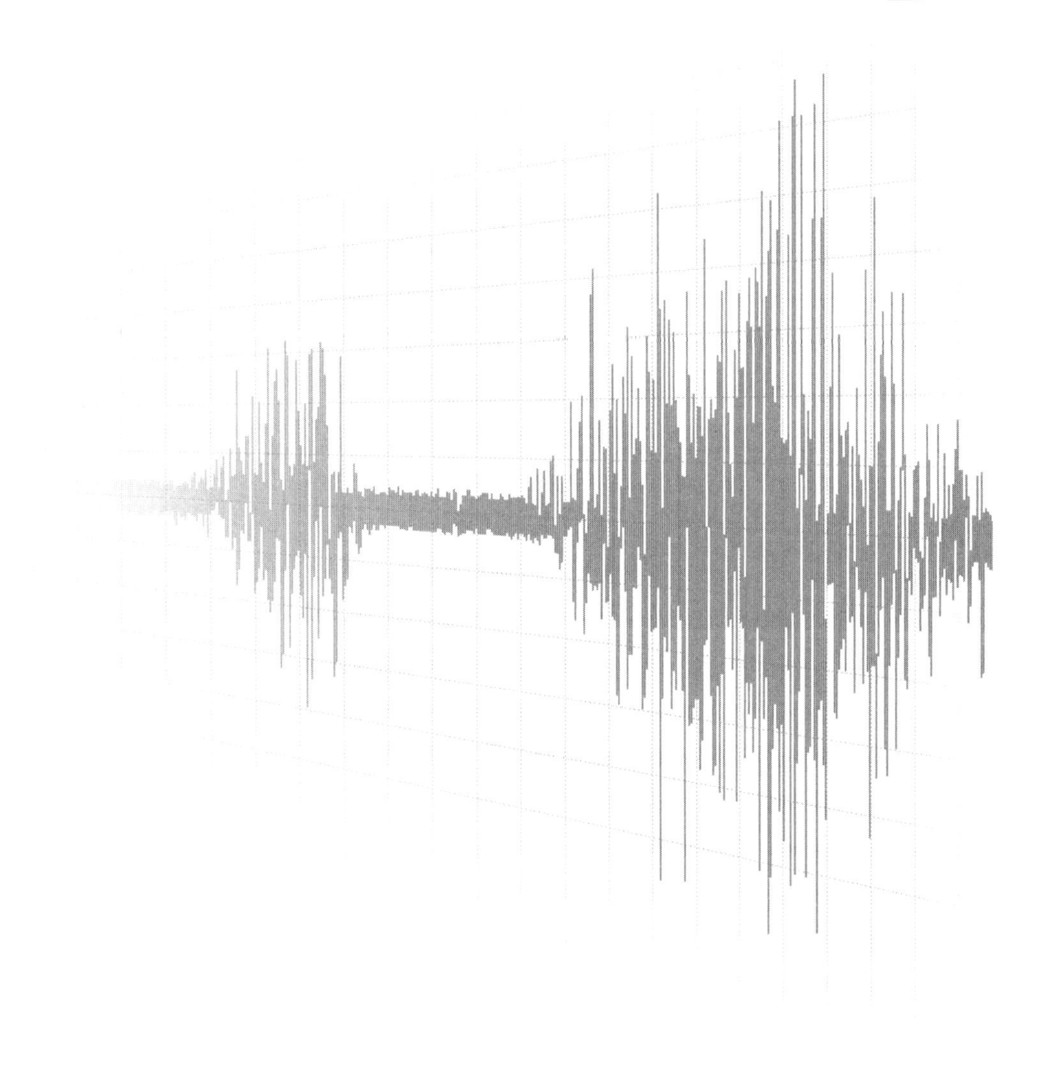

In the previous chapters, we have discussed the basic physiological characteristics of the electromyogram, important bioelectricity and instrumentation issues, and a variety of procedures used to analyze the EMG signal. In this chapter, we begin our discussion of how we use this information. This chapter discusses two applications that have resulted in many research publications. The first section details the analysis of the EMG signal at various muscular forces and the way in which force relates to EMG amplitude. Second, we discuss the issue of muscular fatigue and both amplitude and frequency characteristics of the EMG signal during the fatigue state.

Relationships Between Muscular Force and EMG

There are many cases in which knowledge of the relationship between EMG and force is desired. For example, a seemingly simple contraction involving elbow flexion can actually be somewhat complicated when we examine the characteristics of muscle activation. Elbow flexion is controlled minimally by biceps brachii, brachioradialis, and brachialis. Which of these muscles predominate in muscle activation? How does the muscle activation strategy change as the characteristics of initial position, muscle fiber characteristics, movement velocity, and other variables change?

Another example concerns the design of a proportional control prosthetic limb. It would certainly be important to understand the magnitude of limb movement that would be appropriate given the extent of EMG input (Parker and Scott 1986). In some prosthetic limb systems, the EMG signals from the flexor digitorum superficialis have been used to determine the magnitude of finger movements, while signals from biceps and triceps muscles are frequently used to characterize arm movements (Patterson and Anderson 1999; Zecca et al. 2002). If the relationship between force and EMG amplitude is simply linear, a direct regression equation yields a relatively simple technique to control prosthetic limb function.

Most of the extant research literature involving the EMG–force relationship has involved isometric contractions. If the EMG–force relationship were better understood under isometric conditions, it could be used in numerous nonisometric applications. For example, it could be used in the gait laboratory to advance our understanding of the forces produced during walking or running activities. Alternatively, ergonomists could assess the load on various muscles by monitoring the EMG activity.

EMG Magnitude and Muscular Force

In earlier chapters, we discussed techniques for assessing both the amplitude (or magnitude) of the EMG signal and the frequency characteristics of the EMG signal. In this section we discuss an application involving EMG magnitude and how it changes with muscular force level.

Studies Using Isometric Contractions

In recordings made from the gastrocnemius muscles, Lippold (1952) was among the first to describe a linear relationship between EMG magnitude and muscular force. Numerous subsequent investigations using isometric contractions have also obtained a linear relationship between muscular force observed and EMG amplitude. For example, a linear relationship has been observed in the biceps (Knowlton et al. 1956; Moritani and deVries 1978), masseter (Kawazoe et al. 1981), plantarflexors (Lippold 1952), and first dorsal interosseous (Milner-Brown and Stein 1975), among other muscles. The relationship between EMG and force also seems to depend on the nature of the muscle studied, since some investigators have reported a linear relationship for the adductor pollicis and first dorsal interosseous and soleus, and a nonlinear relationship for the biceps and deltoid (Lawrence and De Luca 1983).

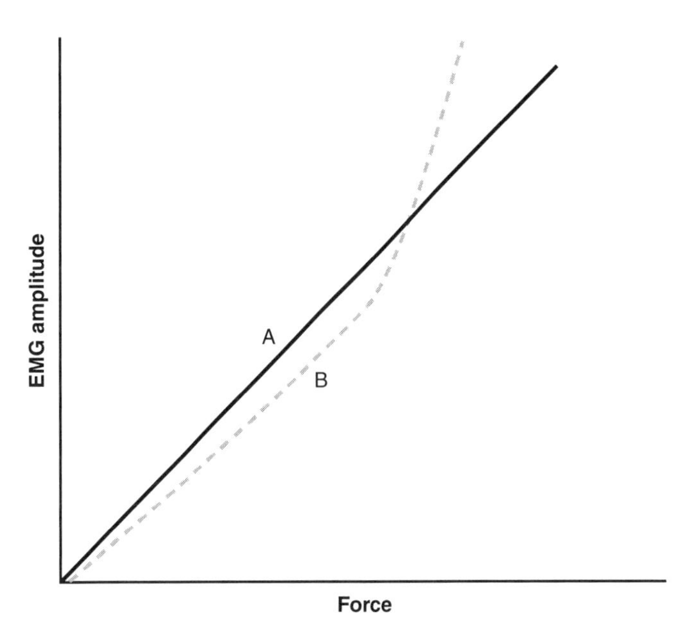

FIGURE 5.1 The relationship between EMG amplitude and muscular force in human muscle. Most studies show a linear relationship between force and EMG amplitude (A). However, nonlinear relationships have also been reported (B).

There have been other, numerous examples of observations of nonlinear relationships between force and EMG amplitude (Alkner et al. 2000; Maton and Bouisset 1977; Metral and Cassar 1981; Thorstensson et al. 1976; Woods and Bigland-Ritchie 1983; Zuniga and Simons 1969). Usually, the low-force and high-force regions have different slopes than the middle portion of the EMG–force curves (figure 5.1). Nightingale has suggested that many studies do not address the EMG–force relationship in the low-force and high-force regions. Otherwise, there might be a greater consensus that the relationship is nonlinear (Nightingale 1960). It is clear that when considering the possible shape of the EMG–force relationship, one needs to consider various features of the movement, such as the type of muscle contraction (isometric vs. dynamic); the size and location of the active muscles; their role as agonists, synergists, or antagonists; air temperature (Bell 1993); and the numerous other physiological and technical factors that affect the electromyogram.

Studies Using Nonisometric Contractions

Some researchers have used nonisometric contractions to assess the force–EMG relationship, although this is problematic since direct assessment of joint force during nonisometric contractions is difficult. Using the elbow extensors, Aoki and colleagues (1986) reported a linear relationship between kinematic variables such as peak velocity and acceleration on the one hand and EMG amplitude on the other. Similar results have been obtained in the elbow flexors (Barnes 1980; Bouisset and Maton 1972; Komi 1973) and plantarflexors (Bigland and Lippold 1954). Others have reported a nonlinear EMG–force relationship during rapid contractions in the first dorsal interosseous (Bronks and Brown 1987). Gerdle and colleagues (1988) reported no relationship between root-mean-square (RMS) amplitude and angular velocity in knee extensor muscles. In cycling, the muscles of the plantarflexors exhibit differential effects, with a linear relationship in the soleus and a nonlinear relationship in the gastrocnemius (Duchateau et al. 1986). When the muscle is allowed to shorten slightly, the EMG–force relationship becomes nonlinear (Currier 1972; Edwards and Lippold 1956), so more linear relationships are observed at longer muscle lengths. This may explain why the slope of the EMG amplitude–force relationship varies with joint angle (Bouisset 1973).

Studies Focusing on Other Factors

Other factors that affect the relationship between EMG and muscular force include the design and placement of the electrodes (Moritani and deVries 1978), muscle length (Inman et al. 1952), speed of contraction (Bouisset and Goubel 1973), and muscle fatigue (Lindstrom et al. 1970). There is also the suggestion that the nonlinearity of the relationship is partially due to the interelectrode distance (Bouisset 1973), with higher amplitudes and a tendency to more nonlinear relationships at greater interelectrode distances. Also, various pathologies can affect the EMG–force relationship (Muro et al. 1982; Tang and Rymer 1981). Subjects producing a sincere effort manifest a nonlinear relationship in the biceps EMG–force curve, while those who are "faking" maximal-force effort produce a linear relationship (Chaffin et al. 1980). Trying to provide a physiological explanation for the numerous equivocal results, Perry and Bekey (1981) suggested that the linear relationships at low-to-moderate force levels are due to recruitment, while the quadratic relationship at higher force level is due to rate coding.

There are some data on the in situ forces produced in both human and animal models (Gregor et al. 1987; Gregor and Abelew 1994; Landjerit et al. 1988). However, a full understanding of the relationship between EMG magnitude and force in situ will require further experimental studies.

Frequency Analyses

The relationship between spectral frequency and muscular force is even more inconsistent than the relationship between EMG amplitude and force. Bilodeau and colleagues (1992) found that the relationship varied by gender and muscle, speculating that differences in skinfold thickness and fiber type between the two subject groups led to differences in the median and mean frequency-versus-force relationships. A somewhat nonlinear relationship was found between both mean and median frequency and force in the anconeus muscle. However, inconsistent relationships were observed in biceps and triceps muscles. In some cases, for example, median frequencies decreased with increasing muscular force. Using the number of turns in the surface electromyogram, Fuglsang-Frederiksen and Mansson (1975) reported that the turns-versus-force relationship became nonlinear in the higher-force regions. Mean power frequency may be more strongly related to fiber type than to the velocity of contraction (Gerdle et al. 1988). Using wavelet analysis techniques, Karlsson and Gerdle (2001) provide some evidence for a linearity between frequency components of the EMG and knee extensor muscular torque. However, others have failed to find a linear relationship between median frequency and force (Onishi et al. 2000).

KEY POINTS

- The relationship between EMG and force is largely dependent on the specific muscle studied, with linear relationships observed in some muscles and nonlinear relationships in other muscles.
- The slope of the EMG amplitude–force relationship varies with joint angle, in part due to the differences in the relationship with muscle length changes.
- The relationship between EMG frequency and force can change considerably depending on factors such as skinfold thickness, fiber type, and specific muscle characteristics.

EMG Analysis During Fatiguing Contractions

The analysis of the electromyogram during fatiguing effort is an area of interest to many researchers. In the field of ergonomics, researchers and practitioners are concerned with workplace fatigue and the extent to which normal everyday activities may result in muscular fatigue that can be documented using EMG. Neck or shoulder pain and back pain are serious problems for many occupations, and there are many applications in ergonomics regarding the use of EMG to explore neck and low back pain.

The EMG signal does change with fatigue (see figure 5.2), and one can find changes in EMG parameters during fatiguing effort in various muscles, including the soleus

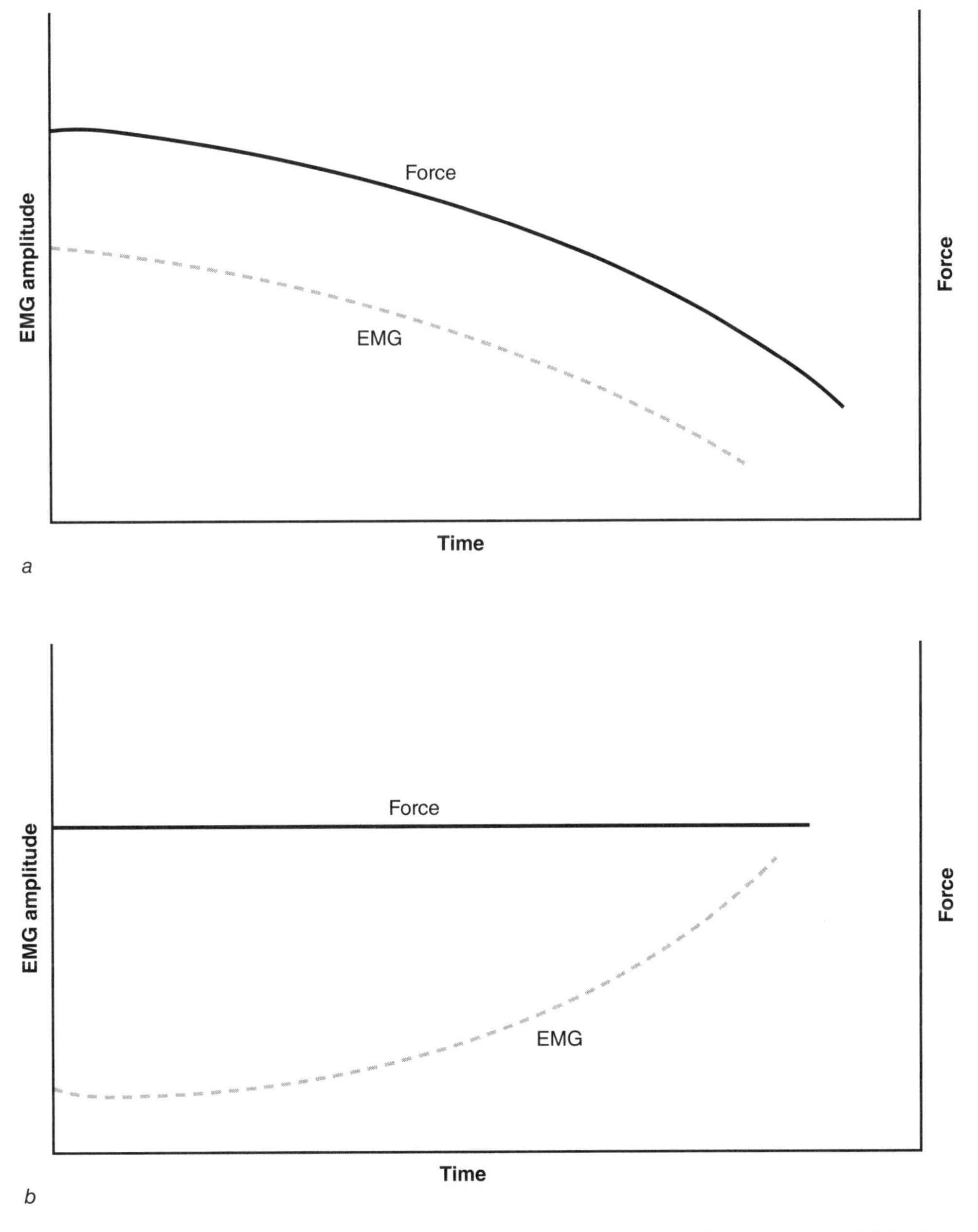

FIGURE 5.2 Changes in EMG amplitude occur during fatiguing isometric contractions. *(a)* During a sustained maximal contraction, EMG amplitude declines soon after contraction onset. *(b)* During a sustained submaximal contraction, EMG amplitude increases as the contraction is maintained.

(Kukulka et al. 1986), masseter (Kroon et al. 1986), forearm musculature (Lind and Petrofsky 1979), biceps brachii (Moritani et al. 1986), tibialis anterior (Reid et al. 1993), and many others. It is possible that the surface EMG signal obtained during fatiguing contractions may become useful for diagnosing neuromuscular diseases such as McArdle's disease, fibromyalgia, and other problems. Future studies will need to provide clinicians with the most useful standardized muscular contractions and the most useful EMG analysis techniques for aiding diagnosis.

EMG Amplitude During Fatigue

In chapter 4, we discussed measures used to describe EMG amplitude. Recall that both the average rectified value (ARV) and the RMS amplitude are used to characterize changes in EMG amplitude during fatiguing contractions.

$$ARV = \frac{1}{N}\sum_{i=1}^{N} | x_i | \quad RMS = \sqrt{\frac{1}{N}\sum_{i=1}^{N} x_i^2}$$

where x_i are the individual EMG values, and N is the total number of samples in the signal.

The changes in EMG amplitude during fatiguing isometric contractions are well documented. During a maximal isometric contraction, EMG amplitude declines (Bigland-Ritchie 1979; Gerdle and Fugl-Meyer 1992; Moritani et al. 1986; Stephens and Taylor 1972). This decline in surface EMG amplitude is likely due to decreases in motor unit firing rate (Bigland-Ritchie et al. 1983), possible neuromuscular propagation failure (Bellemare and Garzaniti 1988), or the slowing of conduction velocity produced by the added K^+ ions and the depletion of Na^+ inside the muscle fiber. These changes in amplitude can be readily measured using the technique of computing the RMS amplitude of the EMG signal.

During sustained submaximal contractions held to fatigue, EMG amplitude is stable initially but then increases (e.g., Krogh-Lund and Jorgensen 1991), and this increased amplitude is likely due to the need for increased motor unit recruitment to maintain the required force (Fuglevand et al. 1993; Krogh-Lund 1993; Maton and Gamet 1989). The observation that maximal EMG activity at the end of a submaximal contraction held to fatigue does not reach the maximal attainable EMG activity observed prior to a fatiguing contraction is evidence of some central fatigue failure (Fuglevand et al. 1993). Changes in RMS amplitude during fatigue in two-joint muscles are different than the changes observed in monoarticular muscles (Ebenbichler et al. 1998), which could suggest different neural control mechanisms for biarticular versus monoarticular muscles. Thus, the electromyographer studying fatigue needs to ensure stable conditions to detect these changes. Skin impedance could change as a result of skin sweating, so proper electrode contact throughout the contraction needs to be monitored. Electrode movement needs to be minimized as well.

Spectral Frequency Characteristics

Considerable attention has been given to understanding the changes in frequency characteristics of the EMG signal during fatiguing effort. Most of this research has utilized the mean power frequency or median power frequency of the EMG signal as discussed earlier. Lindstrom and colleagues (1977) are generally credited with initially defining the method of using the power spectrum to describe changes in the EMG signal with fatigue.

Changes in the mean and median frequency of the EMG signal are both valuable, though they are highly intercorrelated. Consequently, only one of these measures needs be used to describe fatigue changes. These measures can provide insight regarding shape changes in the frequency spectrum during the fatiguing contractions, although information related to median frequency is also available as the skewness measure of the frequency spectrum.

Spectral frequency analysis can be valuable for describing EMG fatigue-related changes that occur during isometric contractions (figure 5.3). As an example, Williams and colleagues (2002) sought to assess changes in mean and median EMG power frequency following fatigue of the elbow flexors. Subjects held a maximal isometric contraction until they could no longer sustain 50% of the maximal force. Spectral analysis revealed that mean and median frequency were about 100 Hz prior to the fatiguing exercise, declining to about 70 Hz after exercise. No difference in the response was exhibited between left and right elbow flexors; neither was a gender difference observed in mean power frequency during fatigue (Bilodeau et al. 2003). Similar changes in mean power frequency have been observed during submaximal isometric contractions: During a 30% maximal voluntary plantarflexion contraction held to fatigue, mean power frequency declined in both the soleus and gastrocnemius muscles (Loscher et al. 1994). The decline in mean and median frequency with fatigue is a common observation, has been seen in a wide variety of muscles including a number of facial muscles (Van Boxtel et al. 1983), and has been verified by many researchers in numerous static and dynamic conditions.

The analysis of median frequency has been useful in the diagnosis of various disorders and in the analysis of ergonomic tasks. Median frequency measures obtained during fatigue in patients presenting with patellofemoral pain syndrome are different in the vastus medialis oblique and vastus lateralis muscles from those observed in healthy subjects (Callaghan et al. 2001). Jensen and colleagues (1993) used EMG analysis to study shoulder muscle fatigue among industrial sewing machine operators. They documented changes in EMG mean power frequency and zero crossing frequency during the workday, which agreed with ratings of perceived exertion over the workday among the workers.

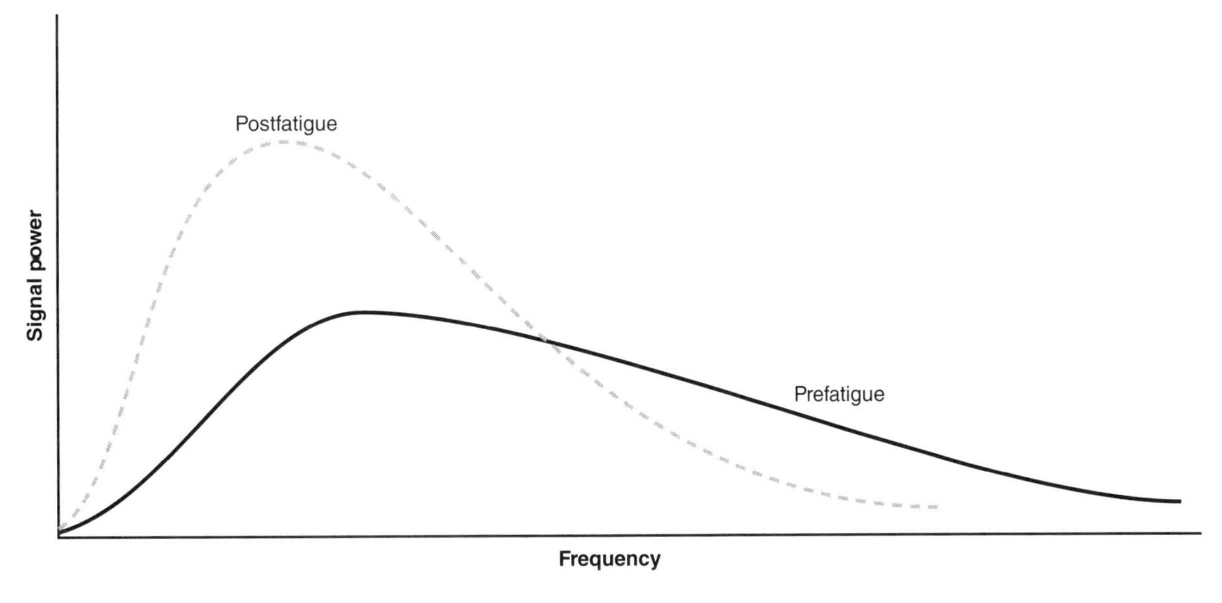

FIGURE 5.3 The EMG frequency spectrum is shifted to the left during fatiguing muscular contractions.

EMG fatigue curves have long been used to help diagnose patients with low back pain (deVries 1968). For example, spectral parameters have been shown to be useful in discriminating patients with low back pain from normal individuals (Klein et al. 1991). Although Roy and colleagues (1989) found a greater reduction of median frequency in a patient population with fatigue, Kramer and colleagues (2005) found that chronic low back pain patients exhibited a *lesser* decline in EMG median frequency during fatigue than normal subjects.

Experimentally, there are certain conditions that affect mean and median frequency. During fatiguing contractions, these conditions need to be maintained at a constant level to minimize the effect of extraneous variables on spectral frequency content. For example, muscle temperature affects frequency content. The frequency spectrum of the EMG signal is shifted toward lower frequencies with decreased muscle temperature (Petrofsky and Laymon 2005) but this effect is attenuated as contraction continues, likely due to the increase in muscle temperature with contraction (Holewijn and Heus 1992). Although it may not be possible to maintain constant muscle temperature throughout a contraction, the initial skin or muscle temperature (preferably both) should be standardized over conditions and across subjects.

Researchers have identified several concerns involving frequency analysis during force-varying, dynamic contractions. There may be problems with computing frequency estimates from EMG signals below 20% to 30% maximal voluntary contraction (MVC), since unphysiologically high frequency values can be obtained (Clancy et al. 2005; Hof 1991). The EMG signal, particularly during dynamic contractions, may be nonstationary, with long periods during which the means and variances are changing over time. Mean frequency is affected by joint angle (MacIsaac et al. 2001b; Matthijsse et al. 1987), for example, so that during dynamic contractions, changes in mean frequency might reflect changes in joint angle as well as fatigue-related factors. The nonstationarities may be caused by alterations in the number of active motor units, which may change during a dynamic contraction. Changes in motor unit wave shapes might also produce nonstationary signals (MacIsaac et al. 2000).

One solution is to analyze the signal during short time periods, considering the EMG signal as short-sense stationary. If force-varying contractions are used, the recommendation is to select intervals of 0.5 to 1 s, zero padding as necessary to compute the Fast Fourier transforms (FFTs) and reporting median or mean frequency results using these intervals. The validity of using short time periods for computing the FFT and reporting frequency estimates is supported by the research literature (MacIsaac et al. 2001a). However, this is an active area of EMG research, and considerably more understanding is needed regarding techniques for analyzing the EMG signal during dynamic, fatiguing contractions.

Details regarding the use of the short-time Fourier transform and its use in force-varying fatiguing contractions can be found in some previous publications (Farina and Merletti 2000; Hannaford and Lehman 1986; MacIsaac et al. 2001a; see also appendix 4.3). Methods are also available to assess EMG characteristics of fatiguing contractions using time-varying autoregressive techniques (Farina and Merletti 2000). However, the short-time Fourier transform methods and the time-varying autoregressive methods produce similar estimates of mean and median EMG frequency (Bower et al. 1984; Clancy et al. 2005).

The mechanism underlying the decrease in spectral frequency during fatiguing contractions is not entirely clear. Muscle fiber conduction velocity may account for part of the decrease in median frequency with fatigue (Lindstrom et al. 1970).

Numerous studies have documented the decline in muscle fiber conduction velocity during fatiguing contractions (Lowery et al. 2002; Mortimer et al. 1970; Schulte et al. 2006; Van Der Hoeven and Lange 1994), and there is some evidence that both conduction velocity and mean frequency decline linearly during fatiguing contractions (Eberstein and Beattie 1985). However, changes in median frequency during fatigue are different in isometric and dynamic tasks that involve the same load. Thus, other factors probably influence fatigue-induced alterations in EMG spectral frequency. Metabolic variables attributable to the difference in muscle blood flow in the two types of muscle contractions are one likely candidate (Masuda et al. 1999).

KEY POINTS

- EMG amplitude changes during fatiguing contractions, with smaller amplitudes during maximal sustained contractions and higher amplitudes during submaximal contractions held to fatigue.
- The mean power frequency of the EMG signal has been reported to decline in both static and dynamic contractions.
- Frequency analysis during fatiguing contractions can be useful in diagnosing injuries, as in patients with low back pain.
- The change in frequency with joint angle makes frequency analysis difficult during dynamic contractions. The nonstationarity of the EMG signal during dynamic contractions is also a serious problem for frequency analysis.
- Short-time FFT methods may be useful for analyzing the EMG signal during fatiguing dynamic contractions.

Advanced EMG Issues During Fatiguing Contractions

In addition to measuring changes in EMG amplitude and frequency characteristics, a number of other techniques are useful in assessing how the muscle environment (including muscle fiber features) and the neural command to the muscle may change during muscle fatigue. As detailed next, these techniques may involve evoked potentials, conduction velocity measurements, or other procedures used in nerve or muscle or in both.

M-Waves During Fatigue

The M-wave or compound muscle action potential (CMAP) is often used as an indicator of neuromuscular propagation failure during fatigue (Stephens and Taylor 1972). Stimulating the motor nerve in the periphery provides for a direct assessment of muscle fiber electrical characteristics. In a study designed to assess neuromuscular propagation, M-wave amplitude in the adductor pollicis muscle decreased during a 90 to 100 s MVC of the thumb adductors. Similar results were obtained by Bellemare and Garzaniti (1988). The declines might be attributed to failure of the muscle fiber action potential to propagate along the muscle fiber membrane, and work by Fuglevand and colleagues (1993) supports this failure of neuromuscular transmission. However, others have failed to find declines in the M-wave during fatiguing contractions (Bigland-Ritchie et al. 1982; Kukulka et al. 1986). An example of the M-wave during a MVC can be seen in figure 5.4.

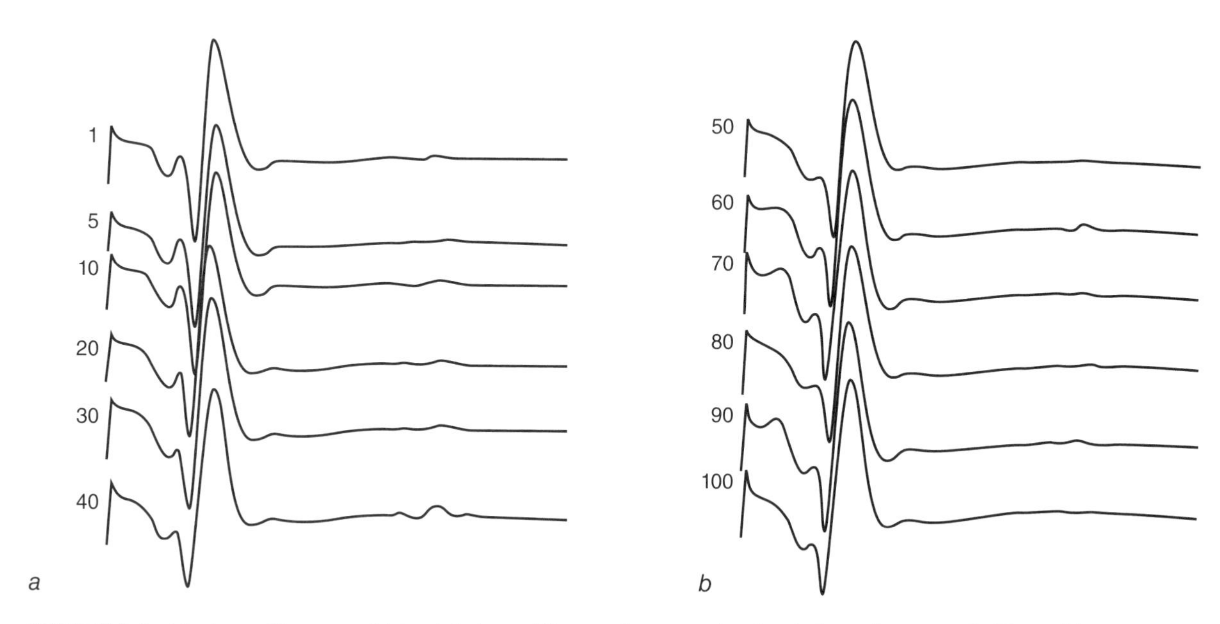

FIGURE 5.4 Maximum M-waves obtained during a 110 s maximum contraction in the soleus muscle. These M-waves were evoked through electrodes applied in the popliteal fossa using single 1 ms square-wave pulses.

Reprinted from *Brain Research* 362(1), C.G. Kulkulka, A.G. Russell, and M.A. Moore, "Electrical and mechanical changes in human soleus muscle during sustained maximum isometric contractions," p. 50, copyright 1986, with permission from Elsevier.

The Importance of Muscle Length

One should note that muscle length is an important determinant of muscle fiber conduction velocity during fatiguing contractions (Arendt-Nielsen et al. 1992). During two different static fatiguing contractions, conduction velocity was measured at 45° and at 90° of knee flexion. Recordings made from the vastus lateralis indicated that conduction velocity decreased with increasing muscle length. Moreover, during fatiguing contractions, the largest increase in RMS amplitude and the fastest decrease in mean power frequency and conduction velocity were observed at 90° of knee flexion.

Muscle length also affects the frequency characteristics of the EMG signal (Inbar et al. 1987). Decreases in median frequency that occur with fatigue are more pronounced at the shortest muscle lengths (Doud and Walsh 1995). These findings demonstrate how difficult it can be to analyze surface EMG data during dynamic contractions, particularly frequency measures. The safest recommendation is to obtain EMG amplitude and frequency measures at the same muscle length, especially during fatiguing dynamic contractions.

Shifts in Spectral Frequency During Fatigue

The bulk of the evidence supports the notion that muscle fiber conduction velocity alone cannot account for the shift in spectral frequency with fatigue. First, the decline in median frequency occurs at a greater rate than the decline in muscle fiber conduction velocity (Broman et al. 1985b; Krogh-Lund and Jorgensen 1993). Another factor that appears to be related to the change in frequency characteristics of the muscle during fatigue is the increased extracellular K^+ concentration (He et al. 2005). Spectroscopy studies have revealed that neither pH nor lactate is responsible for the changes observed in median frequency (Vestergaard-Poulsen et al. 1995). Part of the decline in the spectral frequency is due to the increase in intramuscular pressure during isometric contractions (Korner et al. 1984).

The mechanisms causing the decline in mean frequency with fatigue have not been fully elucidated. One candidate factor is motor unit synchronization—the simultaneous or near-simultaneous discharge of groups of motor units. However, there is little evidence from motor unit recordings that synchronization increases during fatigue. Other factors such as new motor unit recruitment (during submaximal contractions) may affect median frequency (Krogh-Lund and Jorgensen 1991). Again, it is difficult to identify these other factors that affect EMG frequency characteristics during fatigue without recordings from individual motor units. Needle and wire electrodes can be used successfully to demonstrate decreases in motor unit firing rate with fatigue (Rubinstein and Kamen 2005).

Other EMG-Fatigue Reporting Techniques

A simple time domain–based frequency variable that has been used to describe the decrease in frequency characteristics during fatigue is zero crossings. One simply uses the raw EMG signal to count the number of times the signal crosses zero. The number of zero crossings decreases with fatigue (Kilbom et al. 1992), in a fashion similar to that for median and mean spectral frequency (Hagg 1992).

Another technique that has been employed to analyze the EMG signal during fatigue involves interference pattern analysis (IPA) using so-called spike parameters. Interference pattern analysis is a simplified form of frequency analysis that has been shown to yield results similar to those obtained through spectral analysis (Gabriel et al. 2001). The technique was illustrated in figure 4.10 (p. 118). Each pair of upward and downward deflections that both cross zero and exceed the 95% noise confidence interval defines a *spike*. A *peak* (indicated by the "x") is any pair of upward and downward deflections within a spike that does not constitute a discrete spike. From these EMG points, one can define mean spike amplitude, mean spike frequency, mean spike duration, mean spike slope, and the mean number of peaks per spike. Gabriel and colleagues (2001) found that changes in mean spike amplitude were quite similar to those produced by RMS amplitude. Fatiguing contractions also produced similar declines in mean power frequency computed using spectral analysis and in mean spike frequency computed using IPA. Thus, IPA may be an alternative and simplified technique to describe the EMG signal during fatiguing muscular effort.

There are other techniques that can be used to report EMG frequency characteristics. For example, Roman-Liu and colleagues (2004) were interested in identifying the magnitude of fatigue in ergonomic tasks involving shoulder motion. They computed a fatigue index that includes a time domain–based amplitude factor and a frequency domain–based frequency factor. This may be appropriate for repetitive tasks in which a decrease in a frequency domain variable (like median frequency) occurs concurrently with an increase in EMG amplitude during submaximal effort.

Lin and colleagues (2004) used a similar approach to analyze fatigue occurring in healthy females who typed consecutively for 2 h. They recorded from wrist flexor and extensor muscles for a 2 h period. Using a technique termed JASA (joint analysis of spectra and amplitudes), they documented the decrease in amplitude and increase in frequency during the 2 h interval.

Particularly interesting is the use of so-called time–frequency variables aimed at overcoming the nonstationarity of EMG signal, and this technique may be particularly applicable during nonisometric contractions (Bonato et al. 2001). One of the important parameters of time–frequency analysis is instantaneous median frequency, and this variable has been shown to be useful in monitoring fatigue-related changes in the EMG (Knaflitz and Bonato 1999). During a fatiguing squat maneuver, for example, the instantaneous median frequency changes more for an anterior cruciate ligament (ACL)-deficient patient than for a control subject (Bonato et al. 2001).

Nonlinear techniques involving recurrence quantification analysis have revealed that many subjects exhibit nonlinear responses during fatigue (Ikegawa et al. 2000). However, Ravier and colleagues (2005) explored the use of an EMG fractal indicator and found that it was insensitive to fatiguing contractions.

Grid electrodes are becoming increasingly popular (Farina et al. 2006; Holtermann et al. 2005; Staudenmann et al. 2006) and have been used to assess fatigue in the trapezius muscles. Holtermann and Roeleveld (2006) used a 13 × 10 array electrode to study surface EMG activity in the trapezius muscle during fatiguing voluntary contractions. These electrodes may prove useful for identifying localized areas of muscular fatigue.

Reliability of EMG Measures During Fatiguing Contractions

Moderate reliability of both median frequency and EMG amplitude measures can be obtained from fatiguing contractions produced one week apart (Mathur et al. 2005). Ng and Richardson (1996) also reported moderate reliability scores from endurance tests performed three days apart. Their study involved recordings from the iliocostalis and multifidus muscles. One variable that has been used but is not very reliable is the slope of the median frequency during fatigue (Elfving et al. 1999).

Larsson and colleagues (2003) required subjects to perform 100 maximum concentric knee extension contractions on two separate days. Intraclass correlation (ICC) was used to measure the reliability of RMS amplitude and mean frequency measures. The results showed that for these two variables, the ICCs of three knee extensor muscles were all greater than 0.80. Suitable reliability has also been obtained from 50% MVC isometric contractions (Arnall et al. 2002) and from the analysis of time–frequency measures (Ebenbichler et al. 2002).

Other Issues and Recommendations

One way to measure central drive during fatiguing contractions is to use evoked potentials techniques, transcranial magnetic stimulation (TMS), or twitch interpolation (Biro et al. 2006). From these studies, we know that central drive is diminished following fatiguing contractions (Gandevia 2001). Intracortical inhibition can also be assessed with TMS using a paired-pulse protocol (Kujirai et al. 1993); these results demonstrate that intracortical inhibition is decreased following fatigue (Maruyama et al. 2006).

On theoretical grounds, it is generally recommended that surface electrodes *not* be placed over the innervation zone. This is particularly important when repeated tests are to be conducted, for example, on different days. However, if the electrodes are placed over the innervation zone, the results may be similar to those obtained during a fatiguing contraction when the electrodes are placed proximal or distal to the innervation zone (Malek et al. 2006).

Movement artifact can be a problem in some fatigue studies, particularly during nonisometric contractions. Consequently, both high- and low-pass filters should be used. The high-pass filters should be set at approximately 5 to 10 Hz (to remove movement artifact), while the low-pass filters should be set at about 500 to 1000 Hz to attenuate high-frequency noise.

Some of the same recommendations previously made for recording the EMG signal are appropriate here:

- When recording from multiple muscles or from a small muscle whose activity may be confused with that of neighboring muscles, beware of cross-talk.
- Be cognizant of possible nonstationarity issues, particularly during dynamic contractions.
- Ensure that surface electrodes are placed longitudinally along the muscle fibers, not perpendicular to the muscle fibers.
- Normalization should be considered if comparisons are to be made between muscles.

KEY POINTS

- The changes observed in M-wave amplitude can be useful for understanding the electrophysiological characteristics of muscle during fatigue.
- Muscle length can affect both muscle fiber conduction velocity and the frequency characteristics of the EMG signal, rendering EMG interpretation difficult during dynamic contractions.
- Neither pH nor lactate concentration is a determinant of median frequency changes during fatigue, though the increase in intramuscular pressure, particularly during isometric contractions, decreases spectral frequency.
- Although the mechanism producing the decline in EMG frequency with fatigue has not been identified, both motor unit synchronization and new motor unit recruitment have been suggested as possible factors.
- Interference pattern analysis is an alternative means of describing the frequency characteristics of the EMG signal during fatigue, and this technique produces findings similar to those obtained from traditional spectral analysis.
- Time–frequency analysis and other nonlinear techniques such as fractal analysis may ultimately prove useful in describing the EMG signal during fatiguing contractions.

FOR FURTHER READING

Dimitrova, N.A., and G.V. Dimitrov. 2003. Interpretation of EMG changes with fatigue: facts, pitfalls, and fallacies. *Journal of Electromyography and Kinesiology* 13: 13-36.

Gandevia, S.C. 2001. Spinal and supraspinal factors in human muscle fatigue. *Physiological Reviews* 81: 1725-1789.

Hof, A.L. 1984. EMG and muscle force: an introduction. *Human Movement Science* 3: 119-153.

Hof, A.L. 1997. The relationship between electromyogram and muscle force. *Sportverletz Sportschaden* 11: 79-86.

Perry, J., and G.A. Bekey. 1981. EMG–force relationships in skeletal muscle. *CRC Critical Reviews in Biomedical Engineering* 7: 1-22.

chapter 6

Other EMG Applications

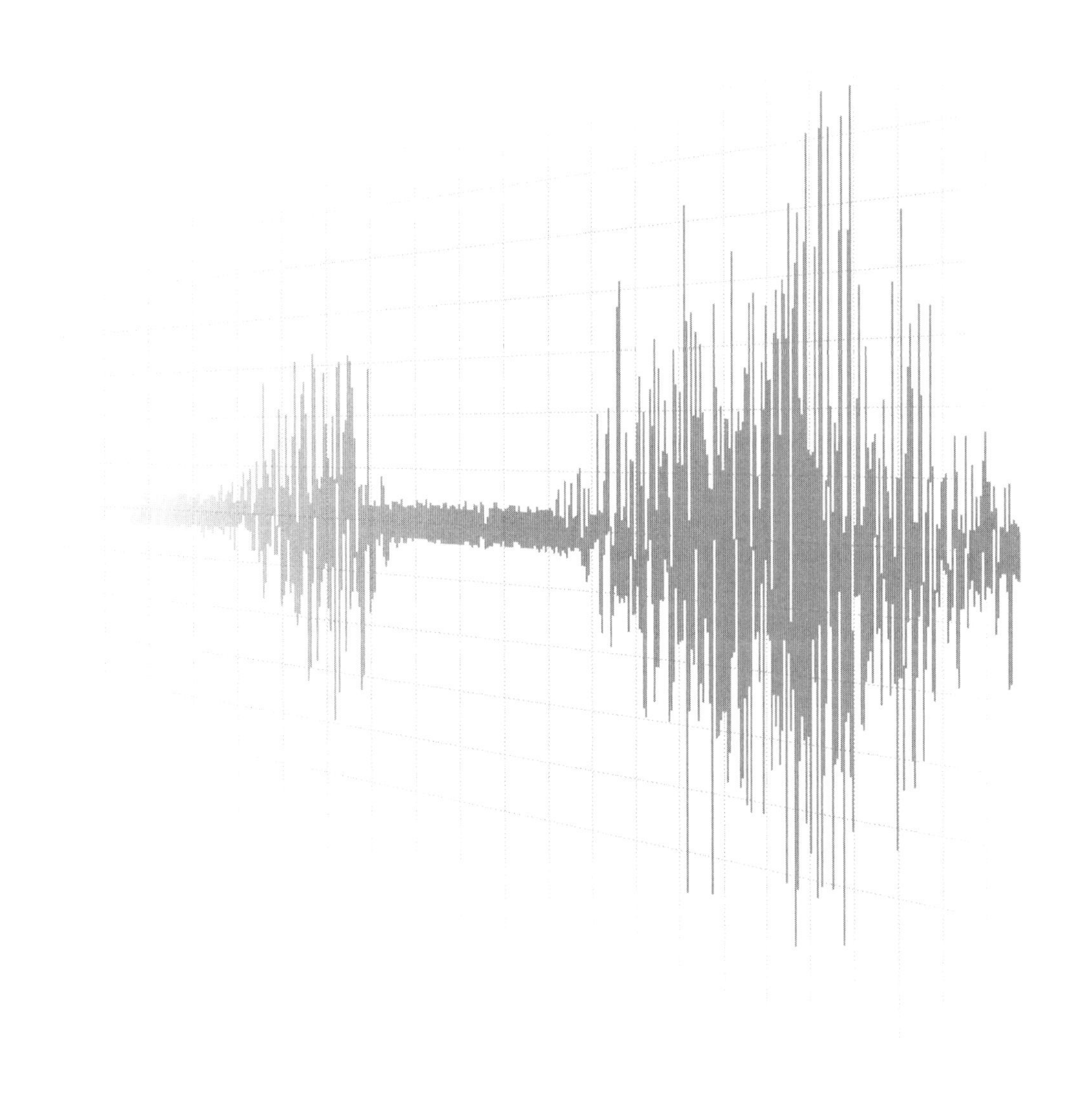

In chapter 5, we discussed some of the major applications of surface EMG techniques, including the assessment of the EMG–force relationship and the use of EMG to study fatigue. However, there are many other research and experimental situations that benefit from the implementation of EMG techniques. In this chapter, we discuss the use of EMG for gait analysis, some of the many stimulus-evoked responses measured using EMG, and the use of EMG for assessing the control of rapid movements.

EMG and Gait

Walking is a cyclical activity that can be impaired by injury or disease. Gait analysis is frequently used to identify the source of an impairment that negatively affects the gait cycle. EMG gait analysis has been applied to sports medicine injuries to determine changes in activation following surgery of the anterior cruciate ligament (Knoll et al. 2004) and total knee replacement (Benedetti et al. 2003), as well as to study muscle activation in patients suffering from patellofemoral injury (Mohr et al. 2003). EMG gait analysis has also been applied to determine the success of operations designed to improve gait in cerebral palsy patients (Cahan et al. 1990; Perry and Hoffer 1977) and to determine the extent of recovery in stroke patients during walking. EMG analysis can supplement other types of gait analysis such as kinematic and kinetic analyses of the walking motion. Numerous reviews have been published detailing the analysis of gait using EMG techniques (Craik and Oatis 1995; Perry 1992; Shiavi 1985; Sutherland 2001; Winter 1991).

Indwelling or Surface Electrodes?

During a walking or running motion, for example, it is certainly more convenient to be able to apply surface electrodes to the muscles of interest than to apply indwelling electrodes. As we have discussed in earlier chapters, recording EMG activity using surface electrodes requires care in electrode application and attention to the recording details (figure 6.1). However, some of the muscles of interest for gait analysis or other areas of investigation may be small or may be sufficiently deep to require wire or indwelling electrode recordings. The amplitude of EMG signals recorded from the skin surface declines very sharply with distance. Moreover, EMG amplitude scores during maximal contractions are smaller in individuals with greater amounts of subcutaneous fat (Nordander et al. 2003). Thus, deeper muscles will require wire electrodes for detection. Of course, the exact take-up area depends greatly on characteristics of the surface electrodes (diameter, interelectrode distance, etc.) and characteristics of the wire electrodes (again, interelectrode distance, amount of exposed wire, and other features). Findings on the question of which electrodes produce more reliable results in superficial muscles are equivocal. Some researchers have found that surface electrodes are more reliable (Kadaba et al. 1985; Komi and Buskirk 1970), and others have suggested that surface and wire electrodes are equally reliable (Giroux and Lamontagne 1990). However, more recently, wire electrodes have been shown to produce results similar to those with surface electrodes

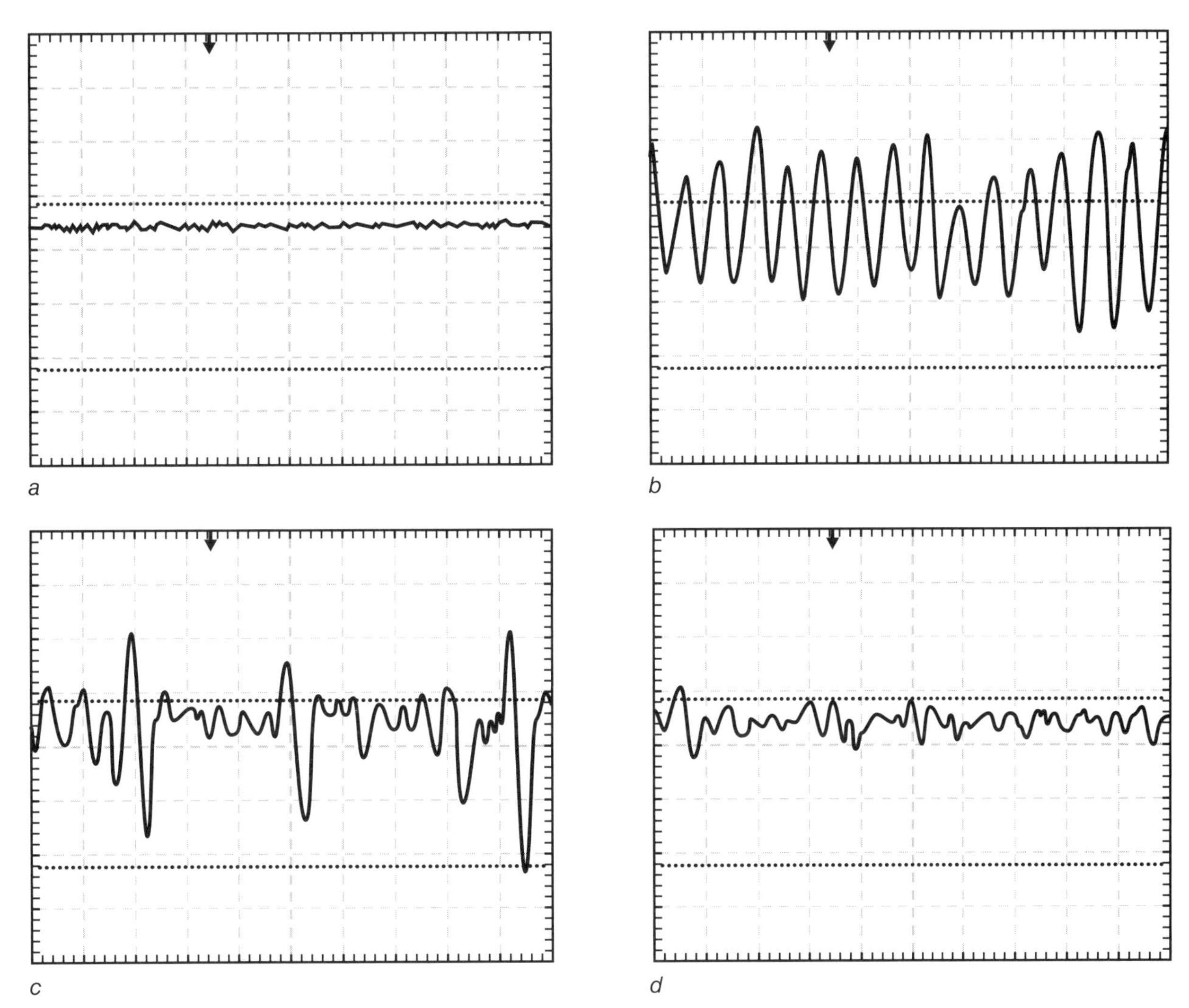

FIGURE 6.1 Examples of various problems that can accompany the application of surface electrodes. *(a)* An electrolyte "bridge" formed by excess electrolyte between two bipolar electrodes can short-circuit the input signal. *(b)* Poor skin preparation and little electrolyte applied can exaggerate line noise. *(c)* Movement artifact causing high-frequency spikes during the movement. *(d)* EMG signal obtained from proper electrode preparation.

(Bogey et al. 2003), and it is possible that more accuracy can be obtained using wire electrodes (Solomonow et al. 1994). Figure 6.2 presents an example of EMG activity recorded using wire electrodes during walking.

Some investigators prefer wire electrodes since the smaller take-up area may present less chance for cross-talk and therefore misinterpretation (Perry et al. 1981). Using double differential surface recordings may also minimize problems produced by cross-talk (Koh and Grabiner 1992). The muscles that may require wire electrodes include the tibialis posterior (Reber et al. 1993), vastus medialis oblique (Mohr et al. 2003), popliteus (Weresh et al. 1994), peroneus brevis (Reber et al. 1993; Walmsley 1977), erector spinae (Thorstensson et al. 1982), flexor hallucis longus (Skinner and Lester 1986), and the hip flexors (Andersson et al. 1997), among others. Of course,

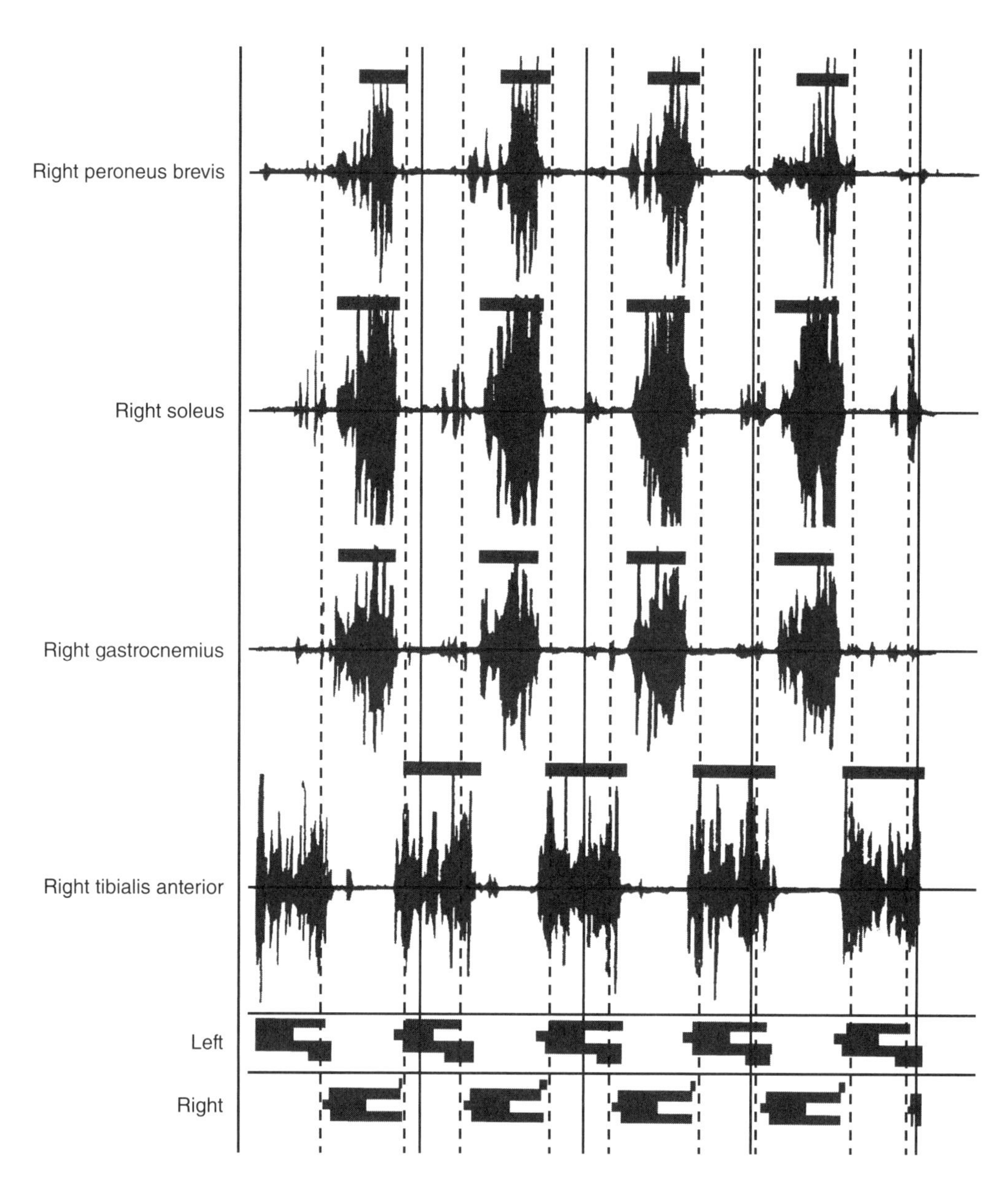

FIGURE 6.2 EMG activity recorded using wire electrodes during walking.

Adapted from *Journal of Biomedical Engineering* 15(6), J. Perry, E.L. Bontrager, R.A. Bogey, et al., "The Rancho EMG analyzer: A computerized system for gait analysis," p. 488, copyright 1993, with permission from IPEM.

one has to ensure that minimal changes in gait occur across several stride cycles with the use of any electrode (Young et al. 1989).

Normalization

Consider a situation in which a researcher would like to compare the EMG amplitude from a small intrinsic muscle of the hand with that recorded in a large knee extensor muscle when these muscles are involved in a maximal-effort contraction. If the same electrodes and electrode configuration are used, the raw EMG amplitude will likely be greater in the larger muscle, since it reflects the activity of a considerably greater number of muscle fibers. Thus, if the researcher plans to present amplitude information from multiple muscles, then *amplitude normalization* is important. Since the amplitude of the signal can be affected by numerous factors, presenting the EMG amplitude of several muscles without normalizing can lead to spurious conclusions. Normalization can greatly facilitate comparisons between muscles, between recording sessions, or between subjects. One of the most frequently employed normalization procedures is the use of a maximal isometric contraction. Subjects can be asked to perform a maximal isometric contraction, and the EMG activity can be normalized to the maximal EMG activity production during the maximal-effort contraction (Arsenault et al. 1986b). Although EMG activity during submaximal isometric effort has also been used, some researchers believe that the reduced reliability observed renders this normalization technique inferior to some of the other normalization methods (Yang & Winter 1984).

The EMG activity during various phases of the gait cycle can be normalized to the maximum dynamic activity during the gait cycle, a technique termed the *peak dynamic method* (Jacobson et al. 1995; Prilutsky et al. 1998; Wu et al. 2004). Alternatively, the mean of the ensemble average during the dynamic activity can be used (Bulgheroni et al. 1997). Thus, there are several normalization techniques that are useful for EMG signals obtained during gait. The decision about which procedure to select is left to the user, taking into account the characteristics of the subject or patient group; specific muscles being analyzed; and access to the instrumentation and equipment that may be required, such as the facilities appropriate for performing submaximal isometric contractions. Normalizing using dynamic isokinetic contractions is not recommended (Burden et al. 2003).

Finally, the appropriate normalization technique may depend on the nature of the research question being raised (Benoit et al. 2003). For example, Benoit and colleagues (2003) examined the ability to detect EMG differences in injured versus uninjured knee extensor muscles of patients diagnosed with anterior cruciate ligament (ACL) injury. Three normalization techniques were used. In one (MEA), the EMG activity was normalized using the mean amplitude during the gait cycle. In the second technique (MAX), the maximum value during the gait cycle was used, while in the third procedure (MVC) the EMG activity was normalized to that obtained during a maximal voluntary contraction.

There were several differences among the three techniques, with the MAX method identifying differences in the rectus femoris activity between the injured and noninjured limb. The MVC technique identified differences in gastrocnemius activity between the two limbs. When an individual is able to produce a maximal voluntary contraction, the MVC method of normalization has been recommended as the best normalization technique; but again, as already noted, the appropriate normalization procedure to use may depend on the nature of the research question.

Appropriate Quantitative Measures

The amplitude of EMG activity measured during gait is usually obtained from the linear envelope detected signal (Patla 1985; Shiavi et al. 1998; see chapter 4). When computing the linear envelope it is important to take into consideration the sources of variability, ensuring that the signals from the gait task are recorded with minimal noise.

Some low-pass filtering of about 10 Hz is recommended (Shiavi et al. 1998). Reliable information can frequently be obtained from as few as three strides (Arsenault et al. 1986c; Shiavi et al. 1998), although under some conditions there can be considerable interstride variability. If the observed variability among strides is a concern, the use of average scores from at least 10 strides is appropriate (Winter 1991). Averaging the linear envelope over several strides can minimize the high-frequency characteristics of the linear envelope, and this may be undesirable. One solution recommended by Herschler and Milner (1978) is to apply what they term a "smoothing filter" to the multiple envelopes. This technique can be implemented using a low-pass filter, as described in chapter 4. Herschler and Milner's technique optimizes the consistency observed in the linear envelope over several stride cycles. Other techniques have been suggested, such as the use of a Hamming filter with a 32 ms window length (Kadaba et al. 1985) and the use of envelope filters with low- and high-pass cutoff frequencies of 3 and 25 Hz (Kleissen 1990).

Recently, multivariate statistical techniques have also been used in an effort to summarize the information provided from many muscle groups. So far, these techniques have included factor analysis (Ivanenko et al. 2004), cluster analysis (Mulroy et al. 2003; Wootten et al. 1990), and neural network and wavelet methods (Chau 2001). New research may lend insight into what additional information can be provided through the use of these analysis techniques.

EMG Onset–Offset Analysis

One of the important measures in gait analysis involves identification of the onset and offset of EMG activity. During the gait cycle, EMG bursts can occur prematurely or may be prolonged or out of phase, or there may be other abnormalities (Perry 1992). Normally, for example, the vastus lateralis and vastus medialis obliquus (VMO) begin their activity simultaneously. However, in patients with patellofemoral pain, VMO activity is often delayed (Cowan et al. 2001). Careful selection of the proper technique to identify EMG onset and offset, then, can be very important in gait analysis. In chapter 4, we detailed some of the techniques available for detecting EMG onset and offset. Here we provide some examples regarding the use of these techniques.

Numerous algorithms have been used to identify the onset of EMG activity during gait, perhaps beginning with visual inspection or the "traumatic ocular" statistic (e.g., Andersson et al. 1997). Using visual inspection, one simply "eyeballs" the point at which the EMG signal exceeds baseline. For gait analysis specifically, others have recommended more quantitative techniques for determining onset and offset of the signal. Perry (1992) suggests that it is appropriate to use an EMG amplitude that is above 5% of the maximum effort obtained from a manual muscle test. Others have used a level above that obtained when the individual is lying supine (Bogey et al. 2000), a level that is some multiple of the standard deviation estimate above a baseline trial (Allison 2003; Morey-Klapsing et al. 2004; Wu et al. 2004), or other methods (Allison 2003; Hodges and Bui 1996; Micera et al. 2001). In the selection of a technique for determining EMG onset during gait, it is important to identify a valid and reliable procedure, ensuring that the signal is uncontaminated by noise or signals from adjacent muscles. Whichever procedure is chosen, it is highly recommended that one compare the EMG onset obtained using the chosen algorithm with that obtained using visual inspection before conducting extensive analysis.

Gait activity is frequently conducted on a treadmill, facilitating the cable connections that might be difficult with overground locomotion. However, EMG signals have also been obtained during overground locomotion (Dubo et al. 1976; Quanbury et al. 1976; Winter and Quanbury 1975). Advances in commercial telemetry systems have facilitated the acquisition of EMG signals during both treadmill and overground locomotion. However, the comparison of EMG results obtained from treadmill and overground running have been interesting and somewhat equivocal. In some studies, similar EMG patterns have been observed during both overground and treadmill running, even when there are kinematic differences between the two conditions. Ordinarily, we would expect a change in kinematics to alter the EMG signal. One possible explanation for the stability of the EMG pattern and amplitude when comparing running in these two conditions may be that the change in kinematic pattern between the tasks is too subtle to produce a difference in EMG that can be statistically observed. (Nymark et al. 2005; Schwab et al. 1983; Wank et al. 1998).

Visual Presentation of EMG Data During Gait

Numerous visual techniques are available for presenting EMG data during gait. Frequently, simple presentations of the raw data may be an important way to provide information to the reader. Figure 6.3 shows an example of raw EMG activity

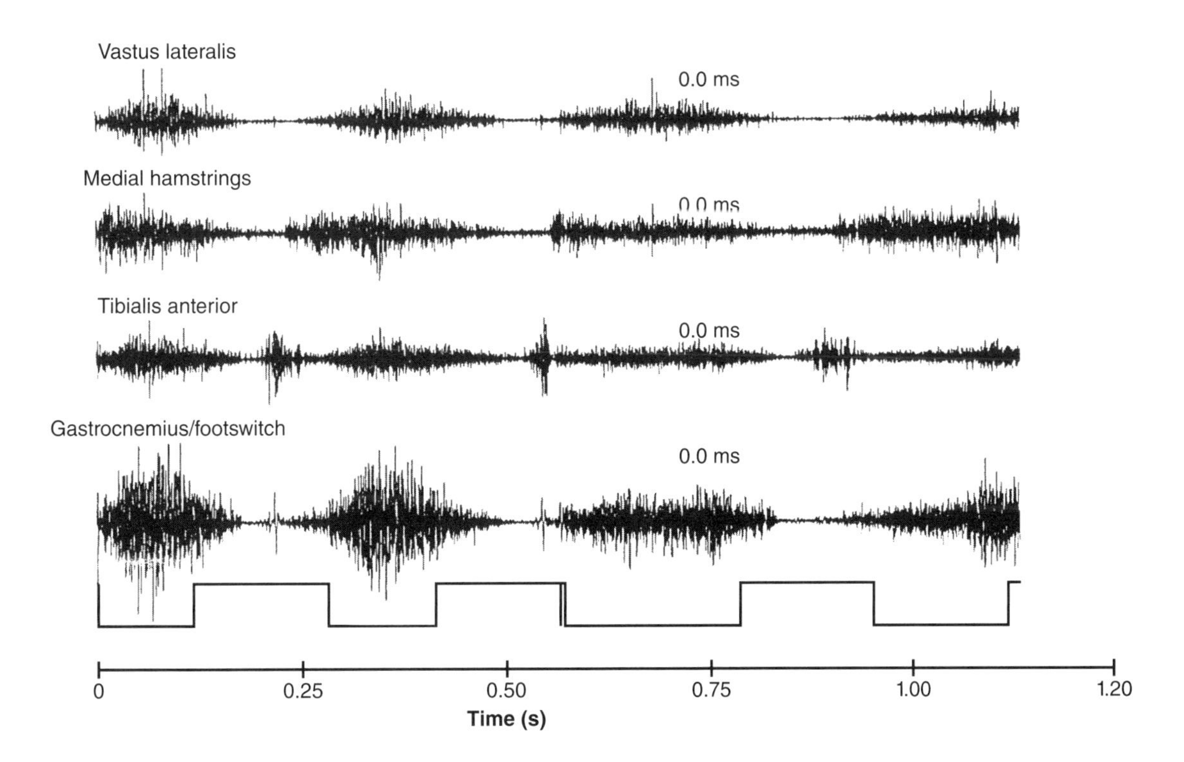

FIGURE 6.3 Raw EMG activity in a cerebral palsy patient, showing extensive intervals of cocontraction.

Adapted, by permission, from R.L. Craik and C.A. Oatis, 1995, *Gait analysis: Theory and application* (St. Louis: Mosby), 310.

during gait in a patient with cerebral palsy. This presentation of the raw EMG activity effectively demonstrates the level of cocontraction that can be present in cerebral palsy patients.

Using the computed onset and offset durations, periods during which each lower limb muscle is active can be presented along a time continuum in which 0 represents the initial heel strike and 100 represents the next heel strike (figure 6.4). The type of presentation shown in figure 6.4 can be effective for providing a demonstration of when muscles are "on" and "off," though the figure may not illustrate periods during which there is minor activity, perhaps reflecting secondary postural activities.

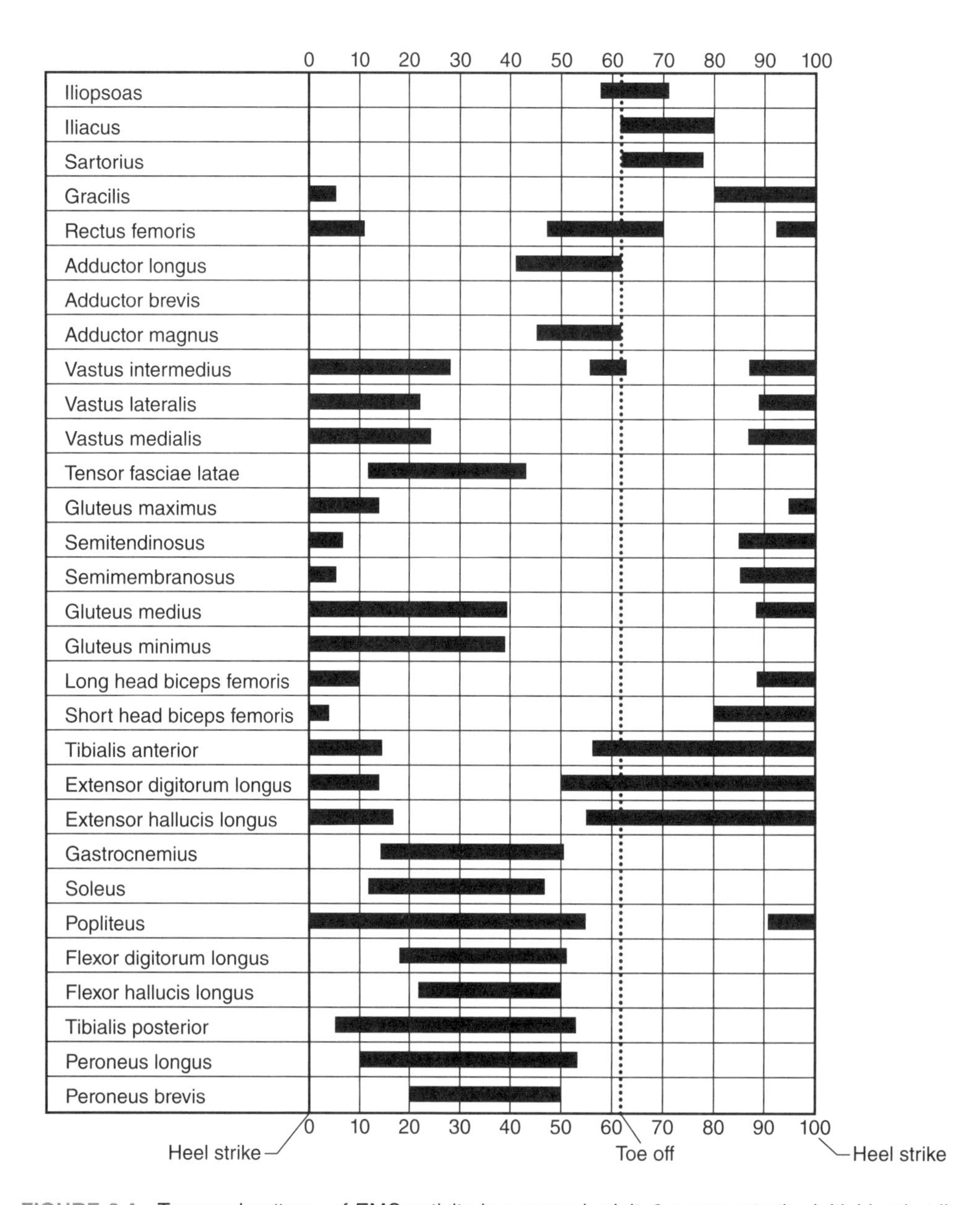

FIGURE 6.4 Temporal patterns of EMG activity in a normal adult: 0 represents the initial heel strike, and 100 represents the next heel strike.

Adapted, by permission, from R.L. Craik and C.A. Oatis, 1995, *Gait analysis: Theory and application* (St. Louis: Mosby), 312.

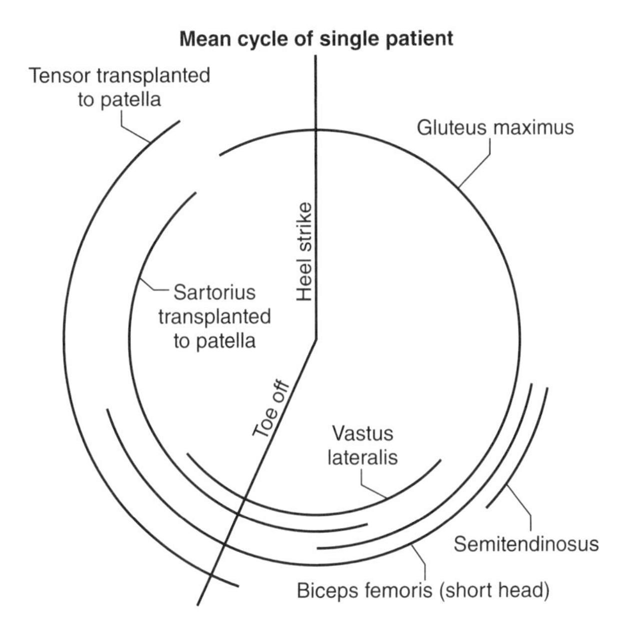

FIGURE 6.5 Temporal presentation of EMG data during the gait cycle using polar plots.

From D.H. Sutherland, 1960, "Electromyographic study of transplanted muscles about the knee in poliomyelitic patients," *Journal of Bone and Joint Surgery* 42-A: 926. Adapted with permission from The Journal of Bone and Joint Surgery, Inc.

Gait EMG data can be presented using a type of *polar plot* (figure 6.5). Polar plots are useful for cyclical activities, like running or walking or cycling, in which the movement repeats. Here, the "toe-off" and "heel strike" points are used to provide movement references to the figure.

It may be also desirable to display the variability in the cycle-to-cycle EMG activity. For example, many stride cycles may be similar to one another, while a few might present atypical EMG bursts. One way to show this variability is to present a "raster" type of plot, in which both the raw EMG signal and the corresponding linear envelopes are displayed over a series of consecutive cycles (figure 6.6).

Gait analysis often includes kinematic and kinetic analyses also, and it may be useful to present some other characteristics of the movement as well as the EMG signal. For example, one can present an illustration of the lower limb's kinematics during the gait cycle in addition to the EMG activity (figure 6.7).

Other Gait EMG Issues

As one might expect, the EMG activity from each muscle can depend on walking speed (Hof et al. 2002; Shiavi et al. 1987). Consequently, it is important that walking speed be controlled for each subject. Also, there are considerable intersubject differences in the normal gait profile, so it can be difficult to establish a "normal" gait pattern (Arsenault et al. 1986b; Dubo et al. 1976; Winter and Yack 1987). Thus, in using EMG activity in gait analysis, it may be important to identify gross feature changes between subjects, rather than subtle differences between subjects.

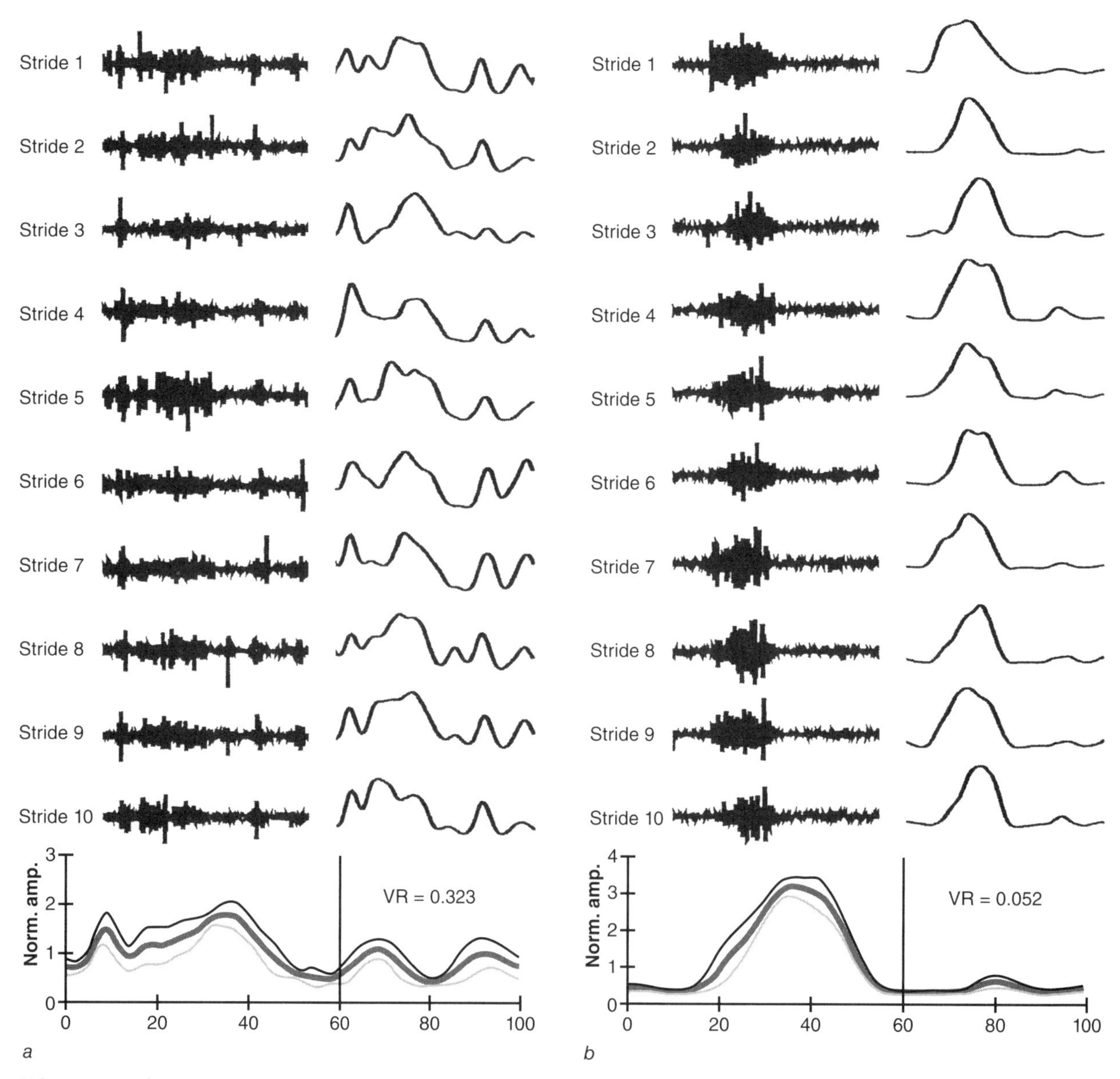

FIGURE 6.6 Simultaneous presentation of raw EMG and corresponding linear envelopes over 10 strides from *(a)* a hemiparetic patient and *(b)* a healthy subject.

Reprinted from *Gait & Posture* 18(1), I.-S. Hwang, H.-M. Lee, R.-J. Cherng, and J.-J. Chen, "Electromyographic analysis of locomotion for healthy and hemiparetic subjects—Study of performance variability and rail effect on treadmill," p. 4, copyright 2003, with permission from Elsevier.

More recently, it has been recognized that there can be important variations from stride to stride within an individual subject. These stride-to-stride variations can be quantified and presented using ensemble averaging plots in which the standard deviations at each time point are presented along with the ensemble average (figure 6.8).

Usually researchers choose to study either left- or right-side muscles. However, it appears that there may be some asymmetry in the EMG response observed during gait (Arsenault et al. 1986a; Ounpuu and Winter 1989; Perttunen et al. 2004; Sadeghi et al. 2000). Consequently, in nondisabled individuals, researchers intending to assess

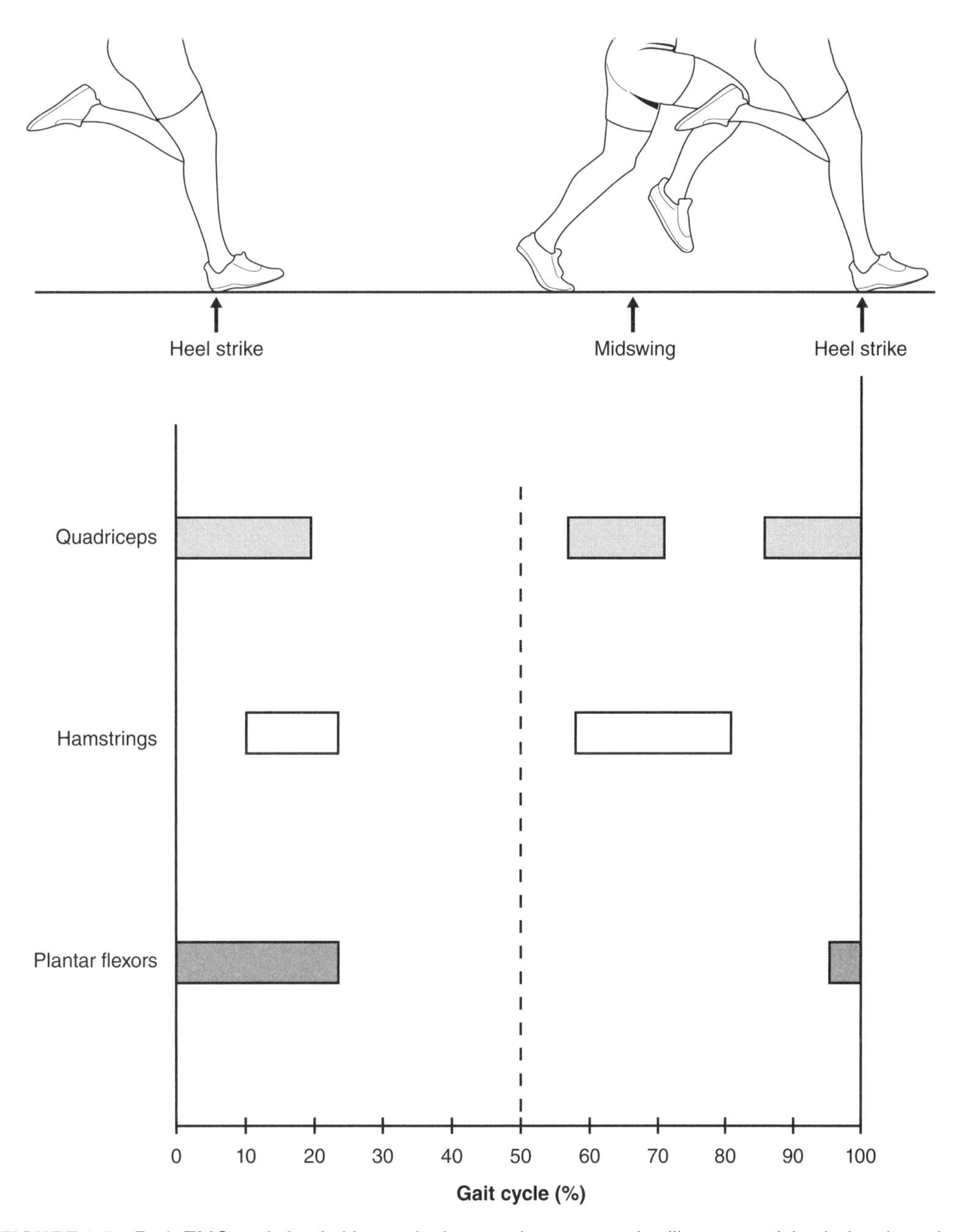

FIGURE 6.7 Both EMG and simple kinematic data can be presented to illustrate activity during the gait cycle. In this example, the right leg is being analyzed. Toe-off occurs at the dashed line.

Adapted, by permission, from G.H. Schwab, D.R. Moynes, F.W. Jobe, and J. Perry, 1983, "Lower extremity electromyographic analysis of running gait," *Clinical Orthopedics and Related Research* 176: 168.

a treatment or intervention using one limb may want to incorporate data from the homologous limb in a control group rather than using the contralateral limb as a control.

Reliability of the EMG Signal During Gait

When several strides are available for analysis, it is often tempting to use the stride cycle that seems most "typical." However, unless one analyzes several strides, the between-stride variability cannot be ascertained. In nondisabled individuals, as few

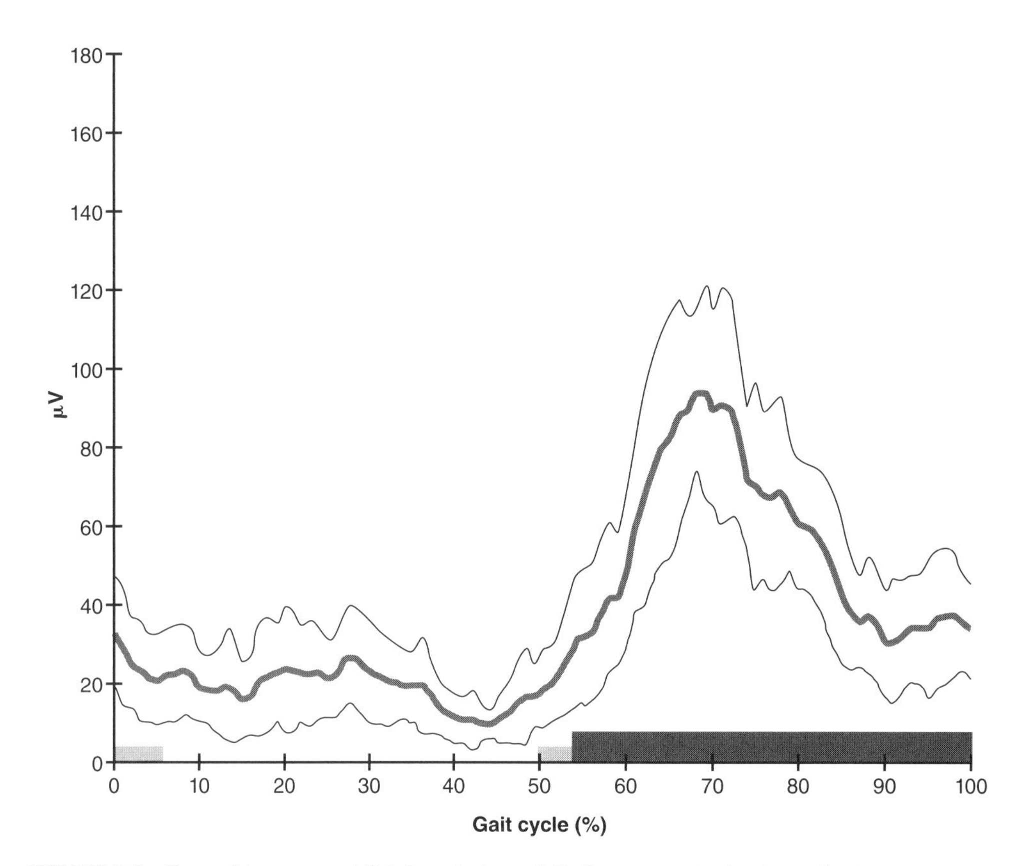

FIGURE 6.8 Ensemble-averaged tibialis anterior activity from a cerebral palsy patient.

Adapted from *Gait & Posture* 18(2), D. Roetenberg, J.H. Buurke, P.H. Veltink, et al., "Surface electromyography analysis for variable gait," p. 117, copyright 2003, with permission from Elsevier.

as three strides may be sufficient to yield reliable EMG data (Arsenault et al. 1986c). However, in patient populations, the number of strides required for ensemble averaging may be considerably greater. Good between-day and within-day reliability can be obtained using surface electrodes (Kadaba et al. 1989). Day-to-day changes in electrode location can contribute to variability in the measured EMG amplitude (Campanini et al. 2006). There is evidence that within-session reliability is greater for adults than for children (Granata et al. 2005). Sometimes the stride-to-stride variability can be indicative of a disease process (Lewek et al. 2006). Again, some care must be taken to ensure that the same walking speed is used for multiple strides, since EMG activity varies with stride speed (Shiavi and Griffin 1983).

An example of the use of EMG for gait analysis is instructive for understanding some of these principles. Stackhouse and colleagues (2007) examined gait initiation in children with cerebral palsy. In addition to ground reaction forces obtained using force plates, EMG characteristics of the anterior tibialis, soleus, gastrocnemius, vastus lateralis, rectus femoris, and medial hamstrings were assessed using commercial surface electrodes with an electrode placement system for gait studies described by Winter and Yack (1987). In order to minimize cross-talk activity in the soleus muscle from the gastrocnemius, the investigators monitored the soleus EMG signal during open kinetic chain knee flexion. The EMG signals were filtered with a low-pass cutoff of 350 Hz, and the gains of each amplifier were individually set to

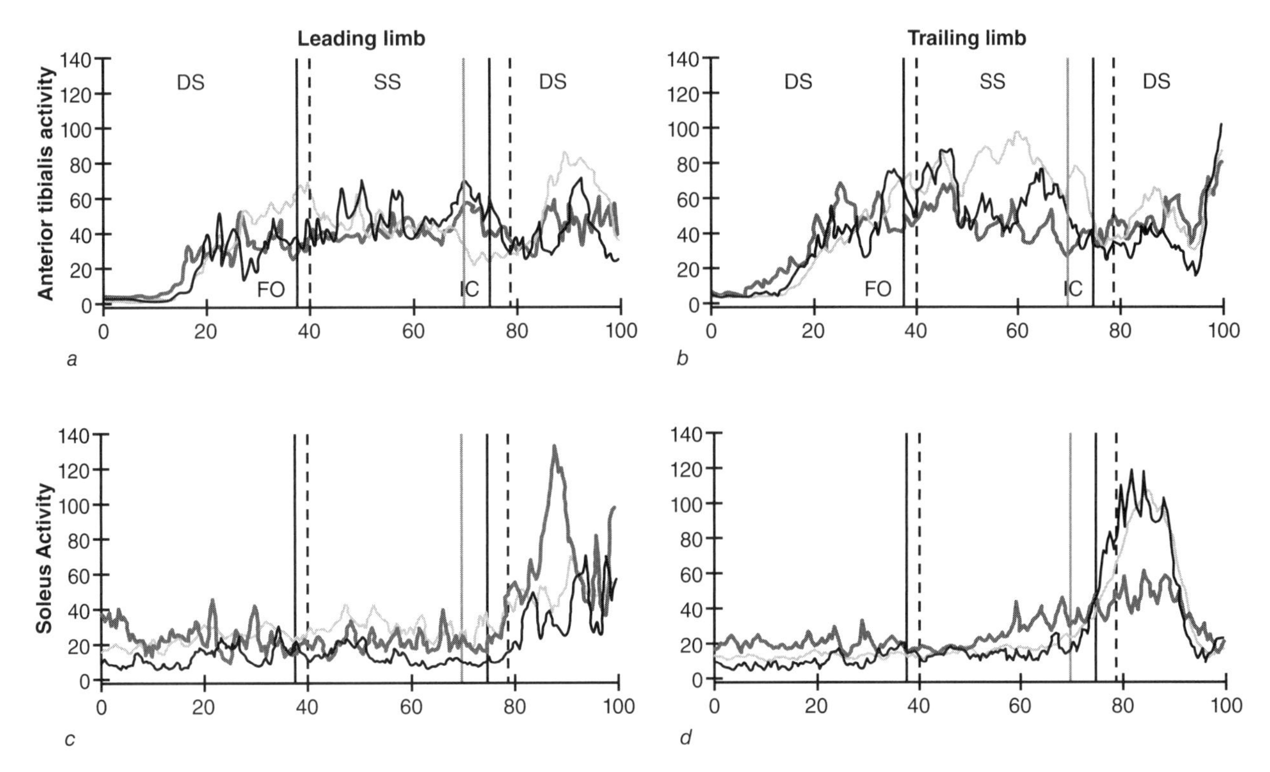

FIGURE 6.9 EMG activity during a gait initiation task in cerebral palsy patients. *(a, b)* Anterior tibialis EMG activity; *(c, d)* soleus EMG activity. DS = double support phase; SS = single support phase; FO = leading limb foot-off; IC = leading limb initial foot contact.

Adapted from *Gait & Posture* 26(2), C. Stackhouse, P.A. Shewokis, S.R. Pierce, et al., "Gait initiation in children with cerebral palsy," p. 306, with permission from Elsevier.

obtain optimal signal resolution. Baseline EMG activity was recorded during sitting, and these baseline signals were used to remove baseline noise during the motor task. A 40 ms moving root-mean-square (RMS) window (Polcyn et al. 1998) was used to process and display the data. The EMG signals were normalized to the peak RMS score obtained in one of the gait initiation conditions.

In figure 6.9 we can see the results. DS and SS are the double and single support phases. FO indicates foot-off, and IC indicates foot contact. The three lines indicate different subject groups. The investigators were particularly interested in the characteristics of gait initiation; this can be seen in the anterior tibialis EMG activity prior to FO. Note how the normalization of the EMG amplitude values for each muscle facilitates the comparison across different muscles and different subject groups.

KEY POINTS

- The nature of the research question being raised in a study may require the use of wire electrodes as well as surface electrodes.
- EMG amplitude normalization may be required to accurately assess muscle activity during gait. Several procedures are available to ensure that the signal is adequately normalized with no loss in reliability.

- The linear envelope is frequently used to display and quantify the amplitude of the signal during gait. Low-pass filtering, which is recommended, optimizes the consistency over several stride cycles.
- Since EMG activity during gait occurs in discrete bursts, careful selection of an appropriate algorithm for determining EMG onset and offset times is important. Several algorithms are available for determining EMG activation timing.
- The reliability of EMG signals during gait is generally good. However, such factors as changes in gait speed, differences in electrode location, and stride-to-stride variability can affect reliability.

EMG Activation Timing

The problem of identifying EMG onset and offset is not restricted to gait analysis. Sometimes a burst of impulses may be present and it may be desirable to identify the onset time and duration of each burst. Such bursts might be recorded using fine-wire, needle, or surface EMG techniques. The bursts may occur during rapid movements, and the electromyographer might be interested in identifying activity in agonist or antagonist muscles. As we have seen, such bursts may occur during gait-related cyclic activities such as walking or running, but they may also occur during activities like cycling or rowing. EMG offset times might need to be identified as well in order to determine the duration of a burst. A variety of procedures are available for detecting EMG burst start and stop times. We began a preliminary discussion of EMG activation timing in chapter 4, and in this section we present some specific applications involving EMG onset and offset detection. Some excellent reviews have previously been published (Delcomyn and Cocatre-Zilgien 1992; Hodges and Bui 1996).

Threshold Detection

We previously discussed threshold detection, and this remains one of the most frequently used algorithms for measuring burst onset. This procedure is highly effective and can be readily implemented using computerized processing. The investigator simply declares a threshold above which EMG activity is defined to begin. For example, Baum and Li (2003) conducted a study designed to assess the effect of cycling frequency and load on EMG characteristics. Using surface electrodes, they recorded EMG signals from several lower limb muscles while subjects bicycled on a stationary ergometer. The data were sampled at 960 samples per second, full-wave rectified, and then smoothed with a low-pass, fourth-order Butterworth filter at 7 Hz. From the resulting linear envelope, the maximum EMG signal occurring during the cycle was obtained. Then, a threshold value of 10% of the maximum value was used to compute onset and offset times, although in some cases a 20% criterion was used, apparently because considerable activity was recorded *between* bursts. Thus, one way to obtain onset and offset times is to declare a threshold amplitude criterion and compute onsets and offsets from this criterion amplitude (figure 6.10).

It is also possible to compute the mean baseline EMG amplitude and declare EMG onset and offset based on a variability measure around the baseline. For example, Neptune and colleagues (1997) sought to better understand neural coordination strategies resulting from changes in cadence during cycling. They recorded surface EMG activity from several lower limb muscles during cycling, sampled the data at

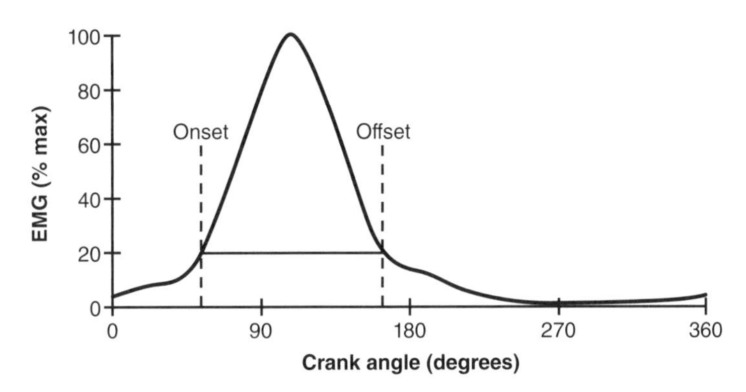

FIGURE 6.10 EMG activation timing can be obtained from relative EMG amplitude measures.

Reprinted from *Journal of Electromyography and Kinesiology* 13(2), B.S. Baum and L. Li, "Lower extremity muscle activities during cycling are influenced by load and frequency," p. 185, copyright 2003, with permission from Elsevier.

850 samples per second, and used a high-pass cutoff frequency of 12 Hz. Maximal voluntary contractions were conducted on and off the bicycle so that the maximum EMG amplitude could be obtained. These values were used to normalize the EMG amplitudes recorded during the cycling tests.

So that EMG onset for each burst could be computed, subjects were placed in a supine position and resting activity was recorded for 10 s. Any EMG activity that was 3 standard deviations in amplitude above the baseline and lasted for at least 50 ms defined the onset of the burst. This task was completed using an automated program. However, the investigators also conducted a visual examination and changed the threshold "when necessary." Neptune and colleagues (1997) used a table to report the average crank angle when burst onset was identified.

The idea of identifying the average EMG activity at rest and using a variance measure to detect burst onset has been implemented in other situations. Bennell and colleagues (2006) assessed the effect of patellar taping on EMG onset of the vastus muscles by identifying the time at which the EMG activity exceeded baseline activity by 2 standard deviations for 25 ms. Chang and colleagues (2007) used the 2 standard deviation criterion to obtain onsets during pediatric gait studies. Numerous other studies have used the criterion of 2 or 3 standard deviations above baseline (Muller and Redfern 2004).

More Complex Techniques

Some applications may require more complex techniques to assess burst characteristics. For example, Santello and McDonagh (1998) needed to identify EMG onset times in lower limb muscles following self-initiated jumps from different heights. These researchers needed an algorithm that would distinguish between a brief burst immediately after takeoff and late bursts occurring just before touchdown. An EMG onset identification algorithm was designed that achieved this goal about 95% of the time, as determined by visual inspection. In some cases, brief, premature EMG bursts seemed to precede the main EMG activity, requiring manual intervention.

First, the EMG signal was sampled at 2000 samples per second. Then the raw EMG signal was full-wave rectified, and a continuous integration of all EMG data points was performed. The integrated EMG (IEMG) and the duration of the fall were then normalized, and the normalized IEMG was plotted against the normalized fall time. The EMG onset time was defined as the point at which the distance between the normalized IEMG slope and a reference line passing through 0 was the greatest. This definition tends to minimize the identification of spurious EMG bursts.

KEY POINTS

- Algorithms based on threshold detection are frequently used to identify EMG onset and offset times.
- The comparison of results obtained from any algorithm with the results obtained by visual examination, at least at the beginning of any study, is recommended.
- More complex techniques for assessing EMG activation timing may be helpful in situations in which the onset or offset is frequently difficult to identify.

Evoked Potentials

Useful EMG information is frequently obtained by stimulation of muscle or sensory or motor nerves or by presenting a visual, auditory, or other stimulus that results in a time-locked EMG response. These *evoked potentials* provide valuable information to EMG clinicians and researchers in motor control.

M-Waves

If the motor nerve is stimulated directly, a response called the *M-wave* can be recorded directly from the muscle (figure 6.11). The M-wave, also called the compound muscle action potential (CMAP), can be produced by stimulation of the motor nerve at some site proximal to the muscle or by application of a single square-wave electrical stimulus over the muscle. This stimulus activates motoneurons directly.

By far the most prevalent measure of M-wave amplitude is M_{max}, or the maximum-size M-wave that can be elicited. M_{max} is a measure of the maximal electrical activity that can be produced in the muscle and is partly related to muscle size (Wee 2006). The correct procedure to record M_{max} is to administer a stimulus with an intensity

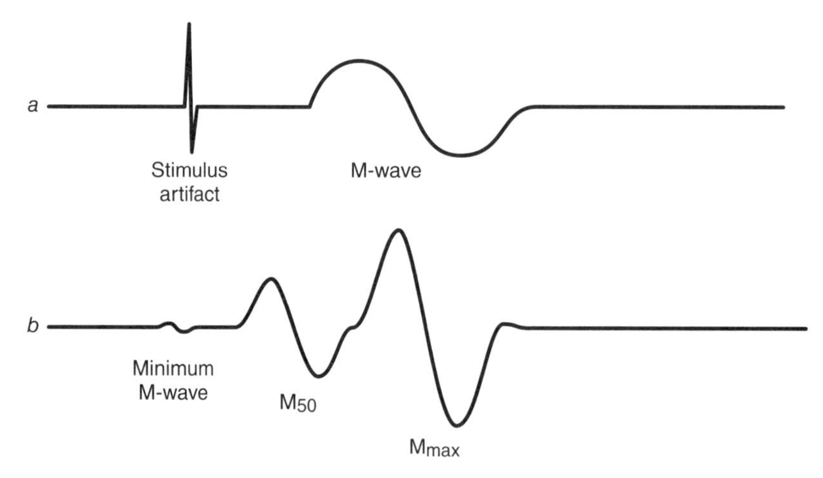

FIGURE 6.11 *(a)* An M-wave is produced in muscle following stimulation of the motor nerves innervating the muscle. *(b)* Varying the intensity of the stimulation alters the amplitude of the M-wave response. M_{50} corresponds to an M-wave that is 50% as large as M_{max}.

sufficient to activate all motoneurons. The failure to activate all motoneurons will result in an M-wave that is not maximal. To ensure that the M-wave is absolutely maximal, the correct protocol is to find the stimulus intensity that yields the highest-amplitude M-wave and then increase the intensity by a safety margin of approximately 30% to ensure that an absolute maximal M-wave will be evoked by the stimulus. Although it is frequently thought that electrical stimulation activates the largest motoneurons first, this is not always the case with skin-based stimulation (Knaflitz et al. 1990); so the failure to use an adequate stimulus intensity may result in the failure to activate either large or small motoneurons whose muscle fibers participate in the M-wave.

M-waves are typically evoked by stimulation of the motor nerve to a muscle; the subsequent EMG response in that muscle is then recorded. For example, it is possible to elicit the M-wave in the soleus muscle by stimulating the posterior tibial nerve behind the knee in the popliteal fossa; the M-wave response is recorded using surface electrodes placed over the soleus muscle. It is also possible to stimulate the muscle directly by placing the stimulator over the motor point (e.g., Marqueste et al. 2003). However, the stimulus artifact may be greater; and depending on the distance between the motor point and the recording zone, the artifact may interfere with the M-wave recording (figure 6.12). Techniques are available that can help reduce the stimulus artifact (Hines et al. 1996; Knaflitz and Merletti 1988). Most frequently, a stimulus duration of 0.1 ms is used. When stimulating over an area in which two or more motor nerves are present, one must take care to ensure that the M-wave is not contaminated by activity from multiple muscles.

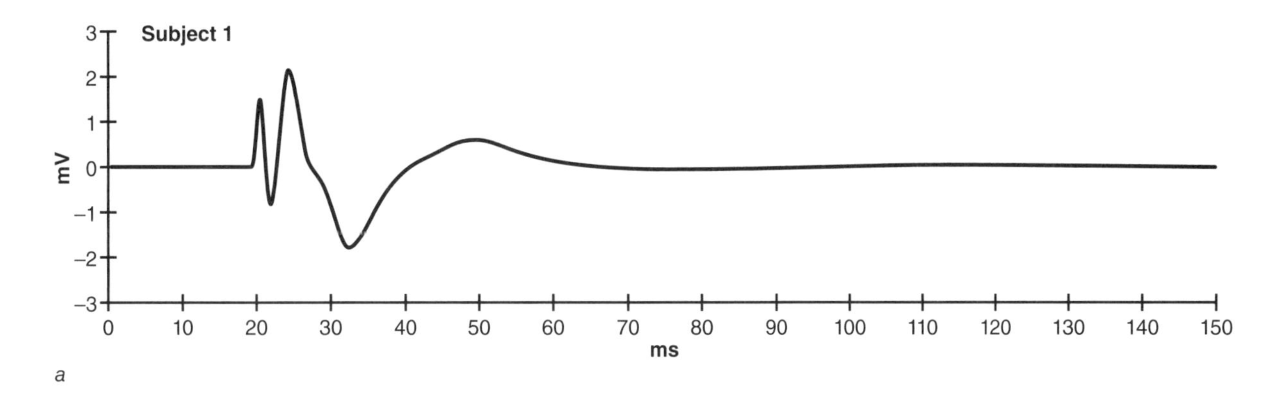

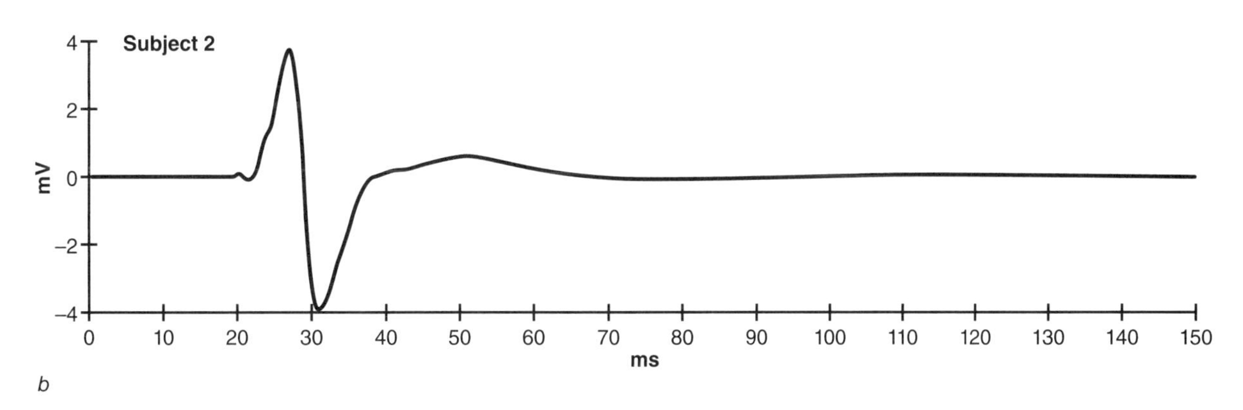

FIGURE 6.12 M-waves recorded by stimulation of the brachial nerve to the biceps at the axilla in two different subjects. The peak-to-peak M-wave response is greater in subject 2 than in subject 1, but the stimulus artifact is greater in subject 1 than in subject 2.

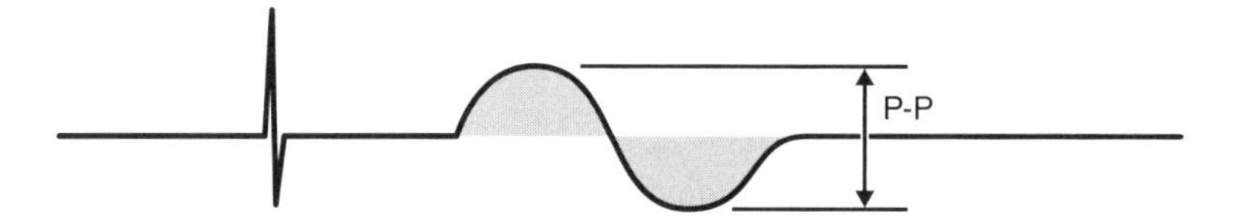

FIGURE 6.13 The M-wave can be analyzed via computation of the peak-to-peak (P-P) amplitude or computation of the area under the M-wave response (shaded area).

There are several M-wave variables that can be used for analysis. Amplitude variables are most prevalent; these include peak-to-peak (P-P) amplitude and M-wave area (figure 6.13). Some observers have also measured the latency of the M-wave—the interval between the stimulus and the onset of the response (Linnamo et al. 2001). However, M-wave latency is quite constant and not sensitive to very much change. The duration of the M-wave has also been used as a criterion measure. For example, M-wave duration decreases during both eccentric and concentric contraction, and this may be related to changes in muscle fiber conduction velocity during the contraction (Linnamo et al. 2001). There is some suggestion that M-wave area may be a more reliable variable than M-wave amplitude, since the authors of one study reported that M-wave area is not affected by repeated stimulation (Aiello et al. 1986). In studies designed to measure either the M-wave P-P amplitude or area, it would be prudent to ensure that the stimulation rate is less than three stimuli per second.

With brief exercise, the M-wave can be enhanced (Hicks et al. 1989). Cupido and colleagues (1996) found that a minute of stimulation at 10 Hz or 20 Hz was capable of doubling the size of the M-wave. With prolonged exercise, the M-wave can be either enhanced or depressed, depending on the nature of the muscle contraction (Cupido et al. 1996; Lentz and Nielsen 2002). Also, with prolonged muscle stimulation causing fatigue, the amplitude of the M-wave decreases (Tanino et al. 2003). In general, fatigue depresses the M-wave (Arnaud et al. 1997), although one study showed that running a marathon had no effect on M-wave characteristics (Millet et al. 2002). The M-wave does not seem to be affected by several weeks of training using functional electrical stimulation (Marqueste et al. 2003).

M-waves can also be produced by stimulation of the peripheral nerve using magnetic stimulation (Al-Mutawaly et al. 2003). Magnetic stimulation has the advantage that it is generally more comfortable than electrical stimulation. Biphasic magnetic pulses generally produce greater M-wave amplitudes than do monophasic pulses.

Large stimulus artifacts can also accompany the M-wave and hamper the ability to identify the response. Stimulus artifacts can sometimes be minimized with various procedures. First, ensure that the recording electrodes are secure, that they are making good contact, and that the voluntary activity during a muscle contraction is free of line frequency (typically 50 or 60 Hz) interference. Moving the ground electrode to another site such as between the stimulating and recording electrodes, improving the ground electrode contact, or using a larger ground may also reduce the artifact. Stimulus artifacts tend to be lower with shorter-duration stimuli. There is some anecdotal suggestion that the placement of the stimulating electrode over an obvious superficial vein may produce larger artifacts. Reduce any moisture on the skin between stimulating and recording electrodes by cleaning the skin area with alcohol to reduce skin-conducted artifacts. When the active and reference stimulating electrodes are separate contacts, changing the position of one stimulating electrode relative to the other can reduce artifact (Kornfield et al. 1985).

Many of the same factors that affect the surface electromyogram during voluntary contraction affect the M-wave. For example, the application of ice decreases the amplitude of the M-wave (Basgoze et al. 1986). Both the maximum M-wave and the maximum Hoffman reflex (H-reflex) decline with age (Scaglioni et al. 2002, 2003). The M-wave depends partly on muscle length and joint angle. Maffiuletti and Lepers (2003) found that the M-wave measured in the rectus femoris was 19% higher when the subject was supine than when measured in the seated position. Similar results have been reported for the soleus M-wave (Allison and Abraham 1995).

H-Reflexes

Hoffmann (1918) was the first to describe a phenomenon we now know as the Hoffmann reflex or simply the **H-reflex.** Measuring the H-reflex is a powerful technique for indirectly measuring motoneuron excitability, though it is also affected by presynaptic inhibition. The H-reflex is elicited by delivery of a submaximal stimulus to the peripheral nerve (figure 6.14). The stimulus activates Ia afferent fibers and impulses from these Ia afferents synapse onto motoneurons. The resultant response can be recorded in the muscle as the H-reflex or H-wave. In this way, the H-reflex is produced by the stimulation of muscle afferents bypassing the muscle spindle. These afferent impulses activate motoneurons producing the EMG response. Low-intensity stimuli may recruit a few Ia afferent fibers and produce a small H-reflex response. As the stimulus intensity increases, the size of the H-reflex increases until motor nerves are activated by the electrical stimulus. At higher stimulus intensities, the **antidromic** impulses from motoneurons collide with the **orthodromic** response produced by

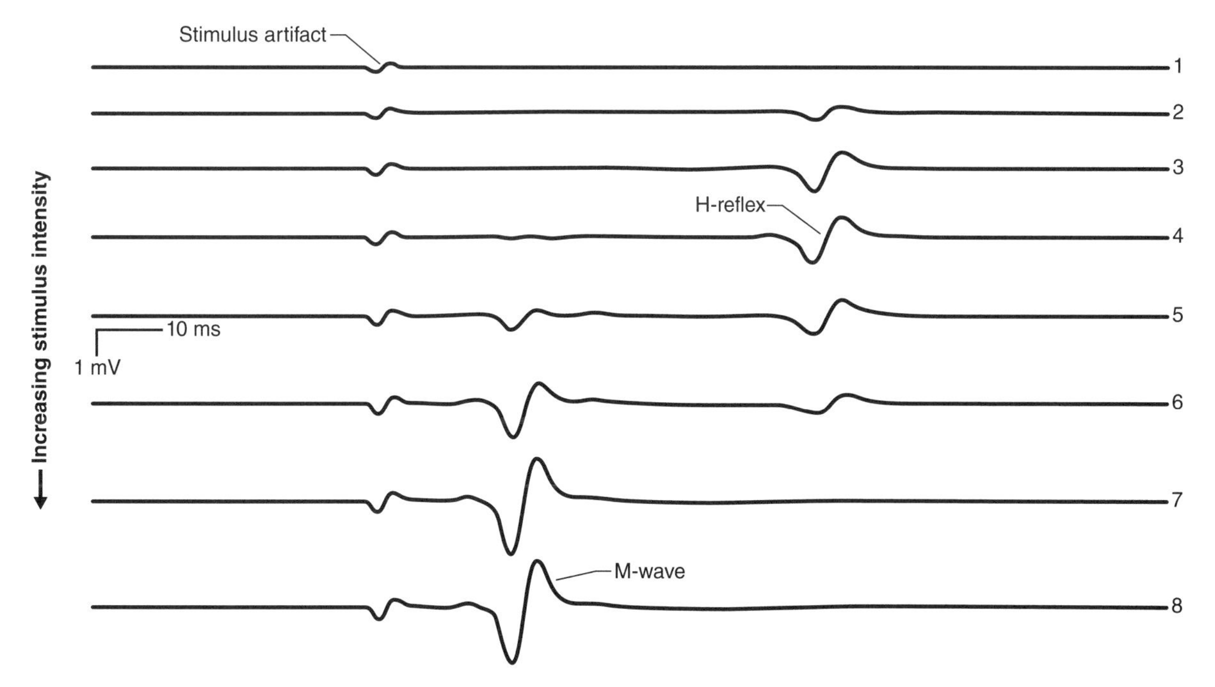

FIGURE 6.14 H-reflexes and M-waves are frequently elicited from the soleus muscle. A very low-intensity stimulus applied to the mixed nerve evokes no response (trace 1). As stimulus intensity increases, the H-reflex response can be observed. Continuing to increase intensity results in direct activation of motoneurons, and a shorter-latency M-wave can be recorded (trace 5). Eventually, the higher-intensity stimuli result in extinction of the H-reflex response due to collision between α-motoneuron and Ia afferent action potentials (trace 7).

activating Ia afferents, and this results in smaller H-reflexes. During supramaximal stimulation producing the largest possible M-waves, the H-reflex is generally absent.

It is commonplace to report the magnitude of the H-reflex in relative values rather than reporting the absolute value (e.g., the value in millivolts) of the response. Most typically, the magnitude of the maximal M-wave is measured first, and the H-reflex magnitude is then normalized to the maximal M-wave response. In this way, the size of the response produced by the H-reflex can be reported with respect to the maximal possible muscle response. In the soleus muscle, for example, H/M ratios of about 0.5 are typical. In H-reflex studies, it is important to measure the M-wave amplitude regularly, since M-wave values can change during a test session (Crone et al. 1999). Test–retest reliability of the H-reflex is high (Crayton and King 1981; Hopkins et al. 2000; Hopkins and Wagie 2003), although it is best to average several responses within a block of trials to obtain suitable reliability (McIlroy and Brooke 1987).

As with measurement of M-waves, sufficient time (at least 5 s) should be placed between repeated stimuli, since the size of the H-reflex is reduced with repetitive activation—a phenomenon termed H-reflex depression (Crone and Nielsen 1989; Floeter and Kohn 1997; Ishikawa et al. 1966; Rossi-Durand et al. 1999). Tetanic stimuli delivered immediately prior to testing can enhance the H-reflex (Blom et al. 1964). A stimulus duration of about 1 ms is recommended (Hugon 1973; Panizza et al. 1989).

Although the H-reflex is most commonly assessed in the soleus muscle, it is possible to obtain H-reflexes in the knee extensors by stimulating the femoral nerve in the inguinal region (Aiello et al. 1982; Mongia 1972). H-reflexes can also be elicited in the tibialis anterior (Brooke et al. 1997; Ellrich et al. 1998), the abductor pollicis brevis (Burke et al. 1989), the abductor hallucis (Ellrich et al. 1998), the masseter (Godaux and Desmedt 1975), and in the flexor carpi radialis (Brooke et al. 2000; Jabre 1981), in which good reliability has been demonstrated (Christie et al. 2005). It is also possible to obtain an H-response in the gastrocnemius (Mongia 1972; Nadeau and Vanden Abeele 1988), though care must be taken to ensure that the responses do not represent cross-talk from the soleus muscle (Perot and Mora 1993).

For measuring soleus H-reflexes, the subject is placed in a prone or semirecumbent position (figure 6.15). It appears that similar values are obtained with the subject in

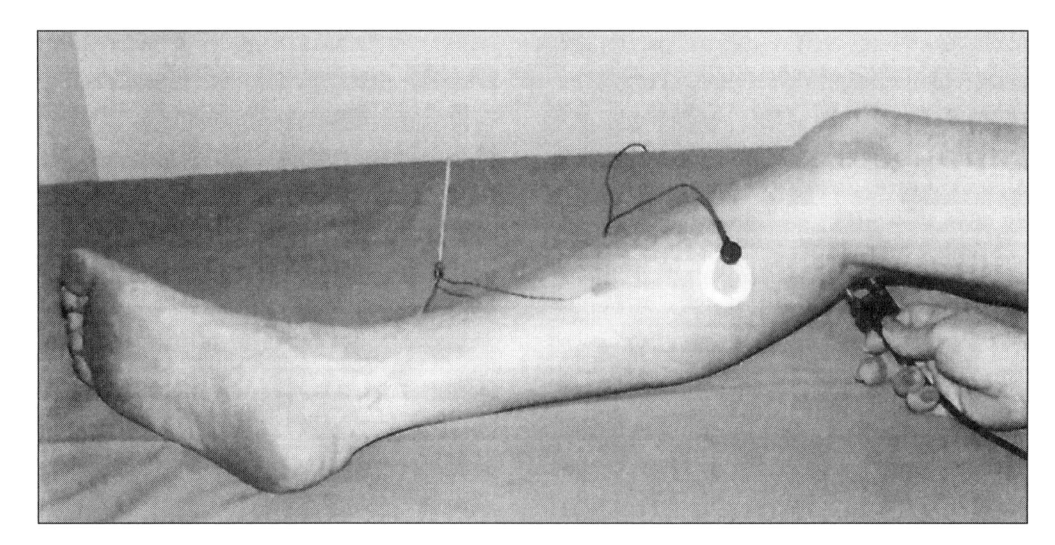

FIGURE 6.15 H-reflexes can be elicited from the soleus muscle by stimulation of the popliteal nerve.

Reprinted from J. Kimura, 2001, *Electrodiagnosis in diseases of nerve and muscle: Principles and practice,* 3rd ed. (New York: Oxford University Press), 469. By permission of Oxford University Press, Inc.

either position (Al-Jawayed et al. 1999). However, inasmuch as passive movement affects the H-reflex (Brooke et al. 1997), the subject should remain still during testing. The neck position should be standardized since it also affects the H-wave through the tonic neck reflex (Hayes and Sullivan 1976; Rossi et al. 1986). Also, the background EMG must be kept constant since light muscle activity can increase H-reflex amplitude (Verrier 1985). Some excellent reviews of H-reflex recording techniques have been published (Braddom and Johnson 1974; Fisher 1992; Hugon 1973; Misiaszek 2003; Pierrot-Deseilligny and Mazevet 2000; Schieppati 1987; Zehr 2002).

Many corollary studies and techniques have been developed using the basic H-reflex protocol. For example, recurrent inhibition from *Renshaw cells* can be measured using a paired or conditioning stimulation technique (Barbeau et al. 2000). It is also possible to measure presynaptic inhibition (Frigon et al. 2004; Hultborn et al. 1987), reciprocal inhibition (Baret et al. 2003; Day et al. 1984), and inhibition or facilitation (or both) from other spinal pathways in neighboring or contralateral muscle groups (Cavallari et al. 1985; Robinson et al. 1979).

Magnetic stimulators can also be used to elicit H-reflexes. In fact, a modification of the H-reflex technique using magnetic stimulation has been applied to elicit H-reflexes in some hand muscles (Mazzocchio et al. 1995).

The proportion of type I motor units activated by the H-reflex stimulus is likely greater than the proportion of type II motor units (Messina and Cotrufo 1976). This is congruent with the idea stated earlier that small tonic motoneurons are responsible for spinal monosynaptic reflexes (Homma and Kano 1962). In part, this explains why the type I–heavy soleus is a good model for measuring H-reflexes.

Light exercise depresses the H-reflex, producing a type of relaxation effect (deVries et al. 1981). The H-reflex can also be depressed by aspirin (Eke-Okoro 1982), pain (Ellrich and Treede 1998), muscle vibration (Martin et al. 1986), and static stretching and proprioceptive neuromuscular facilitation (PNF) techniques (Etnyre and Abraham 1986; Guissard et al. 2001). On the other hand, H-reflexes are enhanced by alcohol and caffeine (Eke-Okoro 1982; Walton et al. 2003) and by vestibular input (Rossi and Nuti 1988). The magnitude of the response can be modified by training using transcendental meditation (Wallace et al. 1983), and the H-reflex is modified during movement (Brooke et al. 2000; Llewellyn et al. 1990; Yang and Whelan 1993) and during the preparation for movement (Michie et al. 1976). Using the H-reflex it has been demonstrated that long-term adaptations in motoneuron excitability are possible with training (Carp and Wolpaw 1995; Trimble and Koceja 1994). The H-reflex is smaller in ballet dancers (Koceja et al. 1991) and in older adults (Tsuruike et al. 2003) than in other groups. Thus, there are many applications of H-reflex techniques in human performance research.

V-Waves

The *V-wave* is considered a measure of the level of central drive to the motoneurons during a maximal contraction. While subjects exert a maximal isometric contraction, a supramaximal nerve stimulus is delivered. This results in a response termed the V-wave. V-waves were originally described by Upton and colleagues (1971). Later reports indicated that resistance exercise training might potentiate the V-wave (Aagaard et al. 2002; Sale et al. 1983). Recent studies indicate that the magnitude of the V-wave is similar regardless of the type of muscle action (Duclay and Martin 2005) and that it varies with the intensity of the contraction (Pensini and Martin 2004).

F-Waves

Like the H-reflex response, the *F-wave* is a late response that is observed in muscle following motor nerve stimulation. The mechanism for F-wave generation involves antidromic reactivation or "backfiring" of motoneurons. The motoneuron action potential travels antidromically to the spinal cord and is then "reflected" back along the motoneuron axon pathway to the muscle. Thus, the F-wave differs from the H-reflex in that it is not produced by activation of the Ia afferents. Indeed, F-waves can be produced in the absence of afferent innervation (McLeod and Wray 1966).

F-waves can be elicited using supramaximal-intensity stimuli with the subject relaxed, yielding a response that has a latency similar to that of the H-reflex. This is a convenient method to use to determine motor conduction time to the spinal cord and back to the recording site, allowing determination of a central latency.

H-reflexes cannot be observed in all muscles, and the F-wave is easier to elicit in those muscles in which an H-response is absent. In studies that have used both H-reflexes and F-waves to measure excitability changes, similar responses have been observed (Walk and Fisher 1993). The effects of conditioning stimuli on F-waves are similar to those produced on the H-reflex (Mastaglia and Carroll 1985). However, when H-reflexes can be elicited, they may be a better measure of excitability. In fact, some suggest that it may be inappropriate to use the F-wave as a measure of excitability (Espiritu et al. 2003). F-waves may have the greatest value in assessing inhibition (Lin and Floeter 2004). The time course of inhibition and facilitation observed through H- and F-responses may differ (Inghilleri et al. 2003).

As with the M-wave, the amplitude of the F-response is usually measured as P-P amplitude, or area under the F-response curve. Sometimes the variability of the F-response is measured (e.g., Espiritu et al. 2003), or the delay or latency of the response (Fisher 1982). Note that a significant number of trials may be necessary to obtain reliable results (Chroni et al. 1996; Gill et al. 1999). In addition to the role of F-waves in electrodiagnosis, F-waves have been combined with motor-evoked potentials (MEPs) obtained during transcranial magnetic stimulation to determine whether excitability changes occur in the motor cortex or at spinal sites under various conditions (Chen et al. 1999; Mercuri et al. 1996; Sohn et al. 2001).

The F-wave is increased in amplitude under conditions in which spinal excitability is increased (Mercuri et al. 1996; Sica et al. 1976). As one might expect, the latency of the F-response varies with limb length or subject height (Fisher 1982; Nobrega et al. 2004). Single motor unit studies reveal that the F-response can be observed in about 25% of motoneurons, though some subjects may have no F-waves (Yates and Brown 1979). A frequent observation is that the F-wave is quite variable. In one study, 100 successive stimulations were delivered to the ulnar nerve, and the F-wave was observed in the hypothenar muscles only 30% of the time among the first 10 stimuli (Barron et al. 1987).

The amplitude of the F-wave is small, perhaps about 1% to 4.5% of the maximum M-wave (Eisen and Odusote 1979). Nevertheless, the F-wave response has been quite useful in diagnosing various disorders, including Guillain-Barré syndrome, Charcot-Marie-Tooth disease, polyneuropathies, and other disorders.

Peripheral Nerve Conduction Velocity

It is also possible to use EMG techniques to measure sensory or motor nerve conduction velocity. For assessing motor nerve conduction velocity, the motor nerve is stimulated at two sites and recordings are made from the muscle. For example, ulnar nerve conduction velocity can be determined at several locations by stimulation at the wrist, the elbow, or the axilla. Recordings can be made from the first dorsal interosseous or the abductor digiti minimi muscle. Measurement of the latency between stimulation and initial onset of the resultant M-wave allows computation of motor nerve conduction velocity. The proximal latency is subtracted from the distal latency and the result is divided by the measured distance between the two stimulation sites, yielding motor nerve conduction velocity (figure 6.16). Values of 50 to 60 m/s are common (Ma and Liveson 1983). Magnetic stimulators can also be used to elicit the motor response (Benecke 1996). A number of good references are available that provide conduction velocity techniques and normal conduction velocity values (Aminoff 1998; Kimura 2001; Oh 2003).

Other Evoked Potentials

There are many other types of potentials that can be evoked by electrical or magnetic stimulation. For example, with *transcranial magnetic stimulation (TMS),* the motor cortex can be activated and responses can be observed using surface electrodes placed

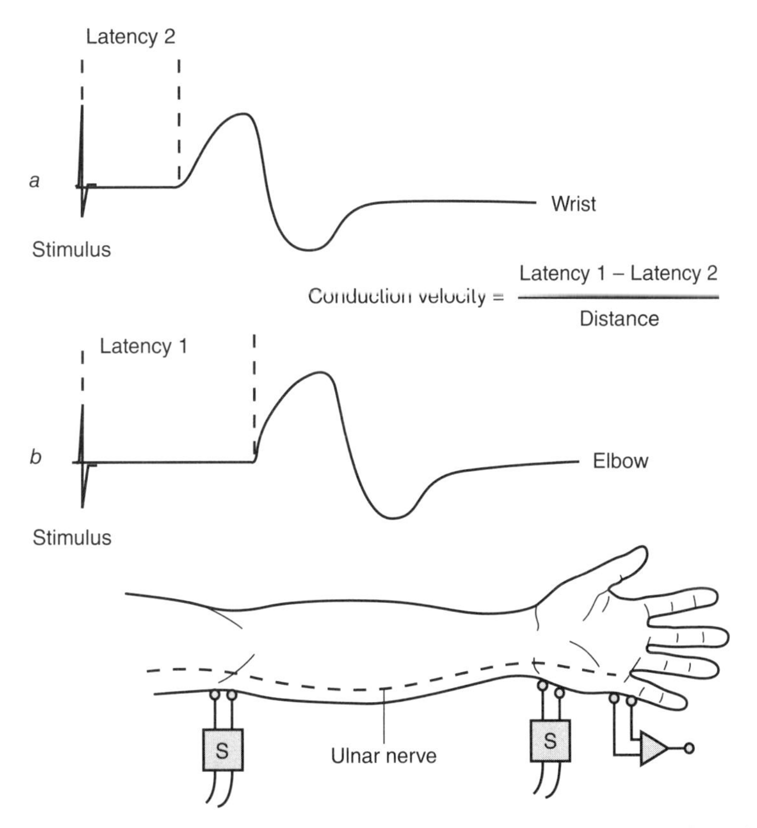

FIGURE 6.16 Motor nerve conduction velocity can be readily assessed using surface electrodes following peripheral nerve stimulation at two sites. Here, the ulnar or median nerve can be stimulated at the wrist or the elbow and the response observed in a hand muscle.

over the stimulated muscle. The response is called the *motor-evoked potential (MEP),* and this can be valuable in a wide variety of motor control experiments, such as those involving fatigue (Gandevia 2001), aging (Peinemann et al. 2001), or the timing of motor activity (Nikolova et al. 2006). Some excellent reviews are available for the user interested in these advanced techniques (Cracco et al. 1999; Hallett 1996; Terao and Ugawa 2002).

KEY POINTS

- The M-wave, also termed the compound muscle action potential (CMAP), is a standard evoked response obtained by recording an EMG response from the muscle following stimulation of the corresponding motor nerve.
- Either P-P amplitudes or the area under the response can be used to quantify the size of evoked responses such as M-waves, H-reflexes, and MEPs.
- The amplitude of the M-wave is sensitive to temperature, active muscle contraction, fatigue, and a brief period of stimulation.
- H-reflexes can frequently be used to quantify the motoneuron excitability state, although changes in H-reflex amplitude can also reflect changes in presynaptic inhibition.
- Recurrent inhibition, presynaptic inhibition, and Ib inhibition can also be measured using variants of the H-reflex recording technique.
- Other evoked responses such as V-waves (recorded during maximal-effort contractions) and F-waves (recorded by supramaximal motoneuron stimulation and observation of the evoked response at approximately H-reflex latency) may also be useful to assess excitability of the neuromuscular system.
- Paired stimulation techniques can be used to measure sensory or motor nerve conduction velocities.

Ballistic Movements

Ballistic movements are rapid movements in either the upper limb or the lower limb. These movements may be performed as rapidly as possible to a target (Gabriel and Boucher 1998) or at various predefined displacements (Brown and Cooke 1981). Research questions may involve the effect of short-term disuse on movement velocity and EMG characteristics (Vaughan 1989) or the effect of subject characteristics such as gender (Ives et al. 1993), training status (Lee et al. 1999), or pathology (Berardelli et al. 1996); and there are certainly a number of studies aimed at understanding the neural control of rapid movements that may require measuring EMG activity from the associated muscles (MacKinnon and Rothwell 2000).

The basic principles of recording EMG activity during rapid movement are similar to those we have discussed previously. The EMG activity of upper limb muscles is probably most frequently recorded in tasks requiring elbow flexion or elbow extension. Similar to the problems encountered with recording EMG during gait, movement artifact in the EMG signal is a potential problem during rapid movements, and this needs to be considered in applying the electrodes. The researcher needs to ensure that the electrodes will not partially detach during the movement.

Proper high-pass filtering is also important. A setting of 10 to 20 Hz is appropriate for a high-pass filter (Merletti et al. 1999), though many other filter setting suggestions have been made (e.g., Potvin and Brown 2004) and the exact choice may be a matter of decision by the researcher considering the characteristics of the project. When cup electrodes are used, braiding the wires between electrodes and amplifier can frequently reduce line noise.

An examination of some signals can reveal some of the typical problems. During a rapid single-limb movement to a target, the well-known triphasic burst pattern occurs (Zehr and Sale 1994): A burst of agonist activity begins in advance of movement onset, and then a burst of EMG activity occurs in the antagonist muscle, followed by a second burst in the agonist. A linear envelope is frequently convenient (figure 6.17), as it minimizes the low-amplitude baseline activity that may precede the movement; and the "traumatic ocular" analysis allows the researcher to more quickly comprehend the dynamics of the EMG response. Furthermore, if many trials are performed over several conditions (such as many subjects or numerous days or other treatment conditions), one can observe the variability of the response by plotting the standard deviation of the EMG amplitude activity at each analysis time point (e.g., figure 6.17*a,* middle frame). Problems that might occur during the recording, such as cross-talk or electrode contact issues, are readily identifiable.

One can appreciate the advantage of computing and visually assessing the linear envelope by comparing figure 6.17*b,* which plots the rectified EMG signal, with the linear envelope seen in figure 6.17*a.* There are some cases (stroke would be one example) in which no triphasic burst pattern can be observed and both agonist and antagonist muscles may display somewhat constant coactivated activity (figure 6.17*c*). Finally, since the distance between electrodes for agonist and antagonist muscles may be fairly small in these types of experiments, researchers wishing to assess the coordination among these muscles need to constantly consider the possibility of volume-conducted cross-talk, an example of which can be seen in figure 6.17*d.* Here, a portion of the biceps signal, which is slightly time shifted (due to delays in volume conduction), can be observed in the triceps. This figure is from a multiday study. In this subject, triceps activity on other days was quite small during biceps activation, lending credence to the conclusion that cross-talk was evident during this experimental session.

KEY POINTS

- Very rapid movements, such as those produced during ballistic contractions, require close attention to proper EMG recording and analysis procedures.
- To ensure quality EMG signals in the absence of artifact or noise, electrodes need to be applied to minimize changes in the skin–electrode interface; and braiding the wires between electrodes may be useful to reduce line noise.
- Averaging the responses over several trials may be required to obtain suitable reliability.
- In the conduct of studies involving agonist–antagonist recordings, careful electrode selection and placement are important, and the analysis needs to consider the possibility that volume-conducted activity is recorded as cross-talk.

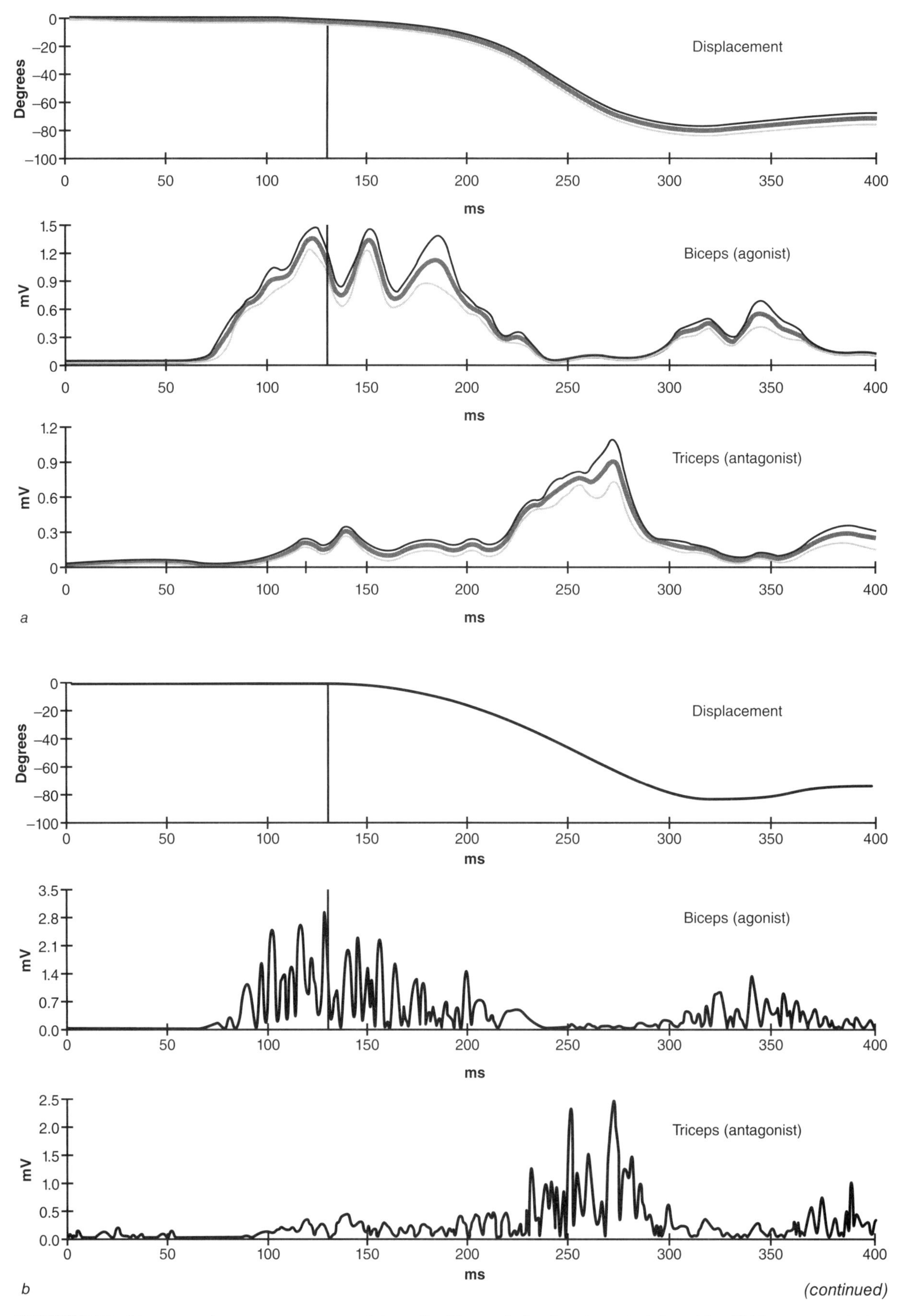

(continued)

FIGURE 6.17 Linear envelopes are one appropriate method for analyzing the characteristics of the EMG response during rapid movement. *(a)* A normal triphasic burst pattern. For each signal, the middle line is the mean linear amplitude value while the upper and lower lines indicate the standard deviation. *(b)* The rectified EMG signal (as discussed in chapter 4) from the same data as in *a*. *(c)* An example in which no coordinated triphasic EMG pattern is evident. *(d)* An example in which volume-conducted biceps activity is recorded as cross-talk in the triceps.

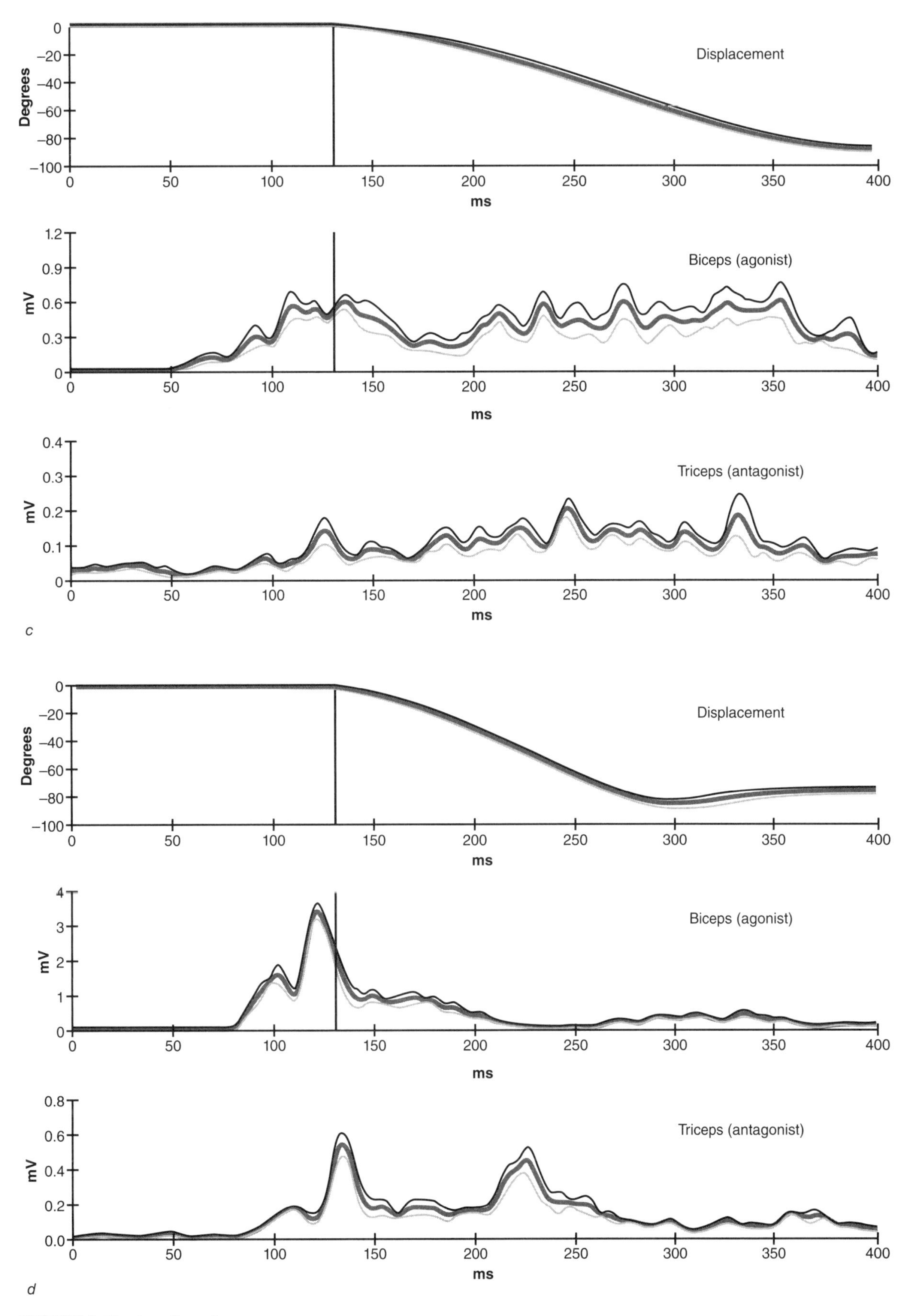

FIGURE 6.17 *(continued)*

FOR FURTHER READING

Fisher, M.A. 1992. AAEM minimonograph #13. H reflexes and F waves: physiology and clinical indications. *Muscle & Nerve* 15: 1223-1233.

Kobayashi, M., and A. Pascual-Leone. 2003. Transcranial magnetic stimulation in neurology. *Lancet* 2: 145-156.

Perry, J. 1992. Dynamic electromyography. In *Gait analysis: normal and pathological function.* Thorofare, NJ: Slack.

Rechtien, J.J., J.B. Gelblum, A. Haig, and A.J. Gitter. 1996. Technology assessment: dynamic electromyography in gait and motion analysis. *Muscle & Nerve* 19: 396-402.

Shiavi, R. 1985. Electromyographic patterns in adult locomotion: a comprehensive review. *Journal of Rehabilitation Research and Development* 22: 85-98.

Soderberg, G.L., and L.M. Knutson. 2000. A guide for use and interpretation of kinesiologic electromyographic data. *Physical Therapy* 80: 485-498.

Appendix 2.1

Calculation of Electric Fields

The depolarization and repolarization phases of the muscle fiber action potential constitute a discrete positive and negative charge system termed a *dipole*. The dipole creates an electric field that is the vector sum of the two separate charges. This is known as the principle of superposition. To map the electric field around the dipole, a small positive charge ($+q_0$) may be placed at different points within the vicinity around the dipole. The net effect of the dipole at each point in space is simply the sum of the two individual charges, determined separately. An example is provided here to enable the reader to understand how the electric fields are constructed as they form the basis of the potential recorded at the electrode. The following is not necessarily meant to be a physiologically relevant example, but merely to provide insight into the mechanics of such calculations.

Consider two equal and opposite charges that are 50 µC in magnitude placed 0.50 m apart. The goal is to calculate the electrical field at position A, which is 0.30 m directly above $+Q_2$. The physical system is depicted in figure 2.1.1. The electric field at position A due to $-Q_1$ is a vector that has both x- and y-components. The same is true for the electric field at position A due to $+Q_2$. The x- and y-components of the electric fields due to the $-Q_1$ and $+Q_2$ at position A must first be resolved separately. The sum of the vectors in the x- and y-directions is then used to calculate the resultant magnitude and direction of the electric field at position A (E_A). Recall that the equation for the electric field is

$$E = \frac{F}{q_0}.$$

Expanding for Coulomb's law:

$$E = k\frac{q_0 Q}{r^2} \cdot \frac{1}{q_0}$$

The equation simplifies so that the electric field depends only on the magnitude of the charge Q producing the field and radial distance r:

$$E = k\frac{Q}{r^2}$$

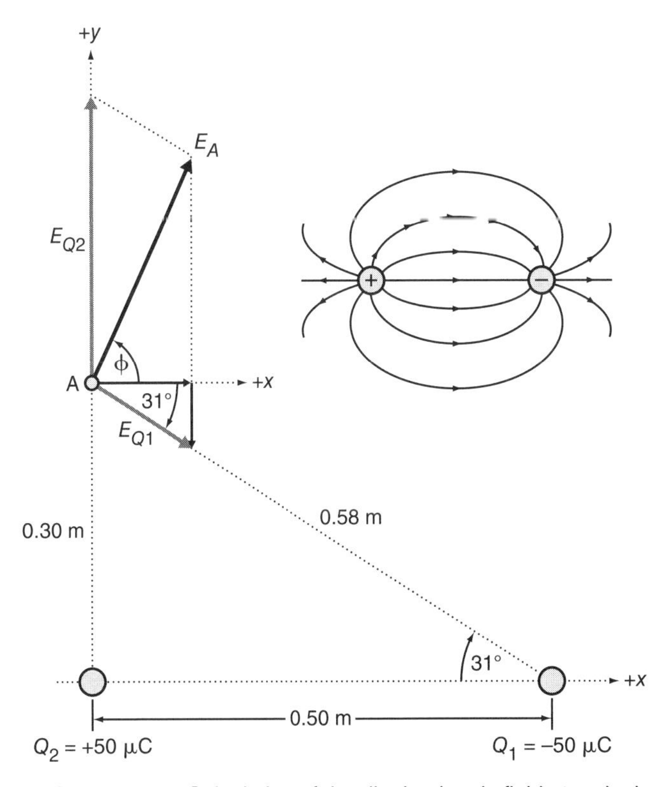

FIGURE 2.1.1 Calculation of the dipole electric field at a single point in space (position A). The inset illustrates electric field lines for the dipole. To construct the lines, many different points in space are evaluated and the electric field vector heads are then joined.

It should be recognized that the points between $-Q_1$, $+Q_2$, and position A form a right-angle triangle so that the angle projecting from $-Q_1$ to position A is

$$\theta = \tan^{-1}\left(\frac{0.30 \text{ m}}{0.50 \text{ m}}\right) = 31°.$$

The hypotenuse of the system is also the radial distance (r_1) between Q_1 and position A:

$$r_1 = \sqrt{0.30^2 + 0.50^2} = 0.583 \text{ m}$$

The electric field at position A due to charge $-Q_1$:

$$E_{Q1} = k\frac{Q_1}{r_1^2} = 9.0 \times 10^9 \frac{\text{Nm}^2}{\text{C}^2} \times \frac{50 \times 10^{-6}\text{C}}{(0.58 \text{ m})^2} = 1.34 \times 10^6 \frac{\text{N}}{\text{C}}$$

This vector acts at a 31° angle with respect to the horizontal so it can be resolved into x- and y-components. The x-component is positive while the y-component is negative:

$$E_{Q1_x} = 1.34 \times 10^6 \frac{\text{N}}{\text{C}} \times \cos(31°) = 1.15 \times 10^6 \frac{\text{N}}{\text{C}}$$

$$E_{Q1_y} = -\left(1.34 \times 10^6 \frac{\text{N}}{\text{C}} \times \sin(31°)\right) = -6.90 \times 10^5 \frac{\text{N}}{\text{C}}$$

The electric field at position A due to charge $+Q_2$:

$$E_{Q2} = k\frac{Q_2}{r_2^2} = 9.0 \times 10^9 \frac{\text{Nm}^2}{\text{C}^2} \times \frac{50 \times 10^{-6} C}{(0.30 \text{ m})^2} = 5.00 \times 10^6 \frac{\text{N}}{\text{C}}$$

This vector acts along the y-axis with respect to position A, so there is no x-component and the full magnitude is then given along the positive y-axis.

$$E_{Q2_x} = 0$$

$$E_{Q2_y} = 5.00 \times 10^5 \frac{\text{N}}{\text{C}}$$

Sum the x- and y-components of the electric fields due to the two charges:

$$\sum E_x = E_{Q1_x} + E_{Q2_x} = 1.15 \times 10^6 \frac{\text{N}}{\text{C}}$$

$$\sum E_y = E_{Q1_y} + E_{Q2_y} = 4.31 \times 10^6 \frac{\text{N}}{\text{C}}$$

Magnitude and direction of the electric field at position A due to the two charges are

$$E_A = \sqrt{\left(\sum E_x\right)^2 + \left(\sum E_y\right)^2} = 4.46 \times 10^6 \frac{\text{N}}{\text{C}};$$

$$\phi = \tan^{-1}\left(\frac{\sum E_y}{\sum E_x}\right) = 75.1°.$$

Thus, the magnitude of the electric field at position A is 4.46×10^6 N/C and it acts at a 75.1° angle relative to the horizontal. The inset of figure 2.1.1 illustrates that electric field lines are constructed by evaluating many different positions around the two charges and then connecting the vector heads pointing in the field direction.

Appendix 2.2

Calculating the Electric Potential at a Point

The following example illustrates how an electric potential at a specific point is calculated due to a dipole. The electric potential is that recorded by a single electrode. The formula used in the calculation is based on the physical system depicted in figure 2.2.1. The starting point is to determine the electric potential at a specific point (r_a) due to a single charge ($+Q$). This is accomplished using a much smaller charge ($+q_0$). Since only differences in electric potential between two points can be physically measured, a second point must be selected to serve as a reference. By convention, a second point located infinitely far away ($r_b = \infty$) from $+Q$ is selected so that the reference potential is zero ($V_b = 0$). This is analogous to the positive test charge ($+q_0$) resting on the negatively charged plate, where it no longer has potential energy (see figure 2.3, p. 21). Thus, the change in electric potential energy as $+q_0$ moves from r_b to r_a is equal to the negative value of the work performed by the electric field if the $+q_0$ is free to naturally move from points r_a to r_b.

If the expression for potential difference (V_{ba}) is multiplied through by q_0, the result is a return to the difference in electric potential energy (U_{ba}) between two

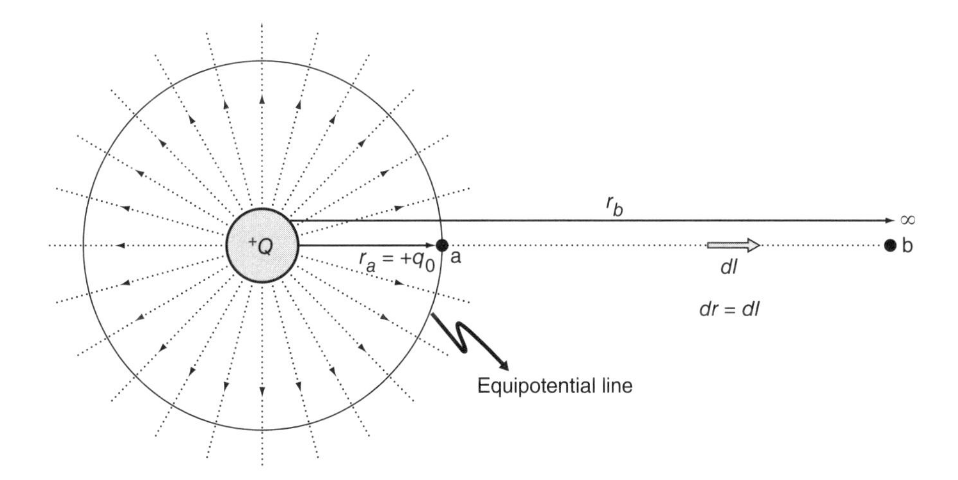

FIGURE 2.2.1 A positive point charge ($+Q$) produces an electric field, depicted by dotted lines extending radially outward from $+Q$. The electric potential (V) at point a is determined by moving a much smaller ($Q \gg q_0$) positive test charge $+q_0$ from point a to b (∞). The integration takes place from point a to b, where dl is an infinitesimal displacement along the field line. The circle (solid line) has a radius r_a, which constitutes an equipotential line that passes through point a.

points: $\Delta U_{ba} = -\Delta W_{ba}$. Where *dl* is an infinitesimal increment in displacement within a nonuniform field around a point charge,

$$U_b - U_a = -\int_a^b F \cdot dl.$$

Knowing that $V = U / q_0$ and $E = F / q_0$, we can divide the expression through by q_0 and obtain a relationship between electric field and potential difference:

$$V_b - V_a = -\int_a^b E \cdot dl.$$

The formula may be expanded if we remember that

$$E = k\frac{Q}{r^2}.$$

At this point we can elaborate further on the proportionality constant defined by coulomb:

$$k = \frac{1}{4\pi\varepsilon_0}$$

where ε_0 is the permittivity of free space ($\varepsilon_0 \approx 8.85 \times 10^{-12}\,\mathrm{C}^2 / \mathrm{Nm}^2$). The permittivity of free space is a measure of how much charges can affect each other in a medium, which is the open air in this case.

The distance between the two charges is expressed as a radial distance when we are working with the electric field (E). The infinitesimal displacement dl between a and b may therefore be replaced by the equivalent infinitesimal radial displacement (dr) between r_a to r_b:

$$V_b - V_a = -\int_a^b \frac{1}{4\pi\varepsilon_0}\frac{Q}{r^2} \cdot dr$$

Moving the constant outside the integrand,

$$V_b - V_a = -\frac{Q}{4\pi\varepsilon_0}\int_a^b \frac{1}{r^2} \cdot dr;$$

$$V_b - V_a = -\frac{Q}{4\pi\varepsilon_0} \cdot \left[-\frac{1}{r}\right]_{r_a}^{r_b};$$

$$V_b - V_a = -\frac{1}{4\pi\varepsilon_0} \cdot \left[\frac{Q}{r_b} - \frac{Q}{r_a}\right].$$

At a radial distance infinitely far away from the point charge ($r_b = \infty$), the electric potential at this point is ($V_b = 0$) and the term inside the brackets that includes (r_b) becomes infinitesimally small. Then, multiplying both sides by (−1), the electric potential due to a point charge is

$$V_a = \frac{1}{4\pi\varepsilon_0}\frac{Q}{r_a}.$$

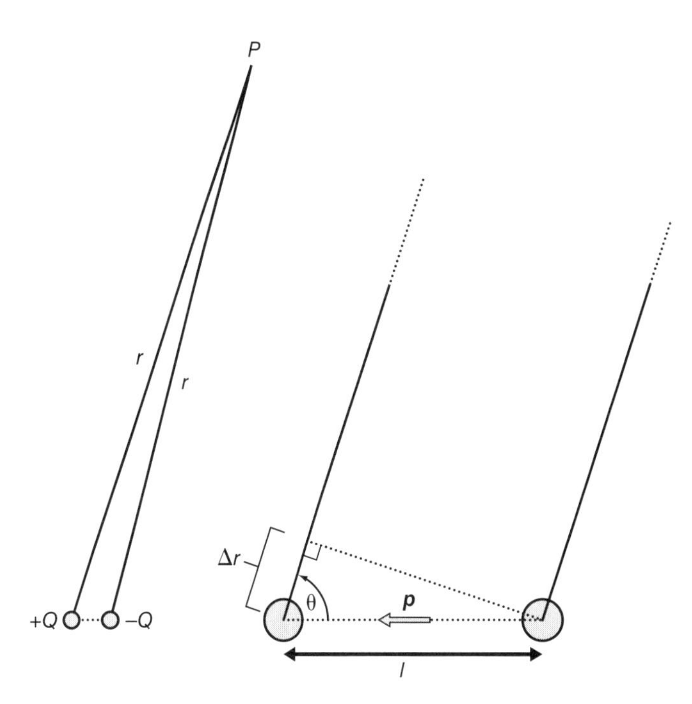

FIGURE 2.2.2 Electric potential (V) at a point (P) in space due to a dipole.

Thus, there is an intimate relationship between the electric field (E) and the potential (V). The electric potential is exactly the same for any position around the charge ($+Q$), as long as the radial distance (r) remains constant. The reason is that electric field strength depends on the distance (r) away from the charge ($+Q$). Electric potential reflects the amount of work required to bring a small charge (q_0) from infinitely far away to a closer point in the electric field, which gets harder to do as the opposing field becomes stronger, closer to the charge ($+Q$). For any location that has the same radial distance (r) from the charge ($+Q$), the electric potential is the same and constitutes an equipotential line (figure 2.2.1).

We have been leading up to the use of the superposition principle to calculate the potential at a particular point (P) in space, far away from an electric dipole. In this case, the point (P) can be called a point electrode with no real surface area. The physical system is depicted in figure 2.2.2. We must still choose a reference point that is infinitely far away from the dipole source where ($r_b = \infty$) and ($V_b = 0$). The basic equation is taken from the previous example except that the electric potential at P follows the principle of superposition, which is the simple sum of the potentials associated with the two charges:

$$V = \sum_{i=1}^{2} V_i = V_{(+)} + V_{(-)}$$

$$V = \frac{1}{4\pi\varepsilon_0}\frac{+Q}{r} + \frac{1}{4\pi\varepsilon_0}\frac{-Q}{r}$$

$$V = \frac{Q}{4\pi\varepsilon_0}\left(\frac{1}{r} - \frac{1}{r+\Delta r}\right)$$

Simplify under one common denominator:

$$V = \frac{Q}{4\pi\varepsilon_0 r} - \frac{Q}{4\pi\varepsilon_0(r+\Delta r)} = \frac{Q(r+\Delta r) - Qr}{4\pi\varepsilon_0(r+\Delta r)}$$

$$V = \frac{Q\Delta r}{4\pi\varepsilon_0 r(r+\Delta r)}$$

Remember that the point electrode is very far away from the dipole source. As a result, the distance from the dipole is much greater than the distance between the two charges ($r \gg l$). There are two important consequences. First, the dipole can form the basis of a right-angle triangle where $\Delta r \approx l\cos\theta$ (figure 2.2.2). Second, the distance r is so great that Δr is negligible by comparison and can therefore be dropped from the denominator:

$$V = \frac{Ql\cos\theta}{4\pi\varepsilon_0 r^2}$$

$$V = \frac{1}{4\pi\varepsilon_0}\frac{\vec{p}\cos\theta}{r^2}$$

The quantity (Ql) is called the dipole moment. The dipole moment may be represented by the vector $\vec{p}$ whose magnitude is given by $|\vec{p}| = Ql$ and whose direction points from $-Q$ to $+Q$. This formula is used extensively in biomedical engineering because on an anatomical scale the distance between the charges of the dipole (i.e., action potential depolarization and repolarization) is very small relative to the distance between the dipole itself and the point electrode (Geddes and Baker 1968).

A physiological application of the preceding equation is based on recording an action potential from a single muscle fiber that is sitting in a small tank filled with extracellular fluid. The following physiological data were used with the dipole formula (Boyd et al. 1978; Fuglevand et al. 1992). The magnitude of each charge on the muscle fiber is 388 nanocoulombs ($nC = 10^{-9}$ C). The charges are 0.5 mm apart on the muscle fiber. A point electrode is then allowed to probe the external medium 360° around the dipole at fixed radial distances, ranging from 1 to 2 mm in 0.05 mm increments. The conductivity of extracellular fluid ($2.44\ \Omega \cdot m^{-1}$) replaces the permittivity of free space (ε_0) in the proportionality constant (k). Recall that conductivity is the reciprocal of resistance and reflects how well current flows through a medium.

Figure 2.2.3 shows a series of equipotential lines. Each equipotential line corresponds to a different radial electrode distance, mapped 360° around the dipole. The

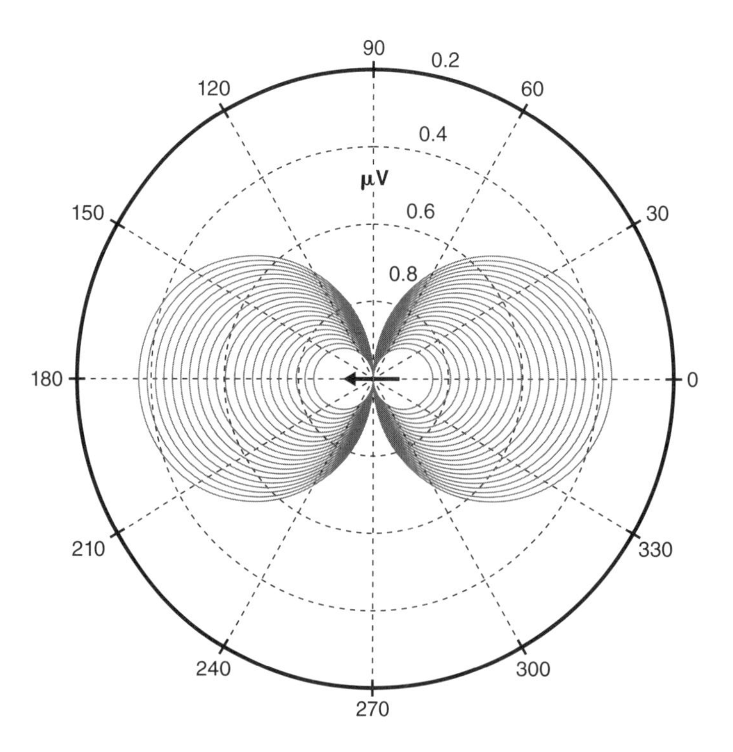

FIGURE 2.2.3 Equipotential lines for a dipole. The dipole moment vector $|\vec{p}|$ depicted in the center of the figure points from $-Q$ to $+Q$.

strongest electric potential is along the dipole axis because either charge can dominate the sum. The electric potential goes toward zero in the middle of the dipole because the positive and negative charges have equal contributions. The greatest equipotential line is associated with the smallest radial electrode distance (1 mm). Each 0.05 mm increment in radial electrode distance results in a decrease in the magnitude of the equipotential lines. The equipotential line with the smallest magnitude is associated with the farthest radial electrode distance (2 mm).

Appendix 2.3

Electric Circuits

Capacitors

Consider two capacitors arranged in series (15 and 5 μF), a third (10 μF) placed in parallel to the other two, and all three connected to a 120 V battery (figure 2.3.1). A worked example is provided to determine the following: (a) the equivalent capacitance of the circuit, (b) the charge on each capacitor, and (c) the voltage across the 15 μF capacitor. The first step is to reduce the series capacitors to a single equivalent:

$$\frac{1}{C_{2,3}} = \frac{1}{C_2} + \frac{1}{C_3}$$

$$C_{2,3} = \frac{C_2 C_3}{C_2 + C_3} = \frac{15\ \mu\text{F} \times 5\ \mu\text{F}}{15\ \mu\text{F} + 5\ \mu\text{F}} = 3.75\ \mu\text{F}$$

The problem can now be treated as capacitors connected in parallel. The single equivalent for the parallel capacitors is $C_{eq} = C_1 + C_{2,3} =$ 13.75 μF. The total charge must first be calculated to determine how it is applied across the circuit: $Q = C_{eq}V =$ 1,650 μC. Since both plates of the first capacitor are connected directly

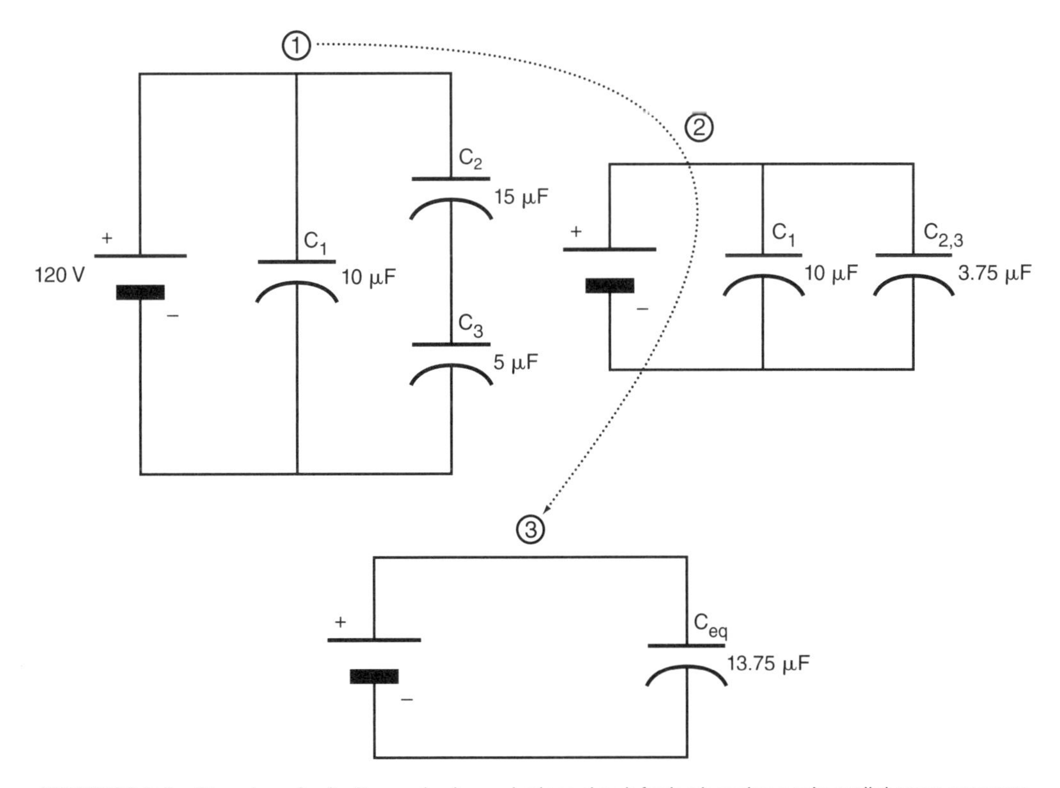

FIGURE 2.3.1 The steps for finding a single equivalent circuit for both series and parallel arrangements of capacitors.

to the battery, the other two capacitors have the remaining charge. The charge on the first capacitor is $Q_1 = C_1V =$ 1,200 μC. The remaining charge for the other two capacitors is $Q - Q_1 =$ 450 μC. As a cross-check, if the voltage is applied to the single equivalent, $Q_{2,3} = C_{2,3}V$ = 450 μC. The voltage across the 15 μF capacitor is $V_2 = Q_2 / C_2 =$ 30 V. The remaining voltage drop is 90 V across the 5 μF capacitor because in the series arrangement, $V = V_1 + V_2$ and the total voltage is 120 V.

Resistors

Consider the electric circuit depicted in figure 2.3.2 that consists of one resistor ($R_1 = 5\ \Omega$) in series with two parallel resistors ($R_2 = 15\ \Omega$ and $R_3 = 10\ \Omega$). Find the equivalent resistance and determine the total current flow if the source is a 12 V battery. Then, determine both the current and potential difference across each resistor. The first step is to simplify the parallel system to determine the equivalent resistor for R_2 and R_3:

$$\frac{1}{R_{2,3}} = \frac{1}{R_2} + \frac{1}{R_3}$$

$$R_{2,3} = \frac{R_2R_3}{R_2 + R_3} = \frac{15\ \Omega \times 10\ \Omega}{15\ \Omega + 10\ \Omega} = 6\ \Omega$$

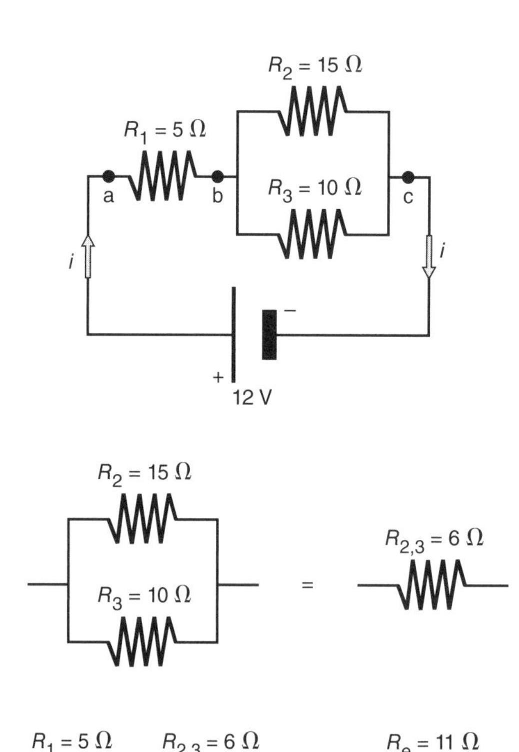

FIGURE 2.3.2 Electric circuit with resistors in series and in parallel.

The second step is to treat the first resistor (R_1) as in series with the equivalent resistor ($R_{2,3}$):

$$R_e = R_1 + R_{2,3} = 5\ \Omega + 6\ \Omega = 11\ \Omega$$

The first resistor receives the total current (i) across the circuit:

$$i_1 = \frac{V}{R_e} = \frac{12\ \text{V}}{11\ \Omega} = 1.1\ \text{A}$$

To find the current and potential difference (voltage drop) across each resistor, we must first determine the potential difference across R_1 because the current (i_1) that passes through R_1 is split across the parallel system (R_2 and R_3). The potential difference (V_{ab}) across R_1 is

$$V_{ab} = i_1 R_1 = 1.1\ \text{A} \times 5\ \Omega = 5.5\ \text{V}.$$

The remaining voltage $(12.0\ \text{V} - 5.5\ \text{V} = 6.5\ \text{V})$ is then applied across the parallel system. Since R_2 and R_3 are in parallel they have the same potential difference ($V_{bc} = 6.5$ V). The current across the parallel resistors is

$$i_2 = \frac{V_{bc}}{R_2} = \frac{6.5\ \text{V}}{15\ \Omega} = 0.43\ \text{A};$$

$$i_3 = \frac{V_{bc}}{R_3} = \frac{6.5\ \text{V}}{10\ \Omega} = 0.65\ \text{A}.$$

As a cross-check, the current passing through the resistors in the parallel system adds up to the total current (1.1 A).

Appendix 2.4

Charging a Capacitor Through a Resistor

The equations that describe the time course of changes in charge on the capacitor plates $q(t)$, the potential difference across the capacitor $V_C(t)$, and the current in the circuit $i(t)$ are used extensively in instrumentation and signal processing. A derivation of the equations is included to facilitate understanding of them. The approach follows the same principles as used for solving static equilibrium problems in mechanics in which the forces must sum to zero. According to the conservation of energy law, the changes in potential within a closed circuit must sum to zero. This is also known as Kirchhoff's loop rule for electric circuits. The potential difference provided by the source ($\mathcal{E}$) driving charges into the circuit is equal to the sum of the potential drop across the resistor ($V_R = iR$) and that at the capacitor ($V_C = q/C$) so that

$$\sum \Delta V = 0;$$

$$\mathcal{E} - iR - \frac{q}{C} = 0.$$

Both i and q are related through the equation $i = dq\,/\,dt$. The goal of the following algebraic steps is to substitute for i and rearrange the formula to obtain an expression for charge (q) as a function of time:

$$\mathcal{E} = R\frac{dq}{dt} + \frac{q}{C}$$

$$-R\frac{dq}{dt} = \frac{q}{C} - \mathcal{E}$$

$$-R\frac{dq}{dt} = \frac{q - C\mathcal{E}}{C}$$

$$-\frac{dq}{dt} = \frac{q - C\mathcal{E}}{C} \cdot \frac{1}{R}$$

$$-\frac{dq}{dt} = \frac{q - C\mathcal{E}}{RC}$$

$$dq(RC) = -dt(q - C\mathcal{E})$$

$$\frac{dq}{q - C\mathcal{E}} = -\frac{dt}{RC}$$

Integrate to solve the differential equation:

$$\int_0^q \frac{dq}{q - C\mathcal{E}} = -\int_0^t \frac{dt}{RC}$$

$$\int_0^q \frac{dq}{q - C\mathcal{E}} dq = -\frac{dt}{RC} \int_0^t dt$$

$$\ln(q - C\mathcal{E}) = -\frac{t}{RC} + K$$

The constant of integration (K) can be evaluated using initial conditions ($t = 0$ and $q = 0$) at the moment before the switch closes: $K = \ln(-C\mathcal{E})$. Substituting back into the original equation:

$$\ln(q - C\mathcal{E}) = -\frac{t}{RC} + \ln(-C\mathcal{E})$$

$$\ln(q - C\mathcal{E}) - \ln(-C\mathcal{E}) = -\frac{t}{RC}$$

Using the basic log rules:

$$\ln\left(\frac{q - C\mathcal{E}}{-C\mathcal{E}}\right) = -\frac{t}{RC}$$

$$\ln\left(-\frac{q}{C\mathcal{E}} + \frac{-C\mathcal{E}}{-C\mathcal{E}}\right) = -\frac{t}{RC}$$

$$\ln\left(1 - \frac{q}{CE}\right) = -\frac{t}{RC}$$

Take the exponent of both sides to remove the log and continue to simplify and solve for q:

$$1 - \frac{q}{C\mathcal{E}} = e^{-\frac{t}{RC}}$$

$$1 - e^{-\frac{t}{RC}} = \frac{q}{C\mathcal{E}}$$

$$\frac{1 - e^{-\frac{t}{RC}}}{\frac{1}{C\mathcal{E}}} = q$$

$$\left(1 - e^{-\frac{t}{RC}}\right)C\mathcal{E} = q$$

$$q = C\mathcal{E}\left(1 - e^{-\frac{t}{RC}}\right)$$

The charge as a function of time is then

$$q(t) = C\mathcal{E}\left(1 - e^{-\frac{t}{RC}}\right).$$

The time course of current $i(t)$ may be obtained by differentiating with respect to the charge:

$$i = \frac{dq}{dt}$$

$$i = \frac{d}{dt}\left\{C\mathcal{E}\left(1 - e^{-\frac{t}{RC}}\right)\right\}$$

Applying the chain rule for differentiating composite functions:

$$i = C\mathcal{E}\left(-e^{-\frac{t}{RC}}\right)\left(-\frac{1}{RC}\right)$$

$$i = \frac{\mathcal{E}}{R}e^{-\frac{t}{RC}}$$

The current as a function of time is then

$$i(t) = \frac{\mathcal{E}}{R}e^{-\frac{t}{RC}}.$$

Discharging a Capacitor Through a Resistor

In this situation the capacitor serves as a *nonrenewable* source. That is, the fully charged capacitor is similar to a battery without an electromotive force to maintain a potential difference and flow of current across the circuit. In the absence of a battery ($\mathcal{E} = 0$), the time course of $q(t)$ can be obtained in a similar manner as previously outlined:

$$\sum \Delta V = 0$$

$$-\frac{q}{C} - iR = 0$$

$$-\frac{q}{C} - R\frac{dq}{dt} = 0$$

$$-R\frac{dq}{dt} = \frac{q}{C}$$

$$\frac{dq}{q} = -\frac{1}{RC}dt$$

Solve the differential equation by integrating starting from initial conditions where at time $(t = 0)$ the charge on the capacitor is $q = Q$, where Q is the maximal value:

$$\int_Q^q \frac{1}{q}dt = -\frac{1}{RC}\int_0^t dt$$

$$\left[\ln(q)\right]_Q^q = \left[-\frac{1}{RC}t\right]_0^t$$

$$\ln(q) - \ln(Q) = -\frac{1}{RC}t - 0$$

$$\ln\left(\frac{q}{Q}\right) = -\frac{t}{RC}$$

$$\frac{q}{Q} = e^{-\frac{t}{RC}}$$

$$q = Qe^{-\frac{t}{RC}}$$

The charge as a function of time is then

$$q(t) = Qe^{-\frac{t}{RC}}.$$

The same is also true for $i(t)$ while the capacitor is discharging. As before, the relationship for $i(t)$ may be obtained by differentiating with respect to the charge:

$$i = \frac{dq}{dt}$$

$$i = \frac{d}{dt}\left(Qe^{-\frac{t}{RC}}\right) = Q\left(e^{-\frac{t}{RC}}\right)\left(-\frac{1}{RC}\right)$$

$$i = -\frac{Q}{RC}e^{-\frac{t}{RC}}$$

The current as a function of time is then

$$i(t) = \frac{Q}{RC}e^{-\frac{t}{RC}}.$$

Appendix 2.5

The Muscle Fiber as an RC Circuit

Resistivity of the myoplasm (ρ_m) and resistance of the fiber membrane (R_m) are expressed per unit length ($\Omega \cdot$ cm), because they decrease over the length of the fiber as more channels are available for current to leak out. Similarly, membrane capacitance (C_m) is expressed per unit area ($\text{F} \cdot \text{cm}^2$). Physiologic values for the electric properties of the frog adductor magnus muscle are taken from Katz (1948) to illustrate the resistive and capacitive properties of muscle fibers. The resistivity of the myoplasm is $\rho_m = 176\ \Omega \cdot \text{cm}$; capacitance of the muscle fiber membrane is $C_m = 6\ \mu\text{F} \cdot \text{cm}^2$; resistance of muscle fiber membrane is $R_m = 1500\ \Omega \cdot \text{cm}^2$; and the radius ($a$) of the muscle fiber is 75 μm.

The axial resistance of a 1 cm muscle fiber is determined not only by its length (l) but also by its cross-sectional area (πa^2) and the per unit length resistivity of the myoplasm (ρ_m). The muscle fiber length (l = 1 cm) was selected for convenience:

$$R = \rho_m \left(\frac{l}{A} \right)$$

$$R = \frac{\rho_m l}{\pi a^2} = \frac{(176\ \Omega \cdot \text{cm})(1\ \text{cm})}{\pi \left(75 \times 10^{-4} \text{cm}\right)^2} = 9.96 \times 10^5\ \Omega$$

This resistance is very large. To provide a frame of reference, the resistance for a typical household extension cord (with the same dimensions made of copper wire), is $4.1 \times 10^{-5} \Omega$. The resistance of copper wire is so small that it is considered negligible in comparison to that of the actual resistor elements within the circuit and is ignored in circuit problems.

The same steps are used to determine the radial resistance of the membrane to leakage current (R'), over a specific length (l = 1 cm) of the muscle fiber. The surface area ($2\pi rl$) of the specific segment length must be linked with the resistance to leakage current per unit surface area of the muscle (R_m):

$$R' = \frac{R_m}{2\pi a l} = \frac{1500\ \Omega \cdot \text{cm}^2}{2\pi \left(75 \times 10^{-4} \text{cm}\right) 1\ \text{cm}} = 3.18 \times 10^4 \Omega$$

The result indicates that resistance in the radial direction is only 3% of that in the axial direction. This means that most of the current leaks out through the membrane to the interstitial fluid within a 1 cm segment of the muscle fiber.

A fundamental quantity in electrophysiology is the *length constant* (λ). It is the segment length at which the axial and radial resistances are equal. At distances greater than λ, the axial resistance is greater than the radial resistance and the majority of the current leaks through the membrane. Typical values for myelinated and unmyelinated axons are 0.05 and 0.7 cm, respectively. Using the space parameter (λ) instead of an arbitrary length (l), we can solve for the length at which the two resistances are equal for the frog adductor magnus:

$$\frac{\rho_m \lambda}{\pi a^2} = \frac{R_m}{2\pi a \lambda}$$

$$\lambda = \sqrt{\frac{R_m a}{2\rho_m}}$$

$$\lambda = \sqrt{\frac{1500\ \Omega \cdot \text{cm}^2 \times 75 \times 10^{-4}\,\text{cm}}{2 \times 176\ \Omega \cdot \text{cm}}} = 0.179\ \text{cm}$$

The muscle fiber membrane is very thin relative to its length, so it basically looks flat over a very short section. The interior and exterior of the membrane can therefore be treated in the same manner as two flat plates wherein the capacitance is proportional to the surface area. We need to know the capacitance per unit area of the membrane (C_m) to determine its value over a 1 cm segment of the muscle fiber with surface area ($2\pi al$):

$$C = C_m(2\pi al)$$

$$C = 6 \times 10^{-6}\,\text{F} \cdot \text{cm}^2 (2\pi\ 75 \times 10^{-4}\,\text{cm} \times 1\ \text{cm})$$

$$C = 283 \times 10^{-6}\,\text{F} = 283\ \text{pF}$$

Capacitance is also a function of the distance (d) between plates. In the case of muscle fibers, this is determined by the distance between the interstitial fluid and the myoplasm (see figure 2.13, p. 43).

The resistive and capacitive functions of the muscle membrane afford a representation of the fiber as an RC circuit (figure 2.5.1). The basic circuit unit includes two resistors; one resistor represents axial resistance to the flow of current through the myoplasm (R), and the other represents radial resistance to leakage current through the membrane (R'). Membrane capacitance (C) and leakage resistance (R') are in parallel with each other, and together they represent the function of the muscle fiber

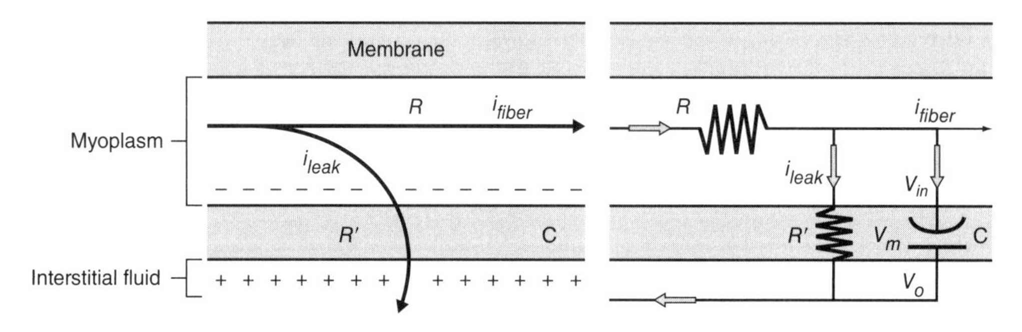

FIGURE 2.5.1 The muscle fiber is modeled as an RC circuit. The entire length of the muscle fiber is then a series of RC circuits linked together. Each successive transmembrane potential (V_m) includes the resistive and capacitive effects of the preceding circuit through which the electromotive force must travel. Thus, there is a progressive decay in both the rate of increase of transmembrane potential and its maximum.

membrane. The electromotive force ($\mathcal{E}$) is an applied stimulus that results in the movement of positive charges within the myoplasm. The entire length of the muscle fiber is then represented as a series of RC circuits linked together.

The electromotive force ($\mathcal{E}$) is a *subthreshold* electrical stimulus applied to the membrane to cause a change in the transmembrane potential (V_m). The transmembrane potential (V_m) is the difference between the electric potential inside and outside the membrane, and it corresponds to the potential difference across the capacitor in the RC circuit. Thus, once the switch closes on the first circuit (the stimulus is applied), the potential difference across the membrane (V_m) will gradually increase to 63% of its maximum charge within $\tau = RC$. Each successive RC circuit unit is farther away from the stimulus and therefore includes the resistors and capacitor from each previous RC unit, in series. As a result, the rate of increase in transmembrane potential (V_m) is attenuated with each additional RC circuit unit, and it takes increasingly longer to reach 63% of the maximum charge.

The leakage resistor (R') ultimately affects the maximum potential (V_m) across the membrane capacitor (C). The current that leaves the first resistor R is split in the parallel system between R' and C. A portion of the total current is in the path between R and R'. The charge and resulting potential difference across the capacitor (V_m) never achieve the predicted maximum in a simple RC circuit because of R'. The potential drop across R and the final potential across C are less than the electromotive force ($\mathcal{E}$). The leakage current R' results in a steady decrease in the maximum potential difference (V_m) across the capacitor of each additional RC pair. This process continues along the fiber until the current is completely dissipated.

The conduction velocity of action potentials depends on how quickly the membrane can be brought to threshold, and how far ahead of the active region the membrane can be brought to threshold by the passive spread of positive charges. The rate at which the membrane can be brought to threshold depends on the membrane's RC time constant ($\tau = RC$). The smaller the RC time constant, the more quickly the membrane will depolarize to threshold and facilitate conduction. The length of the leading edge of the depolarizing current is dependent upon the length constant (λ). The formula for the length constant (λ) shows that the length at which R and R' are equal increases with the square root of the diameter ($a^{1/2}$).

Larger muscle fibers have a greater length constant, which means that the depolarizing current will travel farther forward, passively. Ultimately, this occurs because there is a decrease in axial resistance associated with an increase in fiber diameter. The distance along the muscle fiber at which the potential falls to 37% of its maximum value also increases. The significance of the 37% value should be familiar from the RC time constant. The action potential has a greater forward extent (or leading edge) that brings the membrane area farther ahead of the traveling dipole, closer to threshold (see figure 2.5, p. 26). The membrane area will depolarize more rapidly, decreasing the RC time constant. The result is an overall increase in conduction velocity with an increase in muscle fiber size.

Appendix 3.1

Muscle-Tendon End Effects

The MFAP may be represented as a quadripole (+ – – +). The quadripole is then used to construct two adjacent dipoles (figure 3.1.1). The middle of the negative portion of the MFAP is divided in two, with each portion represented as a separate current sink. The current sink for the first half of the negativity and the weak current source of passive depolarization are the leading dipole (LD). The current sink for the second half

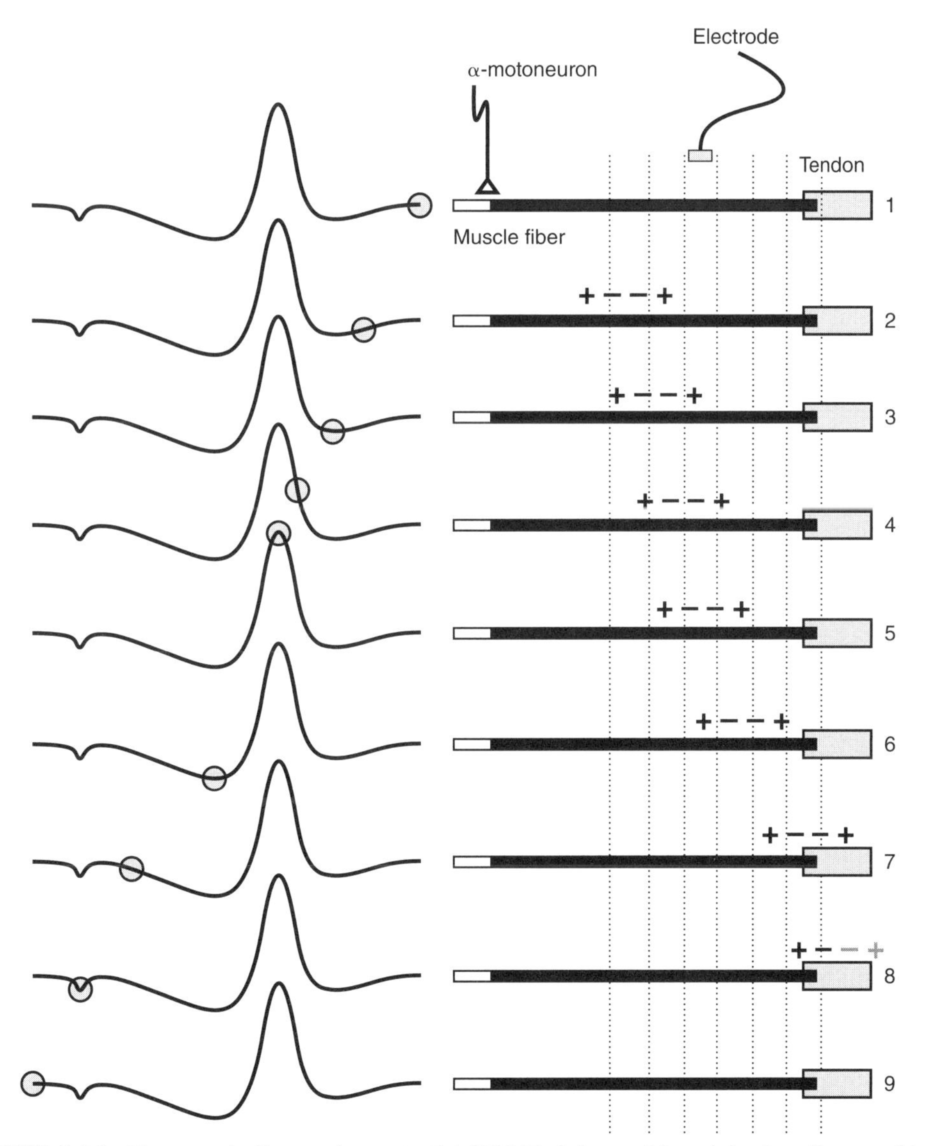

FIGURE 3.1.1 The muscle fiber action potential (MFAP) (left panels) and the traveling quadripole (right panels). The circle on the MFAP shows the potential recorded at the electrode due to the traveling quadripole. The convention for polarity of the MFAP is positive below the horizontal and negative above the horizontal.

of the negativity and the strong current source of repolarization constitute the trailing dipole (TD) (Dumitru 2000; Lateva and McGill 2001; Dimitrova and Dimitrov 2003).

Additional technological developments in recording the MFAP reveal the presence of muscle fiber–tendon end effects. This phenomenon is observed as a minor positivity located between the depolarization phase and the end of the MFAP, depending on the length of the muscle fiber. Figure 3.1.1 shows how the TD–LD relationship relates to muscle fiber–tendon end effects. In each panel, the MFAP is drawn on the left and the muscle fiber–tendon complex on the right. Symmetry is assumed so the α-motoneuron demarcates the innervation zone with the MFAP propagating an equal distance along the muscle fiber in both directions. The vertical lines show which portion of the TD-LD dominates the potential recorded at the electrode as the MFAP travels along the membrane (panels 2-8). The farther away from the electrode, the less influence a particular charge has on the recorded potential. The circle on the MFAP highlights the net potential recorded at the electrode associated with TD–LD position to reveal its evolution. The top and bottom panels depict muscle fiber resting states before and after MFAP generation, panel 1 and panel 9, respectively.

When the MFAP leaves the motor end plate, it can be viewed as two adjacent dipoles traveling together (TD-LD) along the surface of the muscle fiber. As the LD enters the proximity of the electrode, the positive polarity of the leading edge is formed (panels 2 and 3). The two adjacent current sinks approach and pass underneath the electrode, giving rise to the large negativity (panels 4 and 5). As the MFAP continues, the TD has the greatest impact on the potential recorded at the electrode. The positive charge of the TD associated with repolarization dominates the polarity (panel 6). The MFAP continues traveling so that the LD is now over the tendon, leaving behind the TD. Recall that the net potential is determined by the difference ($r_1 - r_2$), which is relatively small for large distances. The potential recorded at the electrode is therefore becoming *less* positive as each charge of the TD has relatively equal influence (panel 7). However, as the negative charge of the TD passes over the tendon, the positive charge now dominates the potential recorded at the electrode. The electric potential is relatively small, and it is registered as a minor positivity toward the tail end of the MFAP (panel 8). The muscle fiber–tendon end effects are evident in MUAPs but are not distinguishable in the more complex interference pattern; they are nonetheless present and contribute to the high-frequency content of the surface electromyographic (sEMG) signal. Muscle fiber–tendon end effects are also prominent in CMAPs recorded with a monopolar electrode configuration but are reduced (not eliminated) by the spatial filtering effects of bipolar recordings (Lateva et al. 1996; Farina et al. 2002; Dimitrova et al. 2002; Dimitrov et al. 2003).

Appendix 4.1

EMG Area and Slope Measurement

The following is a simple algorithm for trapezoidal integration:

$$\int_{t_1}^{t_n} y(t)dt = \sum_{i=1}^{n} \frac{y_{i-1} + y_i}{2} \Delta t$$

where n is the number of data points, y_i is the EMG data value at time t, and Δt is the sampling interval. The units are reported as mV·s. The amplitude scale (mV or µV) depends on the magnitude of EMG, while the time scale may be in seconds or milliseconds depending on the interval of integration.

Numerical integration of the EMG waveform is also used to determine the rate of increase in muscle activity. Gottlieb and colleagues (1989) suggested that the rising phase of the EMG can reasonably be approximated over a short period of time by the function

$$emg(t) = at^n.$$

The constant (a) is unknown and the exponent (n) must be greater than zero. Integrating the linear envelope–detected EMG curve over the short interval (T) results in a quantity that Gottlieb and colleagues (1989) denote as (Q):

$$Q = a\int_0^T t^n dt = a\frac{T^{n+1}}{n+1}$$

Solving for a:

$$a = Q\frac{n+1}{T^{n+1}}$$

A higher low-pass cutoff results in a slope that is continuously changing as a function of time (t). The simplest alternative is to take the slope (m) of the chord between two points over a specific interval (T). The chosen interval starts at EMG onset. The resulting slope calculations are simplified because at $t_1 = 0$ the amplitude of EMG is $emg(t_1) = 0$. At the end of the interval, $t_2 = T$ and the amplitude of EMG is $emg(T)$:

$$m = \frac{\Delta emg}{\Delta t} = \frac{emg(T)}{T}$$

Substituting $emg(t) = at^n$ into this equation:

$$m = \frac{aT^n}{T} = aT^{n-1}$$

If the integration result for a is substituted into the formula for slope:

$$m = \frac{n+1}{T^2} Q$$

This means that integration of the EMG over a short time interval can be used to estimate the average slope over that time interval (T). This result is the same regardless of the power function used to derive it. An interval length of 30 ms was observed to accommodate a variety of contraction conditions. Although the units for the chord slope would be mV/s, it must be remembered that Q_{30} is obtained by numerical integration so the units are still mV·s.

Appendix 4.2

Cross-Correlation Function

The cross-correlation function is formally expressed as

$$R'_{xy}(\tau) = \frac{1}{T}\int_0^T x(t)\,y(t+\tau)\,dt$$

$$R_{xy}(\tau) = \frac{R'_{xy}(\tau)}{\sqrt{R_{xx}(0)\,R_{yy}(0)}}$$

where T is the time period of the signal (Winter et al. 1994). The first expression $R'_{xy}(\tau)$ is calculated using the following steps. Starting with a lag time $\tau = 0$, multiply the two waveforms to obtain the product $x(t)\,y(t)$. Find the area under the product $x(t)\,y(t)$ using trapezoidal integration to obtain the value of the *nonnormalized* cross-correlation at a single lag time $\tau = 0$. Shift $y(t)$ by lag time τ, giving $y(t+\tau)$. Then multiply the shifted $y(t+\tau)$ by $x(t)$, obtaining $x(t)\,y(t+\tau)$. Once again, find the area under the product $x(t)\,y(t+\tau)$ using trapezoidal integration to obtain the value of the nonnormalized cross-correlation at the *next single* lag time τ. Repeat these steps for all successive time lags τ.

The normalized expression $R_{xy}(\tau)$ takes on values between −1.0 and +1.0 when we divide each point in $R'_{xy}(\tau)$ by the square root of the product of the two auto-correlation functions $R_{xx}(0)$ and $R_{yy}(0)$. The auto-correlation functions at lag time $\tau = 0$ are equivalent to

$$R_{xx}(0) = \sum_{i=1}^{N}(x_i - \bar{x})^2;$$

$$R_{yy}(0) = \sum_{i=1}^{N}(y_i - \bar{y})^2.$$

The final result will yield $R_{xy}(\tau)$. The same procedures used for calculating the cross-correlation function are used for the auto-correlation function, except that the signal is cross-correlated with itself.

Appendix 4.3

Calculating Fourier Coefficients

Here we step through the process of calculating the Fourier coefficients for the following square wave (figure 4.3.1*a*). The square wave has a P-P voltage amplitude between $+V$ and V with a period of T, and it is aligned with the y-axis at $T / 2$:

$$f(t) = \begin{cases} -V, & -T/2 \le t < 0 \\ +V, & 0 \le t < T/2 \end{cases}$$

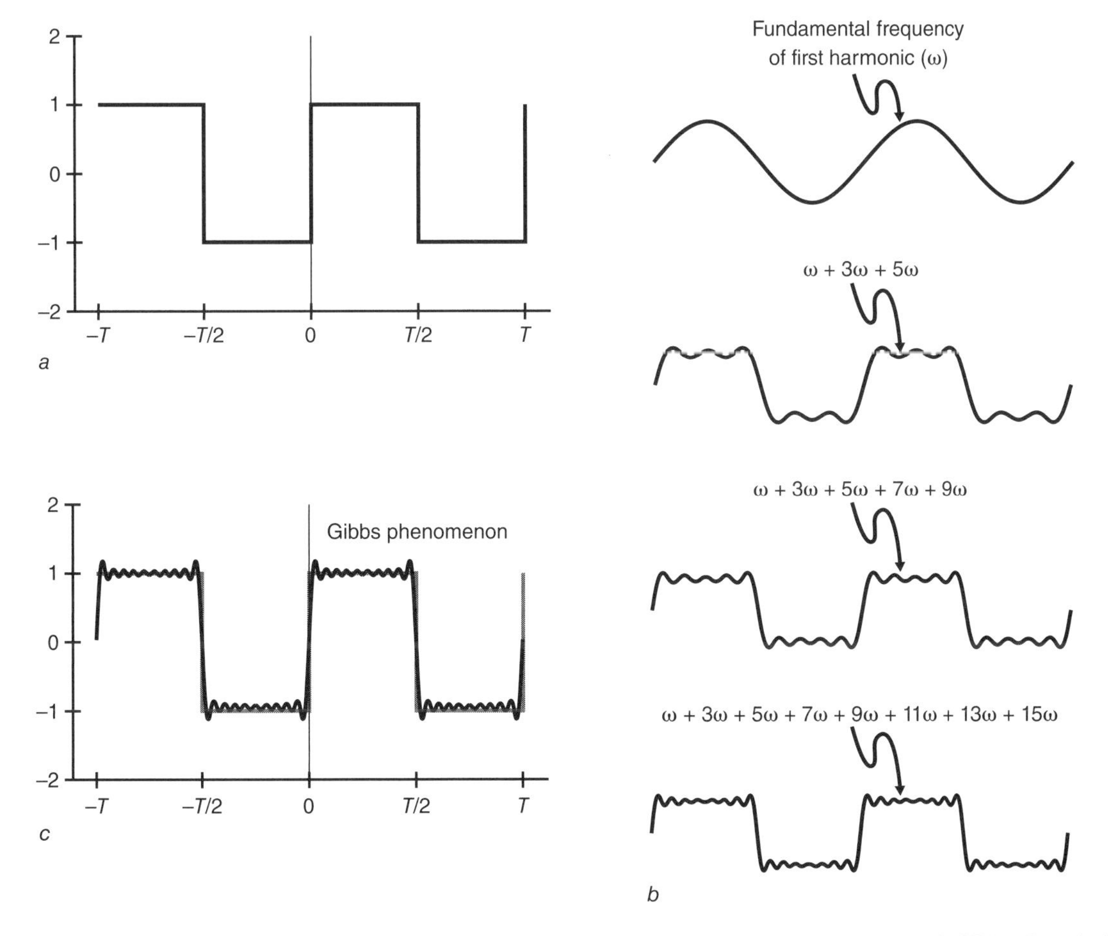

FIGURE 4.3.1 A square wave with a P-P voltage amplitude between $+V$ and $-V$ and a period of T is aligned with the y-axis at $T / 2$ (*a*). The progressive addition of the fundamental frequency (ω) and its higher-frequency harmonics ($n \cdot \omega$, *where* $n = odd$) is used to synthesize a square wave *(b)*. The resulting Fourier synthesis is superimposed on the original square wave to demonstrate the overall fit *(c)*. The increasing oscillation as the synthesized waveform approaches the corner of the square is termed the Gibbs phenomenon.

The derivation of the values for the Fourier coefficients a_0, a_n, and b_n is shown in the following equations. Notice that the DC term is equal to one-half the amplitude of the square wave over time and therefore represents its "average" value.

$$a_0 = \frac{1}{T}\int_{-T/2}^{+T/2} f(t)dt$$

$$a_0 = \frac{1}{T}\left[\int_{-T/2}^{0} -Vdt + \int_{0}^{+T/2} Vdt\right]$$

$$a_0 = \frac{1}{T}\left[[-V]_{-T/2}^{0} + [V]_{0}^{+T/2}\right]$$

$$a_0 = \frac{1}{T}[-VT + VT]$$

$$a_0 = 0$$

The coefficient a_n is zero for all values of n:

$$a_n = \frac{2}{T}\int_{-T/2}^{+T/2} f(t)\cos(n\omega t)dt$$

$$a_n = \frac{2}{T}\left[\int_{-T/2}^{0} -V\cos(n\omega t)dt + \int_{0}^{+T/2} V\cos(n\omega t)dt\right]$$

$$a_n = \frac{2}{T}\left[-V\left[n\omega\sin(n\omega t)\right]_{-T/2}^{0} + V\left[n\omega\sin(n\omega t)\right]_{0}^{+T/2}\right]$$

$$a_n = \frac{2}{T}\left[-\frac{V}{n\omega}[\sin(n\omega t)]_{-T/2}^{0} + \frac{V}{n\omega}[\sin(n\omega t)]_{0}^{+T/2}\right]$$

$$a_n = \frac{2}{T}\left[-\frac{V}{n\omega}\left[0 - \sin\left(-\frac{n\omega T}{2}\right)\right] + \frac{V}{n\omega}\left[\sin\left(\frac{n\omega T}{2}\right) - 0\right]\right]$$

Remembering that $-\sin(-\theta) = \sin(\theta)$:

$$a_n = \frac{2}{T}\left[-\frac{V}{n\omega}\left[\sin\left(\frac{n\omega T}{2}\right)\right] + \frac{V}{n\omega}\left[\sin\left(\frac{n\omega T}{2}\right)\right]\right]$$

$$a_n = 0$$

The coefficient b_n exists only for odd numbers $(2n-1)$:

$$b_n = \frac{2}{T}\int_{-T/2}^{+T/2} f(t)\sin(n\omega t)dt$$

$$b_n = \frac{2}{T}\left[\int_{-T/2}^{0} -V\sin(n\omega t)\,dt + \int_{0}^{+T/2} V\sin(n\omega t)\,dt\right]$$

$$b_n = \frac{2}{T}\left[-V\left[\frac{1}{n\omega}\cos(n\omega t)\right]_{-T/2}^{0} + V\left[\frac{1}{n\omega}\cos(n\omega t)\right]_{0}^{+T/2}\right]$$

$$b_n = \frac{2}{T}\left[\frac{V}{n\omega}\left[\cos(n\omega t)\right]_{-T/2}^{0} - \frac{V}{n\omega}\left[\cos(n\omega t)\right]_{0}^{+T/2}\right]$$

$$b_n = \frac{2}{T}\left[\frac{V}{n\omega}\left[1-\cos\left(-\frac{n\omega T}{2}\right)\right] - \frac{V}{n\omega}\left[\cos\left(\frac{n\omega T}{2}\right)-1\right]\right]$$

Remembering that $\cos(-\theta) = \cos(\theta)$:

$$b_n = \frac{2}{T}\left[\frac{V}{n\omega}\left[1-\cos\left(\frac{n\omega T}{2}\right)\right] - \frac{V}{n\omega}\left[\cos\left(\frac{n\omega T}{2}\right)-1\right]\right]$$

$$b_n = \frac{2}{T}\left[\frac{V}{n\omega}\left[1-\cos\left(\frac{n\omega T}{2}\right)\right] + \frac{V}{n\omega}\left[1-\cos\left(\frac{n\omega T}{2}\right)\right]\right]$$

$$b_n = \frac{2V}{n\omega T}\left[2-2\cos\left(\frac{n\omega T}{2}\right)\right]$$

We substitute a_0, a_n, and b_n into the general formula:

$$f(t) = \frac{a_0}{2} + \sum_{n=1}^{\infty}\left[a_n\cos(n\omega t) + b_n\sin(n\omega t)\right]$$

Since $a_0 = 0$,

$$f(t) = \frac{0}{2} + \sum_{n=1}^{\infty}\left[0\cos(n\omega t) + \frac{2V}{n\omega T}\left[2-2\cos\left(\frac{n\omega T}{2}\right)\right]\sin(n\omega t)\right];$$

$$f(t) = \sum_{n=1}^{\infty}\left[\frac{2V}{n\omega T}\left[2-2\cos\left(\frac{n\omega T}{2}\right)\right]\sin(n\omega t)\right].$$

To simplify the calculation, we take advantage of the fact that $\omega = 2\pi / T$ also gives $\omega T / 2 = \pi$:

$$f(t) = \sum_{n=1}^{\infty}\left[\frac{V}{n\pi}\left[2-2\cos(n\pi)\right]\sin(n\omega t)\right]$$

Expanding the Fourier series for the first five terms reveals that the coefficients for b_n exist only for odd numbers $(2n-1)$ because $\cos(n\pi)$ equals $+1$ for all even n's and -1 for all odd n's.

$$f(t) = \frac{4V}{\pi}\sin(\omega t) + \frac{4V}{3\pi}\sin(3\omega t) + \frac{4V}{5\pi}\sin(5\omega t) + \cdots + \frac{4V}{n\pi}\sin(n\omega t)$$

$$f(t) = \frac{4V}{\pi}\left[\sin(\omega t) + \frac{1}{3}\sin(3\omega t) + \frac{1}{5}\sin(5\omega t) + \cdots + \frac{1}{n}\sin(n\omega t)\right]$$

Part *b* of figure 4.3.1 illustrates Fourier synthesis of a square wave by progressively adding additional terms in the series $(2n - 1)$. The Fourier synthesis is superimposed on the original square wave in part *c*. The squared error between the Fourier series and the square wave minimizes as the series converges; the error tends toward zero as n approaches infinity $(n \to \infty)$. The square wave was specifically chosen to illustrate an additional concept. The corner of the square wave is termed a discontinuity because it is a break point in the function $f(x)$. The Fourier series exhibits an oscillatory error near discontinuities, termed the *Gibbs phenomenon.* Inspection of figure 4.3.1*c* reveals that incorporating more terms in the series does not substantially decrease the error. The oscillatory error just compresses toward the discontinuity.

Glossary

action potential—An electrical signal produced at the membrane of an excitable nerve or muscle cell that is propagated along the membrane to excite adjacent nerve or muscle fiber segments.

active electrode—An EMG system in which a preamplifier is incorporated within the electrode case to increase the magnitude of the signal before it is sent to the amplifier.

afterpotential—A negative phase of the muscle fiber action potential that precedes the return to baseline.

afterwave—Another term for *afterpotential.*

aliasing—The phenomenon in which the frequency of the original analog signal is reduced to the lower-frequency digital signal due to undersampling.

alternating current (AC) —Current that varies consistently between two values in regular fashion.

ampere (A)—The unit of current.

anode—A positively charged terminal.

antidromic—Proceeding in a direction opposite to the usual one. An antidromic impulse is an action potential or series of action potentials transmitted in the direction that is opposite to the normal direction of conduction.

average rectified value (ARV)—The mean amplitude of the absolute value of EMG activity within a defined window of data.

band-pass filter—A filter that attenuates both high- and low-frequency components, leaving only middle-frequency-range signals.

bipolar configuration—The electrode placement configuration in which two active electrodes are placed over or into the muscle. The ground electrode is frequently placed on a nearby bony prominence (surface electrodes) or on the cannula (indwelling electrodes).

capacitor—An arrangement of conductive materials designed to hold a charge.

cathode—A negatively charged terminal.

comb filter—A filter that allows some frequencies to pass while attenuating others.

common mode signal—Any signal that is presented identically and simultaneously by two amplifiers. The constant-frequency power-line electrical noise within the surrounding room is one example.

compound muscle action potential (CMAP)—The muscle EMG response observed when an electrical stimulus is applied to the peripheral motor nerve, resulting in the activation of several motor units. The response is similar in shape to a single motor unit action potential but larger in magnitude. The number of motor units activated by the stimulus is proportional to the stimulation intensity. Also known as *massed action potential (M-wave).*

conductance—The ability to allow the flow of charges.

coulomb (C)—A specific number (6.25×10^{18}) of elementary positive or negative charges.

critically damped—The response to a step increase in voltage is critically damped if the voltage output follows the step increase in voltage without any overshoot, wobbles, or rings before settling down to the input voltage value.

cross-correlation function—A function that describes the degree to which two signals are correlated and any potential time delay that maximizes the correlation.

cross-talk—The presence of volume-conducted potentials from other nearby muscles that are contained in the EMG recording and confound the signal of interest.

current sink—A negative charge that is strong enough to attract positive charges.

current source—A source that provides positive charges.

cutoff frequency—The frequency at which the amplitude of the input signal is reduced, typically by 3 dB. One can obtain this frequency by multiplying the amplitude of the input signal by 0.707. For example, a 1V 1 kHz signal that is input to an amplifier with a 1 kHz cutoff frequency will have an output amplitude of .707 V.

detection volume—See *pickup area.*

deterministic—The behavior of a physical or physiological system is deterministic if it can be described by a mathematical function that predicts specific values of that system in time.

digital filter—A numerical function applied to a signal to alter its frequency content.

dipole—A discrete system of two charges, one positive and the other negative.

direct current (DC)—Current that is a constant value.

double threshold method—A method used to determine the onset or termination (or both) of an EMG burst in which both amplitude and time criteria are applied.

electric current *(i)*—The flow of like charges through a defined surface area.

electric field *(E)*—The electric stress within the vicinity of an electric charge. It acts to create forces between charges.

electric power—The rate at which electric energy is delivered to a circuit.

electromechanical delay (EMD)—The time delay between the onset of muscle electrical activation and the onset of muscular force.

electromotive force ($\mathcal{E}$)—The energy transferred to each charge (joules per coulomb) by a battery.

floating electrode— A surface EMG electrode that is manufactured with the metallic recording surface recessed within a small cup. Electrolyte gel is then used to maintain a conductive bridge with the skin surface. Ag-AgCl plating is frequently used for the metallic recording surface.

folding frequency—See *Nyquist frequency.*

frequency leakage—The false frequencies dispersed across the frequency spectrum due to the inclusion of partial periods within the data analysis window for the Fast Fourier transform.

frequency spectrum—The range of frequencies and their relative contribution that constitute an electrical signal.

ground—The zero electrical potential reference point.

ground loop—The flow of current between equipment in different locations within the same room due to a difference between their respective ground potentials. The result is a noise at power-line frequency.

half-cell potential—The potential difference between the electrolyte at the electrode surface and the surrounding medium.

high-pass filter—A filter that attenuates low-frequency signals.

H-reflex—An evoked response recorded from the skin area overlying the muscle in response to stimulation of the peripheral nerve with sufficient intensity to activate muscle afferents.

impedance (Z)—The total resistance to the flow of current in an alternating current (AC) circuit, which includes both the resistor and capacitor.

innervation ratio—A term that describes the number of muscle fibers per motor unit.

innervation zone—An anatomical location containing a dense cluster of neuromuscular junctions. Motor unit action potentials (MUAPs) originate in this region and then propagate bidirectionally toward the tendon.

leakage current—Current flowing through laboratory equipment due to the capacitive coupling between the chassis and the power-line "hot wire," internal circuitry, and other external cabling.

length constant (λ)—The segment length along the muscle or nerve fiber at which the axial and radial resistances are equal.

linear envelope—A demodulation technique used to extract information from an electrical signal that involves full-wave rectification and low-pass filtering of the signal.

low-pass filter—A filter that attenuates high-frequency signals.

macro-EMG—An EMG technique pioneered by Erik Stålberg. Macro-EMG uses two signals. One signal identifies the activity of an individual muscle fiber, usually from a needle electrode. The second signal obtains a "global" recording of the whole muscle, usually from the needle cannula. The technique provides an estimate of motor unit size.

massed action potential (M-wave)— The muscle EMG response observed when an electrical stimulus is applied to the peripheral motor nerve, resulting in the activation of several motor units. The response is similar in shape to a single motor unit action potential but larger in magnitude. The number of motor units activated by the stimulus is proportional to the stimulation intensity. Also known as *compound muscle action potential (CMAP).*

mean power frequency (MNF)—The frequency at which the mean power of the signal occurs.

mean spike amplitude (MSA)—The average peak-to-peak amplitude of all the EMG activity spikes within a defined window.

mean spike frequency (MSF)—The average number of spikes per second occurring within the EMG interference pattern.

median power frequency (MDF)—The frequency at which the median power of the signal occurs.

membrane potential—The voltage potential difference that exists across a membrane.

miniature end-plate potentials (MEPPs)—Spontaneous high-frequency spikes generated at the muscle fiber that can sometimes be recorded in the electromyogram.

monopolar configuration—An EMG electrode application configuration in which one active electrode is placed over or into the muscle with the reference electrode placed on an electrically neutral site such as the tendon associated with the muscle (surface electrodes) or on the cannula (indwelling electrodes). The ground electrode is typically placed on a nearby bony prominence.

motor point—A focal point on the skin surface where the lowest possible electrical stimulation will produce an observable muscle twitch. This area corresponds to a dense collection of motor end plates.

motor unit (MU)—A motoneuron and all muscle fibers innervated by that motoneuron.

motor unit action potential (MUAP)—An electrical signal produced by the summation of all muscle fiber action potentials belonging to one motor unit.

moving window—A mathematical operation performed on a specified range of data points termed the "window." The operation is applied starting at each data point in the signal and is repeated by moving sequentially point by point along the data stream.

muscle fiber action potential (MFAP)—An action potential produced at the muscle fiber surface.

muscle fiber conduction velocity (MFCV)—The conduction velocity of action potentials propagating along the muscle fiber membrane.

M-wave—See *massed action potential.*

notch filter—A filter that includes a very narrow cutoff at a particular frequency (e.g., 50 or 60 Hz) to remove an unwanted noise source.

Nyquist frequency—The sampling frequency that is twice the highest frequency present in the original analog signal.

orthodromic—The normal and usual direction of action potential conduction.

overdamped— The response to a step increase in voltage is overdamped if the voltage output takes a very long time to achieve the input voltage value.

passive electrode—An EMG system in which the surface electrodes contain no additional electronics to amplify the signal at the skin surface.

phase angle—The angular difference between two sinusoids of the same frequency but different position along the amplitude-versus-angle graph.

pickup area—The volume of muscle tissue from which EMG electrodes can record electrical activity. It is defined as the spherical volume of muscle tissue that has a radius equal to the interelectrode distance. This is also referred to as the *detection volume.*

power spectrum—The distribution of power yielded by the square magnitude of the frequency spectrum. The power (density) spectrum indicates the relative contribution of each frequency to the total signal power.

quantization error (QE)—The rounding error that occurs when a voltage level is presented to an analog-to-digital conversion board that is between two discrete digital conversion values.

RC circuit—An electric circuit that contains both a resistor and a capacitor.

reactive capacitance—Capacitance in an alternating current (AC) circuit.

refractory period—A delayed time interval or a time delay. In muscle and nerve fibers, for example, generation of an action potential results in a refractory period during which no additional action potentials can be generated.

resistance (R)—The ability to impede the flow of charges.

root mean square (RMS)—The square root of the mean of all the squared values of EMG activity within a given window of data.

salt bridge—The phenomenon in which electrolyte gel exists between two recording surfaces across the skin, short-circuiting the electrode circuit.

sarcolemma—The membrane surrounding the muscle fiber.

sarcoplasmic reticulum—A membrane system within the muscle fiber that stores and then sequesters Ca^{++} necessary for the production of muscular force.

selectivity—The capacity to record meaningful muscle activity from a local volume of tissue rather than cross-talk from neighboring muscle fibers.

signal bandwidth—The range of frequencies over which the signal contains components of appreciable magnitude. This is usually defined on the power spectrum as the corner points that encompass the half-power of the signal.

signal transduction—The process through which an electric potential generated by the muscle is converted into an electric signal that is carried through conventional wires to an amplifier.

spatial filter—A filter that produces changes in the amplitude and frequency content of a signal. The interelectrode distance can act as a spatial filter for EMG signals.

stationary—The phenomenon in which the mean and standard deviation of the signal remain constant with the data window of analysis. A signal is stationary if it meets the conditions for stationarity.

stochastic—A signal whose behavior can be described only by a probability density function, as specific values cannot be predicted.

temporal dispersion—The process in which potentials from each muscle or nerve fiber arrive at the recording electrode at slightly different times, broadening the recorded signal. The different latencies can sometimes be attributed to differences in conduction velocity for fibers within a motor unit or peripheral nerve.

terminal wave—A phase of the motor unit action potential that follows the main spike, generally attributed to the termination of the action potential at the neuromuscular junction.

transverse tubular system—A system of channels running perpendicular to the sarcolemma through which the muscle fiber action potential is carried to activate deeper regions of the fiber.

tripole—A discrete system of three charges.

underdamped—The response to a step increase in voltage is underdamped if the voltage output wobbles or rings before settling down to the input voltage value.

variance ratio (VR)—A measure of the variability in shape between successive waveforms. The range is from 0 to 1.0 with a lower number corresponding to a lower variability.

volt (V)—The unit of measure for the electric potential difference between two points within an electric field.

volume conduction—The process through which an action potential is detected through extracellular fluid and tissues.

References

Aagaard, P., E.B. Simonsen, J.L. Andersen, P. Magnusson, and P. Dyhre-Poulsen. 2002. Neural adaptation to resistance training: changes in evoked V-wave and H-reflex responses. *Journal of Applied Physiology* 92: 2309-2318.

Abbs, J.H., V.L. Gracco, and C. Blair. 1984. Functional muscle partitioning during voluntary movement: facial muscle activity for speech. *Experimental Neurology* 85: 469-479.

Aiello, I., G.F. Sau, M. Bissakou, S. Patraskakis, and S. Traccis. 1986. Standardization of changes in M-wave area to repetitive nerve stimulation. *Electromyography and Clinical Neurophysiology* 26: 529-532.

Aiello, I., G. Serra, G. Rosati, and V. Tugnoli. 1982. A quantitative method to analyze the H reflex latencies from vastus medialis muscle: normal values. *Electromyography and Clinical Neurophysiology* 22: 251-254.

Akaboshi, K., Y. Masakado, and N. Chino. 2000. Quantitative EMG and motor unit recruitment threshold using a concentric needle with a quadrifilar electrode. *Muscle & Nerve* 23: 361-367.

Al-Jawayed, I.A., M. Sabbahi, B.R. Etnyre, and S. Hasson. 1999. The H-reflex modulation in lying and a semi-reclining (sitting) position. *Clinical Neurophysiology* 110: 2044-2048.

Alkner, B.A., P.A. Tesch, and H.E. Berg. 2000. Quadriceps EMG/force relationship in knee extension and leg press. *Medicine and Science in Sports and Exercise* 32: 459-463.

Allison, G.T. 2003. Trunk muscle onset detection technique for EMG signals with ECG artefact. *Journal of Electromyography and Kinesiology* 13: 209-216.

Allison, S.C., and L.D. Abraham. 1995. M-wave stability in H-reflex testing: analysis of three rejection criteria. *Electromyography and Clinical Neurophysiology* 35: 165-168.

Al-Mutawaly, N., H. De Bruin, and G. Hasey. 2003. The effects of pulse configuration on magnetic stimulation. *Journal of Clinical Neurophysiology* 20: 361-370.

Aminoff, M.J. 1998. *Electromyography in clinical practice.* 3rd ed. New York: Churchill Livingstone.

An, K-N., W.P. Cooney, E.Y. Chao, L.J. Askew, and J.R. Daube. 1983. Determination of forces in extensor pollicis longus and flexor pollicis longus of the thumb. *Journal of Applied Physiology* 54: 714-719.

Andersen, J.L. 2003. Muscle fibre type adaptation in the elderly human muscle. *Scandinavian Journal of Medicine and Science in Sports* 13: 40-47.

Andersson, E.A., J. Nilsson, and A. Thorstensson. 1997. Intramuscular EMG from the hip flexor muscles during human locomotion. *Acta Physiologica Scandinavica* 161: 361-370.

Andreassen, S., and L. Arendt-Nielsen. 1987. Muscle fibre conduction velocity in motor units of the human anterior tibial muscle: a new size principle parameter. *Journal of Physiology* 391: 561-571.

Andreassen, S., and A. Rosenfalck. 1978. Recording from a single motor unit during strong effort. *IEEE Transactions on Biomedical Engineering* 25: 501-508.

Aoki, F., H. Nagasaki, and R. Nakamura. 1986. The relation of integrated EMG of the triceps brachii to force in rapid elbow extension. *Tohoku Journal of Experimental Medicine* 149: 287-291.

Arendt-Nielsen, L., N. Gantchev, and T. Sinkjaer. 1992. The influence of muscle length on muscle fibre conduction velocity and development of muscle fatigue. *Electroencephalography and Clinical Neurophysiology* 85: 166-172.

Arendt-Nielsen, L., and K.R. Mills. 1988. Muscle fiber conduction velocity, mean power frequency, mean EMG voltage and force during submaximal fatiguing contractions of human quadriceps. *European Journal of Applied Physiology* 58: 20-25.

Arendt-Nielsen, L., and M. Zwarts. 1989. Measurement of muscle fiber conduction velocity in humans: techniques and applications. *Journal of Clinical Neurophysiology* 6: 173-190.

Arnall, F.A., G.A. Koumantakis, J.A. Oldham, and R.G. Cooper. 2002. Between-days reliability of electromyographic measures of paraspinal muscle fatigue at 40, 50 and 60% levels of maximal voluntary contractile force. *Clinical Rehabilitation* 16: 761-771.

Arnaud, S., M.C. Zattara-Hartmann, C. Tomei, and Y. Jammes. 1997. Correlation between muscle metabolism and changes in M-wave and surface electromyogram: dynamic constant load leg exercise in untrained subjects. *Muscle & Nerve* 20: 1197-1199.

Arsenault, A.B., D.A. Winter, and R.G. Marteniuk. 1986a. Bilateralism of EMG profiles in human locomotion. *American Journal of Physical Medicine* 65: 1-16.

Arsenault, A.B., D.A. Winter, and R.G. Marteniuk. 1986b. Is there a "normal" profile of EMG activity in gait? *Medical and Biological Engineering and Computing* 24: 337-343.

Arsenault, A.B., D.A. Winter, R.G. Marteniuk, and K.C. Hayes. 1986c. How many strides are required for the analysis of electromyographic data in gait? *Scandinavian Journal of Rehabilitation Medicine* 18: 133-135.

Babault, N., M. Pousson, A. Michaut, and J. Van Hoecke. 2003. Effect of quadriceps femoris muscle length on neural activation during isometric and concentric contractions. *Journal of Applied Physiology* 94: 983-990.

Baratta, R.V., M. Solomonow, B-H. Zhou, and M. Zhu. 1998. Methods to reduce the variability of the EMG power spectrum estimates. *Journal of Electromyography and Kinesiology* 8: 279-285.

Barbeau, H., V. Marchand-Pauvert, S. Meunier, G. Nicolas, and E. Pierrot-Deseilligny. 2000. Posture-related changes in heteronymous recurrent inhibition from quadriceps to ankle muscles in humans. *Experimental Brain Research* 130: 345-361.

Baret, M., R. Katz, J.C. Lamy, A. Penicaud, and I. Wargon. 2003. Evidence for recurrent inhibition of reciprocal inhibition from soleus to tibialis anterior in man. *Experimental Brain Research* 152: 133-136.

Barkhaus, P.E., and S.D. Nandedkar. 1996. On the selection of concentric needle electromyogram motor unit action potentials: is the rise time criterion too restrictive? *Muscle & Nerve* 19: 1554-1560.

Barnes, W.S. 1980. The relationship of motor-unit activation to isokinetic muscular contraction at different contractile velocities. *Physical Therapy* 60: 1152-1158.

Barron, S.A., J. Mazliah, and E. Bental. 1987. The minimum F-response latency: results from 10,000 stimuli of normal ulnar nerves. *Electromyography and Clinical Neurophysiology* 27: 499-501.

Basgoze, O., K.Y. Gokce, and S. Narman. 1986. Effects of ice on the amplitude of M wave in distal latency. *Electromyography and Clinical Neurophysiology* 26: 729-734.

Basmajian, J.V., H.C. Clifford, W.D. McLeod, and H.N. Nunnally. 1975. *Computers in electromyography.* Boston: Butterworths.

Basmajian, J.V., W.J. Forrest, and G. Shine. 1966. A simple connector for fine-wire EMG electrodes. *Journal of Applied Physiology* 21: 1680.

Basmajian, J.V., and G. Stecko. 1963. A new bipolar electrode for electromyography. *Journal of Applied Physiology* 17: 849.

Baum, B.S., and L. Li. 2003. Lower extremity muscle activities during cycling are influenced by load and frequency. *Journal of Electromyography and Kinesiology* 13: 181-190.

Bazzy, A.R., J.B. Korten, and G.G. Haddad. 1986. Increase in electromyogram low-frequency power in nonfatigued contracting skeletal muscle. *Journal of Applied Physiology* 61: 1012-1017.

Beck, T.W., T.J. Housh, G.O. Johnson, J.P. Weir, J.T. Cramer, J.W. Coburn, and M.H. Malek. 2005. The effects of interelectrode distance on electromyographic amplitude and mean power frequency during isokinetic and isometric muscle actions of the biceps brachii. *Journal of Electromyography and Kinesiology* 15: 482-495.

Bell, D. 1993. The influence of air temperature on the EMG/force relationship of the quadriceps. *European Journal of Applied Physiology* 67: 256-260.

Bellemare, F., and N. Garzaniti. 1988. Failure of neuromuscular propagation during human maximal voluntary contraction. *Journal of Applied Physiology* 64: 1084-1093.

Bendat, J.S., and A.G. Piersol. 1971. *Random data: analysis and measurement procedures.* New York: Wiley.

Benecke, R. 1996. Magnetic stimulation in the assessment of peripheral nerve disorders. *Bailliere's Clinical Neurology* 5: 115-128.

Benedetti, M.G., F. Catani, T.W. Bilotta, M. Marcacci, E. Mariani, and S. Giannini. 2003. Muscle activation pattern and gait biomechanics after total knee replacement. *Clinical Biomechanics (Bristol, Avon)* 18: 871-876.

Bennell, K., M. Duncan, and S. Cowan. 2006. Effect of patellar taping on vasti onset timing, knee kinematics, and kinetics in asymptomatic individuals with a delayed onset of vastus medialis oblique. *Journal of Orthopaedic Research* 24: 1854-1860.

Benoit, D.L., M. Lamontagne, G. Cerulli, and A. Liti. 2003. The clinical significance of electromyography normalisation techniques in subjects with anterior cruciate ligament injury during treadmill walking. *Gait & Posture* 18: 56-63.

Berardelli A., M. Hallett, J.C. Rothwell, R. Agostino, M. Manfredi, P.D. Thompson, and C.D. Marsden. 1996. Single-joint rapid arm movements in normal subjects and in patients with motor disorders. *Brain* 119 (Pt 2): 661-674.

Bigland, B., and O.C.J. Lippold. 1954. The relation between force, velocity and human integrated electrical activity in human muscles. *Journal of Physiology* 123: 214-224.

Bigland-Ritchie, B. 1979. Factors contributing to quantitative surface electromyographic recording and how they are affected by fatigue. *American Review of Respiratory Disease* 119: 95-97.

Bigland-Ritchie, B., R. Johansson, O.C.J. Lippold, S. Smith, and J.J. Woods. 1983. Changes in motoneurone firing rates during sustained maximal voluntary contractions. *Journal of Physiology* 340: 335-346.

Bigland-Ritchie, B., C.G. Kukulka, O.C. Lippold, and J.J. Woods. 1982. The absence of neuromuscular transmission failure in sustained maximal voluntary contractions. *Journal of Physiology* 330: 265-278.

Bilodeau, M., A.B. Arsenault, D. Gravel, and D. Bourbonnais. 1992. Influence of gender on the EMG power spectrum during an increasing force level. *Journal of Electromyography and Kinesiology* 2: 121-129.

Bilodeau, M., M. Cincera, A.B. Arsenault, and D. Gravel. 1997. Normality and stationarity of EMG signals during ramp and step isometric contraction. *Journal of Electromyography and Kinesiology* 7: 87-96.

Bilodeau, M., S. Schindler-Ivens, D.M. Williams, R. Chandran, and S.S. Sharma. 2003. EMG frequency content changes with increasing force and during fatigue in the quadriceps femoris muscle of men and women. *Journal of Electromyography and Kinesiology* 13: 83-92.

Biro, A., L. Griffin, and E. Cafarelli. 2006. Reflex gain of muscle spindle pathways during fatigue. *Experimental Brain Research* 177: 157-166.

Blanksma, N.G., and T.M. van Eijden. 1990. Electromyographic heterogeneity in the human temporalis muscle. *Journal of Dental Research* 69: 1686-1690.

Blijham, P.J., G.J. Hengstman, H.J. Ter Laak, B.G. van Engelen, and M.J. Zwarts. 2004. Muscle-fiber conduction velocity and electromyography as diagnostic tools in patients with suspected inflammatory myopathy: a prospective study. *Muscle & Nerve* 29: 46-50.

Blom, S., K.E. Hagbarth, and S. Skoglund. 1964. Post-tetanic potentiation of H-reflexes in human infants. *Experimental Neurology* 89: 198-211.

Bodine-Fowler, S., A. Garfinkel, R.R. Roy, and V.R. Edgerton. 1990. Spatial distribution of muscle fibers within the territory of a motor unit. *Muscle & Nerve* 13: 1133-1145.

Bogey, R., K. Cerny, and O. Mohammed. 2003. Repeatability of wire and surface electrodes in gait. *American Journal of Physical Medicine* 82: 338-344.

Bogey, R.A., J. Perry, E.L. Bontrager, and J.K. Gronley. 2000. Comparison of across-subject EMG profiles using surface and multiple indwelling wire electrodes during gait. *Journal of Electromyography and Kinesiology* 10: 255-259.

Bonato, P., M.S. Cheng, J. Gonzalez-Cueto, A. Leardini, J. O'Connor, and S.H. Roy. 2001. EMG-based measures of fatigue during a repetitive squat exercise. *IEEE Engineering and Medicine in Biology Magazine* 20: 133-143.

Bouisset, S. 1973. EMG and muscle force in normal motor activities. In *New developments in electromyography and clinical neurophysiology,* ed. Desmedt, J.E. (1: 547-583). Basel: Karger.

Bouisset, S., and F. Goubel. 1973. Integrated electromyographical activity and muscle work. *Journal of Applied Physiology* 35: 695-702.

Bouisset, S., and B. Maton. 1972. Quantitative relationship between surface EMG and intramuscular electromyographic activity in voluntary movement. *American Journal of Physical Medicine* 51: 285-295.

Bower, J.S., T.G. Sandercock, E. Rothman, P.H. Abbrecht, and D.R. Dantzker. 1984. Time domain analysis of diaphragmatic electromyogram during fatigue in men. *Journal of Applied Physiology* 57: 913-916.

Boyd, D.C., P.D. Lawrence, and P.J.A. Bratty. 1978. On modeling the single motor unit action potential. *IEEE Transactions on Biomedical Engineering* 25: 236-243.

Braddom, R.L., and E.W. Johnson. 1974. H reflex: review and classification with suggested clinical uses. *Archives of Physical Medicine and Rehabilitation* 55: 412-417.

Broman, H., G. Bilotto, and C.J. De Luca. 1985a. A note on the noninvasive estimation of muscle fiber conduction velocity. *IEEE Transactions on Biomedical Engineering* 32: 341-344.

Broman, H., G. Bilotto, and C.J. De Luca. 1985b. Myoelectric signal conduction velocity and spectral parameters: influence of force and time. *Journal of Applied Physiology* 58: 1428-1437.

Bronks, R., and J.M. Brown. 1987. IEMG/force relationships in rapidly contracting human hand muscles. *Electromyography and Clinical Neurophysiology* 27: 509-515.

Brooke, J.D., W.E. McIlroy, M. Miklic, W.R. Staines, J.E. Misiaszek, G. Peritore, and P. Angerilli. 1997. Modulation of H reflexes in human tibialis anterior muscle with passive movement. *Brain Research* 766: 236-239.

Brooke, J.D., G. Peritore, W.R. Staines, W.E. McIlroy, and A. Nelson. 2000. Upper limb H reflexes and somatosensory evoked potentials modulated by movement. *Journal of Electromyography and Kinesiology* 10: 211-215.

Brown, S.H., and J.D. Cooke. 1981. Amplitude- and instruction-dependent modulation of movement-related electromyogram activity in humans. *Journal of Physiology* 316: 97-107.

Brown, W.F. 1984. *The physiological and technical basis of electromyography.* Boston: Butterworths.

Buchthal, F., C. Guld, and P. Rosenfalck. 1957. Volume conduction of the spike of the motor unit potential investigated with a new type of multieletrode. *Acta Physiologica Scandinavica* 38: 331-354.

Buchthal, F., P. Pinelli, and P. Rosenfalck. 1954. Action potential parameters in normal human muscle and their physiological determinants. *Acta Physiologica Scandinavica* 32: 219-229.

Buchthal, F., and P. Rosenfalck. 1973. On the structure of motor units. In *New developments in electromyography and clinical neurophysiology,* ed. Desmedt, J.E. Basel: Karger.

Bulgheroni, P., M.V. Bulgheroni, L. Andrini, P. Guffanti, and A. Giughello. 1997. Gait patterns after anterior cruciate ligament reconstruction. *Knee Surgery, Sports Traumatology, Arthroscopy* 5: 14-21.

Burden, A.M., M. Trew, and V. Baltzopoulos. 2003. Normalisation of gait EMGs: a re-examination. *Journal of Electromyography and Kinesiology* 13: 519-532.

Burke, D., R.W. Adams, and N.F. Skuse. 1989. The effects of voluntary contraction on the H reflex of human limb muscles. *Brain* 112: 417-433.

Burke, R.E., P. Rudomin, and F.E. Zajac III. 1970. Catch property in single mammalian motor units. *Science* 168: 122-124.

Cahan, L.D., J.M. Adams, J. Perry, and L.M. Beeler. 1990. Instrumented gait analysis after selective dorsal rhizotomy. *Developmental Medicine and Child Neurology* 32: 1037-1043.

Calder, K., L.A. Hall, S.M. Lester, G.I. Inglis, and D.A. Gabriel. 2005. Reliability of the biceps brachii M-wave. *Journal of NeuroEngineering and Rehabilitation* 2:33. http://www.jneuroengrehab.com/content/2/1/33.

Callaghan, M.J., C.J. McCarthy, and J.A. Oldham. 2001. Electromyographic fatigue characteristics of the quadriceps in patellofemoral pain syndrome. *Manual Therapy* 6: 27-33.

Campanini, I., A. Merlo, P. Degola, R. Merletti, G. Vezzosi, and D. Farina. 2006. Effect of electrode location on EMG signal envelope in leg muscles during gait. *Journal of Electromyography and Kinesiology* 17(4): 515-526.

Carp, J.S., and J.R. Wolpaw. 1995. Motoneuron properties after operantly conditioned increase in primate H-reflex. *Journal of Neurophysiology* 73: 1365-1373.

Cavallari, P., E. Fournier, R. Katz, K. Malmgren, E. Pierrot-Deseilligny, and M. Shindo. 1985. Cutaneous facilitation of transmission in Ib reflex pathways in the human upper limb. *Experimental Brain Research* 60: 197-199.

Chaffin, D.B., M. Lee, and A. Freivalds. 1980. Muscle strength assessment from EMG analysis. *Medicine and Science in Sports and Exercise* 12: 205-211.

Chang, W.N., J.S. Lipton, A.I. Tsirikos, and F. Miller. 2007. Kinesiological surface electromyography in normal children: range of normal activity and pattern analysis. *Journal of Electromyography and Kinesiology* 17: 437-445.

Chau, T. 2001. A review of analytical techniques for gait data. Part 2: Neural network and wavelet methods. *Gait & Posture* 13: 102-120.

Chen, R., B. Corwell, and M. Hallett. 1999. Modulation of motor cortex excitability by median nerve and digit stimulation. *Experimental Brain Research* 129: 77-86.

Christie, A.D., J.G. Inglis, J.P. Boucher, and D.A. Gabriel. 2005. Reliability of the FCR H-reflex. *Journal of Clinical Neurophysiology* 22: 204-209.

Christie, A., and G. Kamen. 2006. Doublet discharges in motoneurons of young and older adults. *Journal of Neurophysiology* 95: 2787-2795.

Chroni, E., N. Taub, and C.P. Panayiotopoulos. 1996. The importance of sample size for the estimation of F wave latency parameters in the peroneal nerve. *Electroencephalography and Clinical Neurophysiology* 101: 375-378.

Clancy, E.A., D. Farina, and R. Merletti. 2005. Cross-comparison of time- and frequency-domain methods for monitoring the myoelectric signal during a cyclic, force-varying, fatiguing hand-grip task. *Journal of Electromyography and Kinesiology* 15: 256-265.

Clancy, E.A., and N. Hogan. 1999. Probability density of the surface electromyogram and its relation to amplitude detectors. *IEEE Transactions on Biomedical Engineering* 46: 730-739.

Clancy, E.A., E.L. Morin, and R. Merletti. 2002. Sampling, noise-reduction and amplitude estimation issues in surface electromyography. *Journal of Electromyography and Kinesiology* 12: 1-16.

Cooper, R. 1963. Electrodes. *American Journal of EEG Technology* 3: 91-101.

Cowan, S.M., K.L. Bennell, P.W. Hodges, K.M. Crossley, and J. McConnell. 2001. Delayed onset of electromyographic activity of vastus medialis obliquus relative to vastus lateralis in subjects with patellofemoral pain syndrome. *Archives of Physical Medicine and Rehabilitation* 82: 183-189.

Cracco, R.Q., J.B. Cracco, P.J. Maccabee, and V.E. Amassian. 1999. Cerebral function revealed by transcranial magnetic stimulation. *Journal of Neuroscience Methods* 86: 209-219.

Craik, R.L., and C.A. Oatis. 1995. *Gait analysis: theory and application.* St. Louis: Mosby.

Cram, J.R., G.S. Kasman, and J. Holtz. 1998. *Introduction to surface electromyography.* Gaithersburg, MD: Aspen.

Crayton, J.W., and S. King. 1981. Inter-individual variability of the H-reflex in normal subjects. *Electromyography and Clinical Neurophysiology* 21: 183-200.

Crone, C., L.L. Johnsen, H. Hultborn, and G.B. Orsnes. 1999. Amplitude of the maximum motor response (Mmax) in human muscles typically decreases during the course of an experiment. *Experimental Brain Research* 124: 265-270.

Crone, C., and J. Nielsen. 1989. Methodological implications of the post activation depression of the soleus H-reflex in man. *Experimental Brain Research* 78: 28-32.

Cruz Martinez, A., and J.M. López Terradas. 1992. Motor unit remodelling in Duchenne muscular dystrophy. Electrophysiological assessment. *Electromyography and Clinical Neurophysiology* 32: 351-358.

Cupido, C.M., V. Galea, and A.J. McComas. 1996. Potentiation and depression of the M wave in human biceps brachii. *Journal of Physiology* 491 (Pt 2): 541-550.

Currier, D.P. 1972. Maximal isometric tension of the elbow extensors at varied positions. 2. Assessment of extensor components by quantitative electromyography. *Physical Therapy* 52: 1265-1276.

Darling, W.G., J.D. Cooke, and S.H. Brown. 1989. Control of simple arm movements in elderly humans. *Neurobiology of Ageing* 10: 149-157.

Daube, J.R. 1991. AAEM minimonograph #11: needle examination in clinical electromyography. *Muscle & Nerve* 14: 685-700.

Day, B.L., C.D. Marsden, J.A. Obeso, and J.C. Rothwell. 1984. Reciprocal inhibition between the muscles of the human forearm. *Journal of Physiology* 349: 519-534.

De la Barrera, E.J., and T.E. Milner. 1994. The effects of skinfold thickness on the selectivity of surface EMG. *Electroencephalography and Clinical Neurophysiology* 93: 91-99.

De Luca, C.J. 1979. Physiology and mathematics of myoelectric signals. *IEEE Transactions on Biomedical Engineering* 26: 313-325.

De Luca, C.J., and R. Merletti. 1988. Surface myoelectric signal cross-talk among muscles of the leg. *Electroencephalography and Clinical Neurophysiology* 69: 568-575.

De Luca, C.J., and E.J. Van Dyk. 1975. Derivation of some parameters of myoelectric signals recorded during sustained constant force isometric contractions. *Biophysical Journal* 15: 1167-1180.

Delcomyn, F., and J.H. Cocatre-Zilgien. 1992. Computer method for identifying bursts in trains of spikes. In *Methods in Neurosciences* Vol. 10, *Computer and computations in the neurosciences,* ed. Conn, M. (pp. 228-240). New York: Academic Press.

deVries, H.A. 1968. EMG fatigue curves in postural muscles. A possible etiology for idiopathic low back pain. *American Journal of Physical Medicine* 47: 175-181.

deVries, H.A., R.K. Burke, R.T. Hopper, and J.H. Sloan. 1976. Relationship of resting EMG level to total body metabolism with reference to the origin of "tissue noise." *American Journal of Physical Medicine* 55: 139-147.

deVries, H.A., R.A. Wiswell, R. Bulbulian, and T. Moritani. 1981. Tranquilizer effect of exercise. Acute effects of moderate aerobic exercise on spinal reflex activation level. *American Journal of Physical Medicine* 60: 57-66.

DiFabio, R.P. 1987. Reliability of computerized surface electromyography for determining the onset of muscle activity. *Physical Therapy* 67: 43-48.

Dimitrov, G.V., and N.A. Dimitrova. 1998. Fundamentals of power spectra of extracellular potentials produced by skeletal muscle fibre of finite length. Part I: Effect of fiber anatomy. *Medical Engineering and Physics* 20: 580-587.

Dimitrov, G.V., C. Disselhorst-Klug, N.A. Dimitrova, E. Schulte, and G. Rau. 2003. Simulation analysis of the ability of different types of multi-electrodes to increase selectivity of detection and to reduce cross-talk. *Journal of Electromyography and Kinesiology* 13: 125-138.

Dimitrova, N.A., and G.V. Dimitrov. 2003. Interpretation of EMG changes with fatigue: facts, pitfalls, and fallacies. *Journal of Electromyography and Kinesiology* 13: 13-36.

Dimitrova, N.A., G.V. Dimitrov, and Z.C. Lateva. 1991. Influence of the fiber length on the power spectra of single muscle fiber extracellular potentials. *Electromyography and Clinical Neurophysiology* 31: 387-398.

Dimitrova, N.A., G.V. Dimitrov, and O.A. Nikitin. 2001. Longitudinal variations of characteristic frequencies of skeletal muscle fiber potentials detected by a bipolar electrode or multi-electrode. *Journal of Medical Engineering and Technology* 25: 34-40.

Dimitrova, N.A., G.V. Dimitrov, and O.A. Nikitin. 2002. Neither high-pass filtering nor mathematical differentiation of the EMG signals can considerably reduce cross-talk. *Journal of Electromyography and Kinesiology* 12: 235-246.

Doud, J.R., and J.M. Walsh. 1995. Muscle fatigue and muscle length interaction: effect on the EMG frequency components. *Electromyography and Clinical Neurophysiology* 35: 331-339.

Drake, J.D.M., and J.P. Callaghan. 2006. Elimination of electrocardiogram from electromyogram signals: an evaluation of currently used removal techniques. *Journal of Electromyography and Kinesiology* 16: 175-187.

Dubo, H.I., M. Peat, D.A. Winter, A.O. Quanbury, D.A. Hobson, T. Steinke, and G. Reimer. 1976. Electromyographic temporal analysis of gait: normal human locomotion. *Archives of Physical Medicine and Rehabilitation* 57: 415-420.

Dubowitz, V., and M.H. Brooke. 1973. *Muscle biopsy: a modern approach.* Philadelphia: Saunders.

Duchateau, J., S. Le Bozec, and K. Hainaut. 1986. Contributions of slow and fast muscles of triceps surae to a cyclic movement. *European Journal of Applied Physiology* 55: 476-481.

Duclay, J., and A. Martin. 2005. Evoked H-reflex and V-wave responses during maximal isometric, concentric, and eccentric muscle contraction. *Journal of Neurophysiology* 94: 3555-3562.

Dumitru, D. 2000. Physiologic basis of potentials recorded in electromyography. *Muscle & Nerve* 23: 1667-1685.

Dumitru, D., and J.C. King. 1992. Far-field potentials in circular volumes: evidence to support the leading/trailing dipole model. *Muscle & Nerve* 15: 101-105.

Dwyer, D., J. Browning, and S. Weinstein. 1999. The reliability of muscle biopsies taken from vastus lateralis. *Journal of Science and Medicine in Sport* 2: 333-340.

Ebenbichler, G.R., P. Bonato, S.H. Roy, S. Lehr, M. Posch, J. Kollmitzer, and C.U. Della. 2002. Reliability of EMG time-frequency measures of fatigue during repetitive lifting. *Medicine and Science in Sports and Exercise* 34: 1316-1323.

Ebenbichler, G., J. Kollmitzer, M. Quittan, F. Uhl, C. Kirtley, and V. Fialka. 1998. EMG fatigue patterns accompanying isometric fatiguing knee-extensions are different in mono- and bi-articular muscles. *Electroencephalography and Clinical Neurophysiology* 109: 256-262.

Eberstein, A., and B. Beattie. 1985. Simultaneous measurement of muscle conduction velocity and EMG power spectrum changes during fatigue. *Muscle & Nerve* 8: 768-773.

Edstrom, L., and E. Kugelberg. 1968. Histochemical composition, distribution of fibres and fatiguability of single motor units. Anterior tibial muscle of the rat. *Journal of Neurology, Neurosurgery and Psychiatry* 31: 424-433.

Edwards, R.G., and O.J. Lippold. 1956. The relation between force and integrated electrical activity in fatigued muscle. *Journal of Physiology* 132: 677-681.

Eisen, A., and K. Odusote. 1979. Amplitude of the F wave: a potential means of documenting spasticity. *Neurology* 29: 1306-1309.

Eke-Okoro, S.T. 1982. The H-reflex studied in the presence of alcohol, aspirin, caffeine, force and fatigue. *Electromyography and Clinical Neurophysiology* 22: 579-589.

Elfving, B., D. Liljequist, E. Mattsson, and G. Németh. 2002. Influence of interelectrode distance and force level on the spectral parameters of surface electromyographic recordings from the lumbar muscles. *Journal of Electromyography and Kinesiology* 12: 295-304.

Elfving, B., G. Nemeth, I. Arvidsson, and M. Lamontagne. 1999. Reliability of EMG spectral parameters in repeated measurements of back muscle fatigue. *Journal of Electromyography and Kinesiology* 9: 235-243.

Ellrich, J., H. Steffens, R.D. Treede, and E.D. Schomburg. 1998. The Hoffmann reflex of human plantar foot muscles. *Muscle & Nerve* 21: 732-738.

Ellrich, J., and R.D. Treede. 1998. Convergence of nociceptive and non-nociceptive inputs onto spinal reflex pathways to the tibialis anterior muscle in humans. *Acta Physiologica Scandinavica* 163: 391-401.

English, A.W., and O.I. Weeks. 1989. Electromyographic cross-talk within a compartmentalized muscle of the cat. *Journal of Physiology* 416: 327-336.

English, A.W., S.L. Wolf, and R.L. Segal. 1993. Compartmentalization of muscles and their motor nuclei: the partitioning hypothesis. *Physical Therapy* 73: 857-867.

Espiritu, M.G., C.S. Lin, and D. Burke. 2003. Motoneuron excitability and the F wave. *Muscle & Nerve* 27: 720-727.

Etnyre, B.R., and L.D. Abraham. 1986. H-reflex changes during static stretching and two variations of proprioceptive neuromuscular facilitation techniques. *Electroencephalography and Clinical Neurophysiology* 63: 174-179.

Farina, D., F. Leclerc, L. Arendt-Nielsen, O. Buttelli, and P. Madeleine. 2006. The change in spatial distribution of upper trapezius muscle activity is correlated to contraction duration. *Journal of Electromyography and Kinesiology* 18: 16-25.

Farina, D., and R. Merletti. 2000. Comparison of algorithms for estimation of EMG variables during voluntary isometric contractions. *Journal of Electromyography and Kinesiology* 10: 337-349.

Farina, D., and R. Merletti. 2004. Methods for estimating muscle fibre conduction velocity from surface electromyographic signals. *Medical and Biological Engineering and Computing* 42: 432-445.

Farina, D., R. Merletti, B. Indino, M. Nazzaro, and M. Pozzo. 2002. Surface EMG crosstalk between knee extensor muscles: experimental and model results. *Muscle & Nerve* 26: 681-695.

Feinstein, B., B. Lindegard, E. Nyman, and G. Wohlfart. 1955. Morphologic studies of motor units in normal human muscles. *Acta Anatomica* 23: 127-142.

Fiorito, A., S. Rao, and R. Merletti. 1994. Analogue and digital instruments for non-invasive estimation of muscle fiber conduction velocity. *Medical and Biological Engineering and Computing* 32: 521-529.

Fisher, M.A. 1982. F response latency determination. *Muscle & Nerve* 5: 730-734.

Fisher, M.A. 1992. AAEM minimonograph #13. H reflexes and F waves: physiology and clinical indications. *Muscle & Nerve* 15: 1223-1233.

Floeter, M.K., and A.F. Kohn. 1997. H-reflexes of different sizes exhibit differential sensitivity to low frequency depression. *Electroencephalography and Clinical Neurophysiology* 105: 470-475.

Forsman, M., L. Birch, Q. Zhang, and R. Kadefors. 2001. Motor unit recruitment in the trapezius muscle with special reference to coarse arm movements. *Journal of Electromyography and Kinesiology* 11: 207-216.

Freund, H.J., H.J. Budingen, and V. Dietz. 1975. Activity of single motor units from human forearm muscles during voluntary isometric contractions. *Journal of Neurophysiology* 38: 933-946.

Frigon, A., D.F. Collins, and E.P. Zehr. 2004. Effect of rhythmic arm movement on reflexes in the legs: modulation of soleus H-reflexes and somatosensory conditioning. *Journal of Neurophysiology* 91: 1516-1523.

Fuglevand, A.J., D.A. Winter, A.E. Patla, and D. Stashuk. 1992. Detection of motor unit action potentials with surface electrodes: influence of electrode size and spacing. *Biological Cybernetics* 67: 143-153.

Fuglevand, A.J., K.M. Zackowski, K.A. Huey, and R.M. Enoka. 1993. Impairment of neuromuscular propagation during human fatiguing contractions at submaximal forces. *Journal of Physiology* 460: 549-572.

Fuglsang-Frederiksen, A., and A. Mansson. 1975. Analysis of electrical activity of normal muscle in man at different degrees of voluntary effort. *Journal of Neurology, Neurosurgery and Psychiatry* 38: 683-694.

Funk, D.A., K-N. An, B.F. Morrey, and J.R. Daube. 1987. Electromyographic analysis of muscles across the elbow joint. *Journal of Orthopaedic Research* 5: 529-538.

Gabriel, D.A. 2000. Reliability of SEMG spike parameters during concentric contractions. *Electromyography and Clinical Neurophysiology* 40: 423-430.

Gabriel, D.A. 2002. Changes in kinematic and EMG variability while practicing a maximal performance task. *Journal of Electromyography and Kinesiology* 12: 407-412.

Gabriel, D.A., J.R. Basford, and K.N. An. 2001. Assessing fatigue with electromyographic spike parameters. *IEEE Engineering and Medicine in Biology Magazine* 20: 90-96.

Gabriel, D., and J. Boucher. 1998. Practice effects on the timing and magnitude of antagonist activity during ballistic elbow flexion to a target. *Research Quarterly for Exercise and Sport* 69: 30-37.

Gabriel, D.A., S.M. Lester, S.A. Lenhardt, and E.D.J. Cambridge. 2007. Analysis of surface EMG spike shape across different levels of isometric force. *Journal of Neuroscience Methods* 159: 142-152.

Gabriel, D.A., J.Y. Matsumoto, D.H. Davis, B.L. Currier, and K-N. An. 2004. Multidirectional neck strength and electromyographic activity for normal controls. *Clinical Biomechanics* 19: 653-658.

Gandevia, S.C. 2001. Spinal and supraspinal factors in human muscle fatigue. *Physiological Reviews* 81: 1725-1789.

Gans, C. and F. deVree. 1987. Functional bases of fiber length and angulation in muscle. *Journal of Morphology* 192: 63-85.

Gantchev, N., A. Kossev, A. Gydikov, and Y. Gerasimenko. 1992. Relation between the motor units recruitment threshold and their potentials propagation velocity at isometric activity. *Electromyography and Clinical Neurophysiology* 32: 221-228.

Garland, S.J., and L. Griffin. 1999. Motor unit double discharges: statistical anomaly or functional entity? *Canadian Journal of Applied Physiology* 24: 113-130.

Gates, H.J., and W.J. Betz. 1993. Spatial distribution of muscle fibers in a lumbrical muscle of the rat. *Anatomical Record* 236: 381-389.

Gath, I., and E. Stålberg. 1977. On the volume conduction in human skeletal muscle: in situ measurements. *Electroencephalography and Clinical Neurophysiology* 43: 106-110.

Gath, I., and E. Stålberg. 1982. On the measurement of fibre density in human muscles. *Electroencephalography and Clinical Neurophysiology* 54: 699-706.

Geddes, L.A., and L.E. Baker. 1968. *Principles of applied biomedical instrumentation.* New York: Wiley.

Geddes, L.A., L.E. Baker, and M. McGoodwin. 1967. The relationship between electrode area and amplifier input impedance in recording muscle action potentials. *Medical and Biological Engineering* 5: 561-569.

Gerdle, B., and A.R. Fugl-Meyer. 1992. Is the mean power frequency shift of the EMG a selective indicator of fatigue of the fast twitch motor units? *Acta Physiologica Scandinavica* 145: 129-138.

Gerdle, B., M.L. Wretling, and K. Henriksson-Larsen. 1988. Do the fibre-type proportion and the angular velocity influence the mean power frequency of the electromyogram? *Acta Physiologica Scandinavica* 134: 341-346.

Gerilovsky, L., D. Karadimov, and B. Ianakiev. 1991. Hypoxia reduces the conduction velocity of the excitation along the striated muscles in man. *Electromyography and Clinical Neurophysiology* 31: 203-208.

Gielen, F.L.H., W. Wallinga de Jonge, and K.L. Boon. 1984. Electrical conductivity of skeletal muscle tissue: experimental results from different muscles in vivo. *Medical and Biological Engineering and Computing* 22: 569-577.

Gill, N.W. III, T.M. Ruediger, R.D. Gochis, W.C. Werling, J.H. Moore, S.C. Allison, S. Shaffer, and F.B. Underwood. 1999. Test-retest reliability of the ulnar F-wave minimum latency in normal adults. *Electromyography and Clinical Neurophysiology* 39: 195-200.

Giroux, B., and M. Lamontagne. 1990. Comparisons between surface electrodes and intramuscular wire electrodes in isometric and dynamic conditions. *Electromyography and Clinical Neurophysiology* 30: 397-405.

Glass, G.V., and K.D. Hopkins. 1996. *Statistical methods in psychology and education.* 3rd ed. Boston: Allyn and Bacon.

Godaux, E., and J.E. Desmedt. 1975. Human masseter muscle: H- and tendon reflexes. Their paradoxical potentiation by muscle vibration. *Archives of Neurology* 32: 229-238.

Gondran, C., E. Siebert, S. Yacoub, and E. Novakov. 1996. Noise of surface biopotential electrodes based on NASICON ceramic and Ag-AgCl. *Medical and Biological Engineering and Computing* 34: 460-466.

Gottlieb, G.L., D.M. Corcos, and G.C. Agarwal. 1989. Organizing principles for single-joint movements I. A speed-insensitive strategy. *Journal of Neurophysiology* 62: 342-357.

Granata, K.P., D.A. Padua, and M.F. Abel. 2005. Repeatability of surface EMG during gait in children. *Gait & Posture* 22: 346-350.

Gregor, R.J., and T.A. Abelew. 1994. Tendon force measurements and movement control: a review. *Medicine and Science in Sports and Exercise* 26: 1359-1372.

Gregor, R.J., P.V. Komi, and M. Jarvinen. 1987. Achilles tendon forces during cycling. *International Journal of Sports Medicine* 8: 9-14.

Gruener, R., L.Z. Stern, and R.R. Weisz. 1979. Conduction velocities in single fibers of diseased human muscle. *Neurology* 29: 1293-1297.

Guissard, N., J. Duchateau, and K. Hainaut. 2001. Mechanisms of decreased motoneurone excitation during passive muscle stretching. *Experimental Brain Research* 137: 163-169.

Hagg, G.M. 1992. Interpretation of EMG spectral alterations and alteration indexes at sustained contraction. *Journal of Applied Physiology* 73: 1211-1217.

Håkansson, C. 1956. Conduction velocity and amplitude of the action potential as related to circumference in the isolated fibre of frog muscle. *Acta Physiologica Scandinavica* 39: 291-312.

Hallett, M. 1996. Transcranial magnetic stimulation: a tool for mapping the central nervous system. *Electroencephalography and Clinical Neurophysiology Supplement* 46: 43-51.

Hammelsbeck, M., and W. Rathmayer. 1989. Intracellular Na+, K+ and Cl– activity in tonic and phasic muscle fibers of the crab Eriphia. *Pflugers Archiv* 413: 487-492.

Hannaford, B., and S. Lehman. 1986. Short time Fourier analysis of the electromyogram: fast movements and constant contraction. *IEEE Transactions on Biomedical Engineering* 12: 1173-1181.

Hayashi, K., R.G. Miller, and K.W. Brownell. 1987. Three-dimensional architecture of sarcoplasmic reticulum and T-system in human skeletal muscle. *Anatomical Record* 218: 275-283.

Hayes, K.C., and J. Sullivan. 1976. Tonic neck reflex influence on tendon and Hoffmann reflexes in man. *Electromyography and Clinical Neurophysiology* 16: 251-261.

He, W., M.Z. Wang, and Z.M. Wang. 2005. Effect of change of plasma K+ and pH value induced by exercise on muscle fatigue and surface EMG. *Sichuan Da Xue Bao Yi Xue Ban* 36: 112-114, 118.

Henneman, E., G. Somjen, and D.O. Carpenter. 1965. Excitability and inhibitability of motoneurons of different sizes. *Journal of Neurophysiology* 28: 599-620.

Hermens, H.J., B. Freriks, C. Disselhorst-Klug, and G. Rau. 2000. Development of recommendations for SEMG sensors and sensor placement procedures. *Journal of Electromyography and Kinesiology* 10: 361-374.

Heron, M.I., and F.J. Richmond. 1993. In-series fiber architecture in long human muscles. *Journal of Morphology* 216: 35-45.

Herschler, C., and M. Milner. 1978. An optimality criterion for processing electromyographic (EMG) signals relating to human locomotion. *IEEE Transactions on Biomedical Engineering* 25: 413-420.

Hicks, A., J. Fenton, S. Garner, and A.J. McComas. 1989. M wave potentiation during and after muscle activity. *Journal of Applied Physiology* 66: 2606-2610.

Hines, A.E., P.E. Crago, G.J. Chapman, and C. Billian. 1996. Stimulus artifact removal in EMG from muscles adjacent to stimulated muscles. *Journal of Neuroscience Methods* 64: 55-62.

Hodges, P.W., and B.H. Bui. 1996. A comparison of computer-based methods for the determination of onset of muscle contraction using electromyography. *Electroencephalography and Clinical Neurophysiology* 101: 511-519.

Hof, A.L. 1991. Errors in frequency parameters of EMG power spectra. *IEEE Transactions on Biomedical Engineering* 38: 1077-1088.

Hof, A.L., H. Elzinga, W. Grimmius, and J.P. Halbertsma. 2002. Speed dependence of averaged EMG profiles in walking. *Gait & Posture* 16: 78-86.

Hoffmann, P. 1918. Uber die Beziehungen der Schenreflexe zur willkurlichen Bewegun zum Tonus. *Zeitschrift fur Biologie* 68: 351-370.

Holewijn, M., and R. Heus. 1992. Effects of temperature on electromyogram and muscle function. *European Journal of Applied Physiology* 65: 541-545.

Holtermann, A., and K. Roeleveld. 2006. EMG amplitude distribution changes over the upper trapezius muscle are similar in sustained and ramp contractions. *Acta Physiologica (Oxford)* 186: 159-168.

Holtermann, A., K. Roeleveld, and J.S. Karlsson. 2005. Inhomogeneities in muscle activation reveal motor unit recruitment. *Journal of Electromyography and Kinesiology* 15: 131-137.

Homma, S., and M. Kano. 1962. Electrical properties of the tonic reflex arc in the human proprioceptive reflex. In *A symposium on muscle receptors,* ed. Barker, D. Hong Kong: Hong Kong University Press.

Hong, C.Z., and W.T. Liberson. 1987. Propagation of compound muscle action potentials measured with small surface recording electrodes. *Electromyography and Clinical Neurophysiology* 27: 415-417.

Hopf, H.C., R.L. Herbort, M. Gnass, H. Gunther, and K. Lowitzsch. 1974. Fast and slow contraction times associated with fast and slow spike conduction of skeletal muscle fibres in normal subjects and in spastic hemiparesis. *Zeitschrift fur Neurologie* 206: 193-202.

Hopkins, J.T., C.D. Ingersoll, M.L. Cordova, and J.E. Edwards. 2000. Intrasession and intersession reliability of the soleus H-reflex in supine and standing positions. *Electromyography and Clinical Neurophysiology* 40: 89-94.

Hopkins, J.T., and N.C. Wagie. 2003. Intrasession and intersession reliability of the quadriceps Hoffmann reflex. *Electromyography and Clinical Neurophysiology* 43: 85-89.

Hugon, M. 1973. Methodology of the Hoffmann reflex in man. In *New developments in electromyography and clinical neurophysiology,* ed. Desmedt, J.E. (3: 277-293). Basel: Karger.

Huigen, E., A. Peper, and C.A. Grimbergen. 2002. Investigation into the origin of the noise of surface electrodes. *Medical and Biological Engineering and Computing* 40: 332-338.

Hultborn, H., S. Meunier, C. Morin, and E. Pierrot-Deseilligny. 1987. Assessing changes in presynaptic inhibition of I a fibres: a study in man and the cat. *Journal of Physiology* 389: 729-756.

Hunter, I.W., R.E. Kearney, and L.A. Jones. 1987. Estimation of the conduction velocity of muscle action potentials using phase and impulse response function techniques. *Medical and Biological Engineering and Computing* 25: 121-126.

Ikegawa, S., M. Shinohara, T. Fukunaga, J.P. Zbilut, and C.L.J. Webber. 2000. Nonlinear time-course of lumbar muscle fatigue using recurrence quantifications. *Biological Cybernetics* 82: 373-382.

Inbar, G.F., J. Allin, and H. Kranz. 1987. Surface EMG spectral changes with muscle length. *Medical and Biological Engineering and Computing* 25: 683-689.

Inghilleri, M., C. Lorenzano, A. Conte, V. Frasca, M. Manfredi, and A. Berardelli. 2003. Effects of transcranial magnetic stimulation on the H reflex and F wave in the hand muscles. *Clinical Neurophysiology* 114: 1096-1101.

Inman, V.T., H.J. Ralston, C.M. Saunders, B. Feinstein, and E.W. Wright. 1952. Relation of human electromyogram to muscular tension. *Electroencephalography and Clinical Neurophysiology* 4: 187-194.

Ishikawa, K., K. Ott, R.W. Porter, and D. Stuart. 1966. Low frequency depression of the H wave in normal and spinal man. *Experimental Neurology* 15: 140-156.

Ivanenko, Y.P., R.E. Poppele, and F. Lacquaniti. 2004. Five basic muscle activation patterns account for muscle activity during human locomotion. *Journal of Physiology* 556: 267-282.

Ives, J.C., L. Abraham, and W. Kroll. 1999. Neuromuscular control mechanisms and strategy in arm movements of attempted supranormal speed. *Research Quarterly for Exercise and Sport* 70: 335-348.

Ives J.C., W.P. Kroll, and L.L. Bultman. 1993. Rapid movement kinematic and electromyographic control characteristics in males and females. *Research Quarterly for Exercise and Sport* 64: 274-283.

Jabre, J.F. 1981. Surface recording of the H-reflex of the flexor carpi radialis. *Muscle & Nerve* 4: 435-438.

Jacobson, W.J., R.H. Gabel, and R.A. Brand. 1995. Surface vs. fine-wire electrode ensemble-averaged signals during gait. *Journal of Electromyography and Kinesiology* 5: 37-44.

Jarcho, L.W., C. Eyzaguirre, B. Berman, and J.J. Lilenthal. 1952. Spread of excitation in skeletal muscle: some factors contributing to the form of the electromyogram. *American Journal of Physiology* 163: 446-457.

Jensen, B.R., B. Schibye, K. Sogaard, E.B. Simonsen, and G. Sjøgaard. 1993. Shoulder muscle load and muscle fatigue among industrial sewing-machine operators. *European Journal of Applied Physiology* 67: 467-475.

Johnson, S.W., P.A. Lynn, J.S.G. Miller, and G.A.L. Reed. 1977. Miniature skin-mounted preamplifier for measurement of surface electromyographic potentials. *Medical and Biological Engineering and Computing* 15: 710-711.

Jonas, D., C. Bischoff, and B. Conrad. 1999. Influence of different types of surface electrodes on amplitude, area and duration of the compound muscle action potential. *Clinical Neurophysiology* 110: 2171-2175.

Juel, C. 1988. Muscle action potential propagation velocity changes during activity. *Muscle & Nerve* 11: 714-719.

Kadaba, M.P., H.K. Ramakrishnan, M.E. Wootten, J. Gainey, G. Gorton, and G.V. Cochran. 1989. Repeatability of kinematic, kinetic, and electromyographic data in normal adult gait. *Journal of Orthopaedic Research* 7: 849-860.

Kadaba, M.P., M.E. Wootten, J. Gainey, and G.V. Cochran. 1985. Repeatability of phasic muscle activity: performance of surface and intramuscular wire electrodes in gait analysis. *Journal of Orthopaedic Research* 3: 350-359.

Kadefors, R. 1973. Myoelectric signal processing as an estimation problem. In *New developments in EMG and clinical neurophysiology,* ed. Desmedt, J.E. (vol. 1, pp. 519-539). Basel: Karger.

Kamen, G. 2004. Electromyographic kinesiology. In *Research methods in biomechanics,* ed. Robertston, D.G.E., Caldwell, D.G., Hamill, J., Kamen, G., and Whittlesey, S.N. (pp. 163-181). Champaign, IL: Human Kinetics.

Kamen, G., and A. Roy. 2000. Motor unit synchronization in young and old adults. *European Journal of Applied Physiology* 81: 403-410.

Kamen, G., S.V. Sison, C.C. Du, and C. Patten. 1995. Motor unit discharge behavior in older adults during maximal-effort contractions. *Journal of Applied Physiology* 79: 1908-1913.

Kamibayashi, L.K., and F.J. Richmond. 1998. Morphometry of human neck muscles. *Spine* 23: 1314-1323.

Kaplanis, P.A., C.S. Pattichis, L.J. Hadjileontiadis, and V.C. Roberts. 2009. Surface EMG analysis on normal subjects based on isometric voluntary contraction. *Journal of Electromyography and Kinesiology* 19:157-171.

Karlsson, J.S., B.E. Erlandson, and B. Gerdle. 1994. A personal computer-based system for real-time analysis of surface EMG signal during static and dynamic contractions. *Journal of Electromyography and Kinesiology* 4: 170-180.

Karlsson, S., and B. Gerdle. 2001. Mean frequency and signal amplitude of the surface EMG of the quadriceps muscles increase with increasing torque—a study using the continuous wavelet transform. *Journal of Electromyography and Kinesiology* 11: 131-140.

Karlsson, J.S., N. Östlund, B. Larrson, and B. Gerdle. 2003. An estimation of the influence of force decrease on mean spectral frequency shift of the EMG during repetitive maximal dynamic knee extensions. *Journal of Electromyography and Kinesiology* 13: 461-468.

Katz, B. 1948. The electrical properties of the muscle fibre membrane. *Proceedings of the Royal Society of London (Biology)* 135: 506-534.

Katz, B. 1966. *Nerve, muscle, and synapse.* New York: McGraw-Hill.

Kaufman, K.R., K-N. An, W.J. Litchy, and E.Y.S. Chao. 1991. Physiological prediction of muscle forces—II. Application to isokinetic exercise. *Neuroscience* 40: 793-804.

Kawazoe, Y., H. Kotani, T. Maetani, T. Hamada, and H. Yatani. 1981. Integrated electromyography activity and biting force during rapid isometric contraction of fatigued masseter muscle in man. *Archives of Oral Biology* 26: 795-801.

Keenan, K.G., D. Farina, K.S. Maluf, R. Merletti, and R.M. Enoka. 2005. Influence of amplitude cancellation on the simulated surface electromyogram. *Journal of Applied Physiology* 98: 120-131.

Kilbom, A., G.M. Hägg, and C. Kall. 1992. One-handed load carrying—cardiovascular, muscular and subjective indices of endurance and fatigue. *European Journal of Applied Physiology* 65: 52-58.

Kimura, J. 2001. *Electrodiagnosis in diseases of nerve and muscle: principles and practice.* New York: Oxford.

Klein, A.B., L. Snyder-Mackler, S.H. Roy, and C.J. De Luca. 1991. Comparison of spinal mobility and isometric trunk extensor forces with electromyographic spectral analysis in identifying low back pain. *Physical Therapy* 71: 445-454.

Kleinpenning, P.H., H.J.M. Gootzen, A. Van Oosterom, and D.F. Stegeman. 1990. The equivalent source description representing the extinction of an action potential at a muscle fiber ending. *Mathematical Biosciences* 101: 41-61.

Kleissen, R.F. 1990. Effects of electromyographic processing methods on computer-averaged surface electromyographic profiles for the gluteus medius muscle. *Physical Therapy* 70: 716-722.

Knaflitz, M., and P. Bonato. 1999. Time-frequency methods applied to muscle fatigue assessment during dynamic contractions. *Journal of Electromyography and Kinesiology* 9: 337-350.

Knaflitz, M., and R. Merletti. 1988. Suppression of simulation artifacts from myoelectric-evoked potential recordings. *IEEE Transactions on Biomedical Engineering* 35: 758-763.

Knaflitz, M., R. Merletti, and C.J. De Luca. 1990. Inference of motor unit recruitment order in voluntary and electrically elicited contractions. *Journal of Applied Physiology* 68: 1657-1667.

Knight, C.A., and G. Kamen. 2005. Superficial motor units are larger than deeper motor units in human vastus lateralis muscle. *Muscle & Nerve* 31: 475-480.

Knoll, Z., R.M. Kiss, and L. Kocsis. 2004. Gait adaptation in ACL deficient patients before and after anterior cruciate ligament reconstruction surgery. *Journal of Electromyography and Kinesiology* 14: 287-294.

Knowlton, G.C., T.F. Hines, K.V. Keever, and R.L. Bennett. 1956. Relation between electromyogram voltage and load. *Journal of Applied Physiology* 9: 472-476.

Koceja, D.M., J.R. Burke, and G. Kamen. 1991. Organization of segmental reflexes in trained dancers. *International Journal of Sports Medicine* 12: 285-289.

Koh, T.J., and M.D. Grabiner. 1992. Cross talk in surface electromyograms of human hamstring muscles. *Journal of Orthopaedic Research* 10: 701-709.

Kohlrausch, A. 1912. Uber das electromyogramm roter und weisser musclen. *Archiv fuer Anatomie und Physiologie, Physiologische Abteilung* 283-295.

Komi, P.V. 1973. Relationship between muscle tension, EMG and velocity of contraction under concentric and eccentric work. In *New developments in electromyography and clinical neurophysiology,* ed. Desmedt, J.E. (1: 596-606). Basel: Karger.

Komi, P.V., and E.R. Buskirk. 1970. Reproducibility of electromyographic measurements with inserted wire electrodes and surface electrodes. *Electromyography* 10: 357-367.

Korner, L., P. Parker, C. Almstrom, P. Herberts, and R. Kadefors. 1984. The relation between spectral changes of the myoelectric signal and the intramuscular pressure of human skeletal muscle. *European Journal of Physiology* 52: 202-206.

Kornfield, M.J., J. Cerra, and D.G. Simons. 1985. Stimulus artifact reduction in nerve conduction. *Archives of Physical Medicine and Rehabilitation* 66: 232-234.

Kossev, A., N. Gantchev, A. Gydikov, Y. Gerasimenko, and P. Christova. 1992. The effect of muscle fiber length change on motor units potentials propagation velocity. *Electromyography and Clinical Neurophysiology* 32: 287-294.

Kramer, M., V. Ebert, L. Kinzl, C. Dehner, M. Elbel, and E. Hartwig. 2005. Surface electromyography of the paravertebral muscles in patients with chronic low back pain. *Archives of Physical Medicine and Rehabilitation* 86: 31-36.

Krause, K.H., I. Magyarosy, H. Gall, E. Ernst, D. Pongratz, and P. Schoeps. 2001. Effects of heat and cold application on turns and amplitude in surface EMG. *Electromyography and Clinical Neurophysiology* 41: 67-70.

Krogh-Lund, C. 1993. Myo-electric fatigue and force failure from submaximal static elbow flexion sustained to exhaustion. *European Journal of Applied Physiology* 67: 389-401.

Krogh-Lund, C., and K. Jorgensen. 1991. Changes in conduction velocity, median frequency, and root mean square-amplitude of the electromyogram during 25% maximal voluntary contraction of the triceps brachii muscle, to limit endurance. *European Journal of Applied Physiology* 63: 60-69.

Krogh-Lund, C., and K. Jorgensen. 1993. Myo-electric fatigue manifestations revisited: power spectrum, conduction velocity, and amplitude of human elbow flexor muscles during isolated and repetitive endurance contractions at 30% maximal voluntary contraction. *European Journal of Applied Physiology and Occupational Physiology* 66: 161-173.

Kroon, G.W., M. Naeije, and T.L. Hansson. 1986. Electromyographic power-spectrum changes during repeated fatiguing contractions of the human masseter muscle. *Archives of Oral Biology* 9: 603-608.

Kujirai, T., M.D. Caramia, J.C. Rothwell, B.L. Day, P.D. Thompson, A. Ferbert, S. Wroe, P. Asselman, and C.D.

Marsden. 1993. Corticocortical inhibition in human motor cortex. *Journal of Physiology* 471: 501-519.

Kukulka, C.G., A.G. Russell, and M.A. Moore. 1986. Electrical and mechanical changes in human soleus muscle during sustained maximum isometric contractions. *Brain Research* 362: 47-54.

Lagerlund, T.D. 1996. Electricity and electronics in clinical neurophysiology. In *Clinical neurophysiology,* ed. Daube, J.R. (pp. 3-17). Philadelphia: Davis.

Landjerit, B., B. Maton, and G. Peres. 1988. In vivo muscular force analysis during the isometric flexion on a monkey's elbow. *Journal of Biomechanics* 21: 577-584.

Lang, A.H., and K.M. Vaahtoranta. 1973. The baseline, the time characteristics and the slow after waves of the motor unit potential. *Electroencephalography and Clinical Neurophysiology* 25: 387-394.

Larsson, B., S. Karlsson, M. Eriksson, and B. Gerdle. 2003. Test-retest reliability of EMG and peak torque during repetitive maximum concentric knee extensions. *Journal of Electromyography and Kinesiology* 13: 281-287.

Lateva, Z.C., and K.C. McGill. 1998. The physiological origin of the slow afterwave in muscle action potentials. *Electroencephalography and Clinical Neurophysiology* 109: 462-469.

Lateva, Z.C., and K.C. McGill. 2001. Estimating motor-unit architectural properties by analyzing motor-unit action potential morphology. *Clinical Neurophysiology* 112: 127-135.

Lateva, Z.C., K.C. McGill, and C.G. Burgar. 1996. Anatomical and electrophysiological determinants of the human thenar compound muscle action potential. *Muscle & Nerve* 19: 1457-1468.

Lateva, Z.C., K.C. McGill, and M.E. Johanson. 2002. Electrophysiological evidence of adult human skeletal muscle fibres with multiple endplates and polyneuronal innervation. *Journal of Physiology (London)* 544: 549-565.

Lawrence, J.H., and C.J. De Luca. 1983. Myoelectric signal versus force relationship in different human muscles. *Journal of Applied Physiology* 54: 1653-1659.

Lee, J.B., T. Matsumoto, T. Othman, M. Yamauchi, A. Taimura, E. Kaneda, N. Ohwatari, and M. Kosaka. 1999. Coactivation of the flexor muscles as a synergist with the extensors during ballistic finger extension movement in trained kendo and karate athletes. *International Journal of Sports Medicine* 20: 7-11.

Lentz, M., and J.F. Nielsen. 2002. Post-exercise facilitation and depression of M wave and motor evoked potentials in healthy subjects. *Clinical Neurophysiology* 113: 1092-1098.

Lewek, M.D., J. Scholz, K.S. Rudolph, and L. Snyder-Mackler. 2006. Stride-to-stride variability of knee motion in patients with knee osteoarthritis. *Gait & Posture* 23: 505-511.

Lexell, J. 1995. Human aging, muscle mass, and fiber type composition. *Journals of Gerontology Series A, Biological Sciences and Medical Sciences* 50: 11-16.

Lexell, J., K. Henriksson-Larsen, and M. Sjostrom. 1983. Distribution of different fibre types in human skeletal muscles. 2. A study of cross-sections of whole m. vastus lateralis. *Acta Physiologica Scandinavica* 117: 115-122.

Li, L., and G.E. Caldwell. 1999. Coefficient of cross correlation and the time domain correspondence. *Journal of Electromyography and Kinesiology* 9: 385-389.

Li, W., and K. Sakamoto. 1996a. The influence of location of electrode on muscle fiber conduction velocity and EMG power spectrum during voluntary isometric contraction measured with surface array electrodes. *Applied Human Science* 15: 25-32.

Li, W., and K. Sakamoto. 1996b. Distribution of muscle fiber conduction velocity of m. biceps brachii during voluntary isometric contraction with use of surface array electrodes. *Applied Human Science* 15: 41-53.

Libet, B., and B. Feinstein. 1951. Analysis of changes in electromyograms with changing muscle length. *American Journal of Physiology* 167: 805.

Lin, J.Z., and M.K. Floeter. 2004. Do F-wave measurements detect changes in motor neuron excitability? *Muscle & Nerve* 30: 289-294.

Lin, M.I., H.W. Liang, K.H. Lin, and Y.H. Hwang. 2004. Electromyographical assessment on muscular fatigue—an elaboration upon repetitive typing activity. *Journal of Electromyography and Kinesiology* 14: 661-669.

Lind, A.R., and J.S. Petrofsky. 1979. Amplitude of the surface electromyogram during fatiguing isometric contractions. *Muscle & Nerve* 2: 257-264.

Lindström, L., R. Kadefors, and I. Petersén. 1977. An electromyographic index for localized muscle fatigue. *Journal of Applied Physiology* 43: 750-754.

Lindström, L.H., and R.I. Magnusson. 1977. Interpretation of myoelectric power spectra: a model and its applications. *Proceedings of the IEEE* 65: 653-662.

Lindström, L., R. Magnusson, and I. Petersén. 1970. Muscular fatigue and action potential conduction velocity changes studied with frequency analysis of EMG signals. *Electromyography* 4: 341-356.

Lindström, L., and I. Petersén. 1983. Power spectrum analysis of EMG signals and its applications. In *Computer-aided electromyography,* ed. Desmedt, J.E. (pp. 1-51). Basel: Karger.

Linnamo, V., V. Strojnik, and P.V. Komi. 2001. EMG power spectrum and maximal M-wave during eccentric and concentric actions at different force levels. *Acta Physiologica Pharmacologica Bulgarica* 26: 33-36.

Lippold, O.C.J. 1952. The relation between integrated action potentials in a human muscle and its isometric tension. *Journal of Physiology* 117: 492-499.

Llewellyn, M., J.F. Yang, and A. Prochazka. 1990. Human H-reflexes are smaller in difficult beam walking than in normal treadmill walking. *Experimental Brain Research* 83: 22-28.

Loeb, G.E., and C. Gans. 1986. *Electromyography for experimentalists.* Chicago: University of Chicago Press.

Loscher, W.N., A.G. Cresswell, and A. Thorstensson. 1994. Electromyographic responses of the human triceps surae and force tremor during sustained submaximal isometric plantar flexion. *Acta Physiologica Scandinavica* 152: 73-82.

Lowery, M., P. Nolan, and M. O'Malley. 2002. Electromyogram median frequency, spectral compression and muscle fibre conduction velocity during sustained sub-maximal contraction of the brachioradialis muscle. *Journal of Electromyography and Kinesiology* 12: 111-118.

Lowery, M., and M.J. O'Malley. 2003. Analysis and simulation of changes in EMG amplitude during high-level fatiguing contractions. *IEEE Transactions on Biomedical Engineering* 50: 1052-1062.

Lowery, M.M., N.S. Stoykov, and T.A. Kuiken. 2003. A simulation study to examine the use of cross-correlation as an estimate of surface EMG cross talk. *Journal of Applied Physiology* 94: 1324-1334.

Lynn, P.A., N.D. Bettles, A.D. Hughes, and S.W. Johnson. 1978. Influences of electrode geometry on bipolar recordings of the surface electromyogram. *Medical and Biological Engineering and Computing* 16: 651-660.

Ma, D.M., and J.A. Liveson. 1983. *Nerve conduction handbook.* Philadelphia: Davis.

MacFarlane, W.V., and J.D. Meares. 1958. Intracellular recording of action and after-potentials of frog muscle between 0 and 45° C. *Journal of Physiology* 142: 97-109.

MacIsaac, D., P.A. Parker, and R.N. Scott. 2000. Non-stationary myoelectric signals and muscle fatigue. *Methods in Information Medicine* 39: 125-129.

MacIsaac, D., P.A. Parker, and R.N. Scott. 2001a. The short-time Fourier transform and muscle fatigue assessment in dynamic contractions. *Journal of Electromyography and Kinesiology* 11: 439-449.

MacIsaac, D.T., P.A. Parker, R.N. Scott, K.B. Englehart, and C. Duffley. 2001b. Influences of dynamic factors on myoelectric parameters. *IEEE Engineering and Medicine in Biology Magazine* 20: 82-89.

MacKinnon, C.D., and J.C. Rothwell. 2000. Time-varying changes in corticospinal excitability accompanying the triphasic EMG pattern in humans. *Journal of Physiology* 528 (Pt 3): 633-645.

Maffiuletti, N.A., and R. Lepers. 2003. Quadriceps femoris torque and EMG activity in seated versus supine position. *Medicine and Science in Sports and Exercise* 35: 1511-1516.

Malek, M.H., J.W. Coburn, J.P. Weir, T.W. Beck, and T.J. Housh. 2006. The effects of innervation zone on electromyographic amplitude and mean power frequency during incremental cycle ergometry. *Journal of Neuroscience Methods* 155: 126-133.

Mambrito, B., and C.J. De Luca. 1984. A technique for detection, decomposition and analysis of the EMG signal. *Electroencephalography and Clinical Neurophysiology* 58: 175-188.

Marmarelis, P.Z., and V.Z. Marmarelis. 1978. *Analysis of physiological systems: the white-noise approach.* New York: Plenum Press.

Marqueste, T., F. Hug, P. Decherchi, and Y. Jammes. 2003. Changes in neuromuscular function after training by functional electrical stimulation. *Muscle & Nerve* 28: 181-188.

Martin, B.J., J.P. Roll, and G.M. Gauthier. 1986. Inhibitory effects of combined agonist and antagonist muscle vibration on H-reflex in man. *Aviation, Space and Environmental Medicine* 57: 681-687.

Martin, S., and D. MacIsaac. 2006. Innervation zone shift with changes in joint angle in the brachial biceps. *Journal of Electromyography and Kinesiology* 16: 144-148.

Maruyama, A., K. Matsunaga, N. Tanaka, and J.C. Rothwell. 2006. Muscle fatigue decreases short-interval intracortical inhibition after exhaustive intermittent tasks. *Clinical Neurophysiology* 117: 864-870.

Mastaglia, F.L., and W.M. Carroll. 1985. The effects of conditioning stimuli on the F-response. *Journal of Neurology, Neurosurgery and Psychiatry* 48: 182-184.

Masuda, K., T. Masuda, T. Sadoyama, M. Inaki, and S. Katsuta. 1999. Changes in surface EMG parameters during static and dynamic fatiguing contractions. *Journal of Electromyography and Kinesiology* 9: 39-46.

Masuda, T., T. Kizuka, J.Y. Zhe, H. Yamada, K. Saitou, T. Sadoyama, and M. Okada. 2001. Influence of contraction force and speed on muscle fiber conduction velocity during dynamic voluntary exercise. *Journal of Electromyography and Kinesiology* 11: 85-94.

Masuda, T., H. Miyano, and T. Sadoyama. 1983. The distribution of myoneural junctions in the biceps brachii investigated by surface electromyography. *Electroencephalography and Clinical Neurophysiology* 56: 597-603.

Masuda, T., and T. Sadoyama. 1987. Skeletal muscles from which the propagation of the motor unit action potentials is detectable with a surface electrode array. *Electroencephalography and Clinical Neurophysiology* 67: 421-427.

Masuda, T., and T. Sadoyama. 1989. Processing of myoelectric signals for estimating the location of innervation zones in the skeletal muscles. *Frontiers in Medical and Biological Engineering* 1: 299-314.

Masuda, T., T. Sadoyama, and M. Shiraishi. 1996. Dependence of average muscle fibre conduction velocity on voluntary contraction force. *Journal of Electromyography and Kinesiology* 6: 267-276.

Mathur, S., J.J. Eng, and D.L. MacIntyre. 2005. Reliability of surface EMG during sustained contractions of the quadriceps. *Journal of Electromyography and Kinesiology* 15: 102-110.

Maton, B., and S. Bouisset. 1977. The distribution of activity among the muscles of a single group during isometric contraction. *European Journal of Applied Physiology* 37: 101-109.

Maton, B., and D. Gamet. 1989. The fatigability of two agonistic muscles in human isometric voluntary submaximal contraction: an EMG study. II. Motor unit firing rate and recruitment. *European Journal of Applied Physiology* 58: 369-374.

Matthijsse, P.C., K.M. Hendrich, W.H. Rijnsburger, R.D. Woittiez, and P.A. Huijing. 1987. Ankle angle effects on endurance time, median frequency and mean power of gastrocnemius EMG power spectrum: a comparison between individual and group analysis. *Ergonomics* 30: 1149-1159.

Mazzocchio, R., J.C. Rothwell, and A. Rossi. 1995. Distribution of Ia effects onto human hand muscle motoneurones as revealed using an H reflex technique. *Journal of Physiology* 489 (Pt 1): 263-273.

McGill, K.C., Z.C. Lateva, and S. Xiao. 2001. A model of the muscle action potential for describing the leading edge, terminal wave, and slow afterwave. *IEEE Transactions on Biomedical Engineering* 48: 1357-1365.

McGillem, C.D., and G.R. Cooper. 1984. *Continuous and discrete signal and system analysis.* 2nd ed. New York: Holt, Rinehart, and Winston.

McIlroy, W.E., and J.D. Brooke. 1987. Within-subject reliability of the Hoffmann reflex in man. *Electromyography and Clinical Neurophysiology* 27: 401-404.

McKeon, B., S. Gandevia, and D. Burke. 1984. Absence of somatotopic projection of muscle afferents onto motoneurons of same muscle. *Journal of Neurophysiology* 51: 185-194.

McLeod, J.G., and S.H. Wray. 1966. An experimental study of the F wave in the baboon. *Journal of Neurology, Neurosurgery and Psychiatry* 29: 196-200.

Mercuri, B., E.M. Wassermann, P. Manganotti, K. Ikoma, A. Samii, and M. Hallett. 1996. Cortical modulation of spinal excitability: an F-wave study. *Electroencephalography and Clinical Neurophysiology* 101: 16-24.

Merletti, R., D. Farina, and M. Gazzoni. 2003. The linear electrode array: a useful tool with many applications. *Journal of Electromyography and Kinesiology* 13: 37-47.

Merletti, R., D. Farina, H.J. Hermens, B. Freriks, and J. Harlaar, 1999. European recommendations for signal processing methods for surface electromyography. In *European recommendations for surface electromyography*, ed. Hermens, H.J., B. Freriks, R. Merletti, D.F. Stegeman, J.H. Blok, G. Rau, C. Disselhorst-Klug, and G. Hagg (pp. 57-70). Enschede, Netherlands: Roessingh Research and Development.

Messina, C., and R. Cotrufo. 1976. Different excitability of type 1 and type 2 alpha-motoneurons. The recruitment curve of H- and M-responses in slow and fast muscles of rabbits. *Journal of the Neurological Sciences* 28: 57-63.

Metral, S., and G. Cassar. 1981. Relationship between force and integrated EMG activity during voluntary isometric anisotonic contraction. *European Journal of Applied Physiology* 46: 185-198.

Micera, S., G. Vannozzi, A.M. Sabatini, and P. Dario. 2001. Improving detection of muscle activation intervals: characteristics of novel statistical algorithms designed to overcome the limitations of traditional methods. *IEEE Engineering in Medicine and Biology* 20: 38-46.

Michie, P.T., A.M. Clarke, J.D. Sinden, and L.C. Glue. 1976. Reaction time and spinal excitability in a simple reaction time task. *Physiology and Behavior* 16: 311-315.

Millet, G.Y., R. Lepers, N.A. Maffiuletti, N. Babault, V. Martin, and G. Lattier. 2002. Alterations of neuromuscular function after an ultramarathon. *Journal of Applied Physiology* 92: 486-492.

Milner-Brown, H.S., and R.B. Stein. 1975. The relation between the surface electromyogram and muscular force. *Journal of Physiology (London)* 246: 549-569.

Milsum, J.H., R.E. Kearney, and H.H. Kwee. 1973. Interactive use of laboratory computer for biomechanical studies. In *Biomechanics III: medicine and sport,* ed. Cerquiglini, S., Venerando, A., and Wartenweiler, J. (vol. 8, pp. 84-103). Basel: Karger.

Misiaszek, J.E. 2003. The H-reflex as a tool in neurophysiology: its limitations and uses in understanding nervous system function. *Muscle & Nerve* 28: 144-160.

Misulis, K.E. 1989. Basic electronics for clinical neurophysiology. *Journal of Clinical Neurophysiology* 6: 41-74.

Mitrovic, S., G. Lüder, and H.C. Hopf. 1999. Muscle fiber conduction velocity at different states of isotonic contraction. *Muscle & Nerve* 22: 1126-1128.

Mogk, J.P.M., and P.J. Keir. 2003. Crosstalk in surface electromyography of the proximal forearm during gripping tasks. *Journal of Electromyography and Kinesiology* 13: 63-71.

Mohr, K.J., R.S. Kvitne, M.M. Pink, B. Fideler, and J. Perry. 2003. Electromyography of the quadriceps in patellofemoral pain with patellar subluxation. *Clinical Orthopedics and Related Research* 415: 261-271.

Mongia, S.K. 1972. H reflex from quadriceps and gastrocnemius muscles. *Electromyography and Clinical Neurophysiology* 12: 179-190.

Morey-Klapsing, G., A. Arampatzis, and G.P. Bruggemann. 2004. Choosing EMG parameters: comparison of different onset determination algorithms and EMG integrals in a joint stability study. *Clinical Biomechanics (Bristol, Avon)* 19: 196-201.

Mori, S., and A. Ishida. 1976. Synchronization of motor units and its simulation in parallel feedback system. *Biological Cybernetics* 21: 107-111.

Morimoto, S. 1986. Effect of length change in muscle fibers on conduction velocity in human motor units. *Japanese Journal of Physiology* 36: 773-782.

Morimoto, S., and M. Masuda. 1984. Dependence of conduction velocity on spike interval during voluntary muscular contraction in human motor units. *European Journal of Applied Physiology and Occupational Physiology* 53: 191-195.

Moritani, T., and H.A. deVries. 1978. Reexamination of the relationship between the surface integrated electromyogram and force of isometric contraction. *American Journal of Physical Medicine* 57: 263-277.

Moritani, T., M. Muro, and A. Nagata. 1986. Intramuscular and surface electromyogram changes during muscle fatigue. *Journal of Applied Physiology* 60: 1179-1185.

Mortimer, J.T., R. Magnusson, and I. Petersen. 1970. Conduction velocity in ischemic muscle: effect on EMG frequency spectrum. *American Journal of Physiology* 219: 1324-1329.

Moss, R.F., P.B. Raven, J.P. Knochel, J.R. Peckham, and J.D. Blachley. 1983. The effect of training on resting muscle membrane potentials. In *Biochemistry of exercise,* ed. Knuttgen, H.G., Vogel, J.A., and Poortmans, J. (pp. 806-811). Champaign, IL: Human Kinetics.

Muller, M.L., and M.S. Redfern. 2004. Correlation between EMG and COP onset latency in response to a horizontal platform translation. *Journal of Biomechanics* 37: 1573-1581.

Mulroy, S., J. Gronley, W. Weiss, C. Newsam, and J. Perry. 2003. Use of cluster analysis for gait pattern classification of patients in the early and late recovery phases following stroke. *Gait & Posture* 18: 114-125.

Muro, M., A. Nagata, K. Murakami, and T. Moritani. 1982. Surface EMG power spectral analysis of neuro-muscular disorders during isometric and isotonic contractions. *American Journal of Physical Medicine* 61: 244-254.

Nadeau, M., and J. Vanden Abeele. 1988. Maximal H- and M-responses of the right and left gastrocnemius lateralis and soleus muscles. *Electromyography and Clinical Neurophysiology* 28: 307-311.

Nandedkar, S.D., D.B. Sanders, and E.V. Stålberg. 1985. Selectivity of electromyographic recording techniques: a simulation study. *Medical and Biological Engineering and Computing* 23: 536-540.

Nandedkar, S.D., J.C. Sigl, Y.I. Kim, and E.V. Stålberg. 1984. Radial decline of the extracellular action potential. *Medical and Biological Engineering and Computing* 22: 564-568.

Neptune, R.R., S.A. Kautz, and M.L. Hull. 1997. The effect of pedaling rate on coordination in cycling. *Journal of Biomechanics* 30: 1051-1058.

Newcomer, K.L., T.D. Jacobson, D.A. Gabriel, D.R. Larson, R.H. Brey, and K-N. An. 2002. Muscle activation patterns in subjects with and without low back pain. *Archives of Physical Medicine and Rehabilitation* 83: 816-821.

Ng, J.K., and C.A. Richardson. 1996. Reliability of electromyographic power spectral analysis of back muscle endurance in healthy subjects. *Archives of Physical Medicine and Rehabilitation* 77: 259-264.

Nightingale, A. 1960. The graphic representation of movement. II. Relationship between muscle force and the EMG in the stand-at-ease position. *Annals of Physical Medicine* 5: 187-191.

Nikolova, M., N. Pondev, L. Christova, W. Wolf, and A.R. Kossev. 2006. Motor cortex excitability changes preceding voluntary muscle activity in simple reaction time task. *European Journal of Applied Physiology* 98: 212-219.

Nishizono, H., T. Fujimoto, H. Ohtake, and M. Miyashita. 1990. Muscle fiber conduction velocity and contractile properties estimated from surface electrode arrays. *Electroencephalography and Clinical Neurophysiology* 75: 75-81.

Nobrega, J.A., D.S. Pinheiro, G.M. Manzano, and J. Kimura. 2004. Various aspects of F-wave values in a healthy population. *Clinical Neurophysiology* 115: 2336-2342.

Nordander, C., J. Willner, G.A. Hansson, B. Larsson, J. Unge, L. Granquist, and S. Skerfuing. 2003. Influence of the subcutaneous fat layer, as measured by ultrasound, skinfold calipers, and BMI, on the EMG amplitude. *European Journal of Applied Physiology* 89: 514-519.

Nourbakhsh, M.R., and C.G. Kukulka. 2004. Relationship between muscle length and moment arm on EMG activity of human triceps surae muscle. *Journal of Electromyography and Kinesiology* 14: 263-273.

Nymark, J.R., S.J. Balmer, E.H. Melis, E.D. Lemaire, and S. Millar. 2005. Electromyographic and kinematic nondisabled gait differences at extremely slow overground and treadmill walking speeds. *Journal of Rehabilitation, Research and Development* 42: 523-534.

Nyquist, Henry. 1928. Certain topics in telegraph transmission theory. *Transactions of the American Institute of Electrical Engineers* 47 : 617-644.

Ödman, S., and P. Öberg. 1982. Movement-induced potentials in surface electrodes. *Medical and Biological Engineering and Computing* 20: 159-166.

Oh, S.J. 2003. *Clinical electromyography: nerve conduction studies.* 3rd ed. Philadelphia: Lippincott Williams & Wilkins.

Okada, M. 1987. Effect of muscle length on surface EMG wave forms in isometric contractions. *European Journal of Applied Physiology* 56: 482-486.

Okajima, Y., Y. Tomita, R. Ushijima, and N. Chino. 2000. Motor unit sound in needle electromyography: assessing normal and neuropathic units. *Muscle & Nerve* 23: 1076-1083.

Onishi, H., R. Yagi, K. Akasaka, K. Momose, K. Ihashi, and Y. Handa. 2000. Relationship between EMG signals and force in human vastus lateralis muscle using multiple bipolar wire electrodes. *Journal of Electromyography and Kinesiology* 10: 59-67.

Ounpuu, S., and D.A. Winter. 1989. Bilateral electromyographical analysis of the lower limbs during walking in normal adults. *Electroencephalography and Clinical Neurophysiology* 72: 429-438.

Panizza, M., J. Nilsson, and M. Hallett. 1989. Optimal stimulus duration for the H reflex. *Muscle & Nerve* 12: 576-579.

Parker, P.A., and R.N. Scott. 1986. Myoelectric control of prostheses. *Critical Reviews in Biomedical Engineering* 13: 283-310.

Patla, A.E. 1985. Some characteristics of EMG patterns during locomotion: implications for the locomotor control process. *Journal of Motor Behavior* 17: 443-461.

Patterson, P.E., and M. Anderson. 1999. The use of self organizing maps to evaluate myoelectric signals. *Biomedical Science and Instrumentation* 35: 147-152.

Peinemann, A., C. Lehner, B. Conrad, and H.R. Siebner. 2001. Age-related decrease in paired-pulse intracortical inhibition in the human primary motor cortex. *Neuroscience Letters* 313: 33-36.

Pensini, M., and A. Martin. 2004. Effect of voluntary contraction intensity on the H-reflex and V-wave responses. *Neuroscience Letters* 367: 369-374.

Pernus, F., and I. Erzen. 1991. Arrangement of fiber types within fascicles of human vastus lateralis muscle. *Muscle & Nerve* 14: 304-309.

Perot, C., and I. Mora. 1993. H reflexes in close muscles: cross-talk or genuine responses? *Electroencephalography and Clinical Neurophysiology* 89: 104-107.

Perotto, A.O., D. Morrison, E.F. Delagi, and J. Iazzetti. 2005. *Anatomic guide for the electromyographer.* 4th ed. Springfield, IL: Thomas.

Perry, J. 1992. *Gait analysis: normal and pathological function.* Thorofare, NJ: Slack.

Perry, J., and G.A. Bekey. 1981. EMG-force relationships in skeletal muscle. *Critical Reviews in Biomedical Engineering* 7: 1-22.

Perry, J., C.S. Easterday, and D.J. Antonelli. 1981. Surface versus intramuscular electrodes for electromyography of superficial and deep muscles. *Physical Therapy* 61: 7-15.

Perry, J., and M.M. Hoffer. 1977. Preoperative and postoperative dynamic electromyography as an aid in planning tendon transfers in children with cerebral palsy. *Journal of Bone and Joint Surgery (American)* 59: 531-537.

Perttunen, J.R., E. Anttila, J. Sodergard, J. Merikanto, and P.V. Komi. 2004. Gait asymmetry in patients with limb length discrepancy. *Scandinavian Journal of Medicine and Science in Sports* 14: 49-56.

Petrofsky, J., and M. Laymon. 2005. Muscle temperature and EMG amplitude and frequency during isometric exercise. *Aviation, Space and Environmental Medicine* 76: 1024-1030.

Petrofsky, J.S., and A.R. Lind. 1980. The influence of temperature on the amplitude and frequency components of the EMG during brief and sustained isometric contractions. *European Journal of Applied Physiology* 44: 189-200.

Phanachet, I., T. Whittle, K. Wanigaratne, and G.M. Murray. 2004. Minimal tonic firing rates of human lateral pterygoid single motor units. *Clinical Neurophysiology* 115: 71-75.

Pierrot-Deseilligny, E., and D. Mazevet. 2000. The monosynaptic reflex: a tool to investigate motor control in humans. Interest and limits. *Neurophysiologie Clinique* 30: 67-80.

Podnar, S. 2004. Usefulness of an increase in size of motor unit potential sample. *Clinical Neurophysiology* 115: 1683-1688.

Podnar, S., and M. Mrkaić. 2003. Size of motor unit potential sample. *Muscle & Nerve* 27: 196-201.

Polcyn, A.F., L.A. Lipsitz, D.C. Kerrigan, and J.J. Collins. 1998. Age-related changes in the initiation of gait: degradation of central mechanisms for momentum generation. *Archives of Physical Medicine and Rehabilitation* 79: 1582-1589.

Polgar, J., M.A. Johnson, D. Weightman, and D. Appleton. 1973. Data on fibre size in thirty-six human muscles. An autopsy study. *Journal of the Neurological Sciences* 19: 307-318.

Potvin, J.R. 1997. Effects of muscle kinematics on surface EMG amplitude and frequency during fatiguing dynamic contractions. *Journal of Applied Physiology* 82: 144-151.

Potvin, J.R., and S.H. Brown. 2004. Less is more: high pass filtering, to remove up to 99% of the surface EMG signal power, improves EMG-based biceps brachii muscle force estimates. *Journal of Electromyography and Kinesiology* 14: 389-399.

Prilutsky, B.I., R.J. Gregor, and M.M. Ryan. 1998. Coordination of two-joint rectus femoris and hamstrings during the swing phase of human walking and running. *Experimental Brain Research* 120: 479-486.

Quanbury, A.O., C.D. Foley, D.A. Winter, R.M. Letts, and T. Steinke. 1976. Clinical telemetry of EMG and temporal information during gait. *Biotelemetry* 3: 129-137.

Rababy, N., R.E. Kearney, and I.W. Hunter. 1989. Method for EMG conduction velocity estimation which accounts for input and output noise. *Medical and Biological Engineering* 27: 125-129.

Ravier, P., O. Buttelli, R. Jennane, and P. Couratier. 2005. An EMG fractal indicator having different sensitivities to changes in force and muscle fatigue during voluntary static muscle contractions. *Journal of Electromyography and Kinesiology* 15: 210-221.

Reber, L., J. Perry, and M. Pink. 1993. Muscular control of the ankle in running. *American Journal of Sports Medicine* 21: 805-810.

Redfern, M.S., R.E. Hughes, and D.B. Chaffin. 1993. High-pass filtering to remove electrocardiographic interference from torso EMG recordings. *Clinical Biomechanics* 8: 44-48.

Reid, M.B., G.J. Grubwieser, D.S. Stokic, S.M. Koch, and A.A. Leis. 1993. Development and reversal of fatigue in human tibialis anterior. *Muscle & Nerve* 16: 1239-1245.

Rich, C., and E. Cafarelli. 2000. Submaximal motor unit firing rates after 8 wk of isometric resistance training. *Medicine and Science in Sports and Exercise* 32: 190-196.

Richmond, F.J., and D.G. Stuart. 1985. Distribution of sensory receptors in the flexor carpi radialis muscle of the cat. *Journal of Morphology* 183: 1-13.

Robertson, D.G.E., and J.J. Dowling. 2003. Design responses of Butterworth and critically damped digital filters. *Journal of Electromyography and Kinesiology* 13: 569-573.

Robinson, K.L., J.S. McIlwain, and K.C. Hayes. 1979. Effects of H-reflex conditioning upon the contralateral alpha motoneuron pool. *Electroencephalography and Clinical Neurophysiology* 46: 65-71.

Roeleveld, K., D.F. Stegeman, H.M. Vingerhoets, and A. Van Oosterom. 1997. Motor unit potential contribution to surface electromyography. *Acta Physiologica Scandinavica* 160: 175-183.

Roman-Liu, D., T. Tokarski, and K. Wojcik. 2004. Quantitative assessment of upper limb muscle fatigue depending on the conditions of repetitive task load. *Journal of Electromyography and Kinesiology* 14: 671-682.

Rossi, A., R. Mazzocchio, and D. Nuti. 1986. Tonic neck influences on lower limb extensor motoneurons in man. *Electromyography and Clinical Neurophysiology* 26: 207-216.

Rossi, A., and D. Nuti. 1988. The effects of caloric vestibular stimulation on the soleus alpha motoneurons reinvestigated in man. *Electromyography and Clinical Neurophysiology* 28: 409-413.

Rossi-Durand, C., K.E. Jones, S. Adams, and P. Bawa. 1999. Comparison of the depression of H-reflexes following previous activation in upper and lower limb muscles in human subjects. *Experimental Brain Research* 126: 117-127.

Roy, S.H., C.J. De Luca, and D.A. Casavant. 1989. Lumbar muscle fatigue and chronic lower back pain. *Spine* 14: 992-1001.

Roy, S.H., G. De Luca, M.S. Cheng, A. Johansson, L.D. Gilmore, and C.J. De Luca. 2007. Electro-mechanical stability of surface EMG sensors. *Medical and Biological Engineering and Computing* 45: 447-457.

Rubinstein, S., and G. Kamen. 2005. Decreases in motor unit firing rate during sustained maximal-effort contractions in young and older adults. *Journal of Electromyography and Kinesiology* 15: 536-543.

Rutkove, S.B. 2001. Effects of temperature on neuromuscular electrophysiology. *Muscle & Nerve* 24: 867-882.

Sadeghi, H., P. Allard, F. Prince, and H. Labelle. 2000. Symmetry and limb dominance in able-bodied gait: a review. *Gait & Posture* 12: 34-45.

Sadoyama, T., and T. Masuda. 1987. Changes of the average muscle fiber conduction velocity during a varying force contraction. *Electroencephalography and Clinical Neurophysiology* 67: 495-497.

Sadoyama, T., T. Masuda, and H. Miyano. 1985. Optimal conditions for the measurement of muscle fibre conduction velocity using surface electrode arrays. *Medical and Biological Engineering and Computing* 23: 339-342.

Sadoyama, T., T. Masuda, H. Miyata, and S. Katsuta. 1988. Fibre conduction velocity and fibre composition in human vastus lateralis. *European Journal of Applied Physiology and Occupational Physiology* 57: 767-771.

Saitou, K., T. Masuda, D. Michikami, R. Kojima, and M. Okada. 2000. Innervation zones of the upper and lower limb muscles estimated by using multichannel surface EMG. *Journal of Human Ergology (Tokyo)* 29: 35-52.

Sakamoto, K., and W. Li. 1997. Effect of muscle length on distribution of muscle fiber conduction velocity for M. biceps brachii. *Applied Human Science* 16: 1-7.

Sale, D.G., J.D. MacDougall, A.R. Upton, and A.J. McComas. 1983. Effect of strength training upon motoneuron excitability in man. *Medicine and Science in Sports and Exercise* 15: 57-62.

Santello, M., and M.J. McDonagh. 1998. The control of timing and amplitude of EMG activity in landing movements in humans. *Experimental Physiology* 83: 857-874.

Sbriccoli, P., I. Bazzucchi, A. Rosponi, M. Bernardi, G. De Vito, and F. Felici. 2003. Amplitude and spectral characteristics of biceps brachii sEMG depend upon speed of isometric force generation. *Journal of Electromyography and Kinesiology* 13: 139-147.

Scaglioni, G., A. Ferri, A.E. Minetti, A. Martin, J. Van Hoecke, P. Capodaglio, A. Sartorio, and M.V. Narici. 2002. Plantar flexor activation capacity and H reflex in older adults: adaptations to strength training. *Journal of Applied Physiology* 92: 2292-2302.

Scaglioni, G., M.V. Narici, N.A. Maffiuletti, M. Pensini, and A. Martin. 2003. Effect of ageing on the electrical and mechanical properties of human soleus motor units activated by the H reflex and M wave. *Journal of Physiology (London)* 548: 649-661.

Schieppati, M. 1987. The Hoffmann reflex: a means of assessing spinal reflex excitability and its descending control in man. *Progress in Neurobiology* 28: 345-376.

Schulte, E., D. Farina, R. Merletti, G. Rau, and C. Disselhorst-Klug. 2004. Influence of muscle fibers shortening on estimates of conduction velocity and spectral frequencies from surface electromyographic signals. *Medical and Biological Engineering and Computing* 42: 477-486.

Schulte, E., L.A. Kallenberg, H. Christensen, C. Disselhorst-Klug, H.J. Hermens, G. Rau, and K. Sogaard. 2006. Comparison of the electromyographic activity in the upper trapezius and biceps brachii muscle in subjects with muscular disorders: a pilot study. *European Journal of Applied Physiology* 96: 185-193.

Schwab, G.H., D.R. Moynes, F.W. Jobe, and J. Perry. 1983. Lower extremity electromyographic analysis of running gait. *Clinical Orthopedics and Related Research* 176: 166-170.

Segal, R.L. 1992. Neuromuscular compartments in the human biceps brachii muscle. *Neuroscience Letters* 140: 98-102.

Segal, R.L., P.A. Catlin, E.W. Krauss, K.A. Merick, and J.B. Robilotto. 2002. Anatomical partitioning of three human forearm muscles. *Cells, Tissues, Organs* 170: 183-197.

Segal, R.L., S.L. Wolf, M.J. DeCamp, M.T. Chopp, and A.W. English. 1991. Anatomical partitioning of three multiarticular human muscles. *Acta Anatomica (Basel)* 142: 261-266.

Seki, K., and M. Narusawa. 1996. Firing rate modulation of human motor units in different muscles during isometric contraction with various forces. *Brain Research* 719: 1-7.

Sherrington, C.S. 1906. *The integrative action of the nervous system.* New Haven, CT: Yale University Press.

Shiavi, R. 1985. Electromyographic patterns in adult locomotion: a comprehensive review. *Journal of Rehabilitation Research and Development* 22: 85-98.

Shiavi, R., H.J. Bugle, and T. Limbird. 1987. Electromyographic gait assessment, part 1: adult EMG profiles and walking speed. *Journal of Rehabilitation Research and Development* 24: 13-23.

Shiavi, R., C. Frigo, and A. Pedotti. 1998. Electromyographic signals during gait: criteria for envelope filtering and number of strides. *Medical and Biological Engineering and Computing* 36: 171-178.

Shiavi, R., and N. Green. 1983. Ensemble averaging of locomotor electromyographic patterns using interpolation. *Medical and Biological Engineering and Computing* 21: 537-578.

Shiavi, R., and P. Griffin. 1983. Changes in electromyographic gait patterns of calf muscles with walking speed. *IEEE Transactions on Biomedical Engineering* 30: 73-76.

Sica, R.E.P., O.P. Sanz, and A. Colombi. 1976. Potentiation of the F wave by remote voluntary contraction in man. *Electromyography and Clinical Neurophysiology* 16: 623-625.

Simons, D.G. 2001. Do endplate noise and spikes arise from normal motor endplates? *American Journal of Physical Medicine* 80: 134-140.

Sinderby, C.A., A.S. Comtois, R.G. Thomson, and A.E. Grassino. 1996. Influence of the bipolar electrode transfer function on the electromyogram power spectrum. *Muscle & Nerve* 19: 290-301.

Skinner, S.R., and D.K. Lester. 1986. Gait electromyographic evaluation of the long-toe flexors in children with spastic cerebral palsy. *Clinical Orthopedics and Related Research* 207: 70-73.

Smith, G. 1989. Padding point extrapolation techniques for the Butterworth digital filter. *Journal of Biomechanics* 22: 967-971.

Sohn, Y.H., A. Kaelin-Lang, H.Y. Jung, and M. Hallett. 2001. Effect of levetiracetam on human corticospinal excitability. *Neurology* 57: 858-863.

Sollie, G., H.J. Hermens, K.L. Boon, W. Wallinga-De Jonge, and G. Zilvold. 1985a. The measurement of the conduction velocity of muscle fibres with surface EMG according to the cross-correlation method. *Electromyography and Clinical Neurophysiology* 25: 193-204.

Sollie, G., H.J. Hermens, K.L. Boon, W. Wallinga-De Jonge, and G. Zilvold. 1985b. The boundary conditions for measurement of the conduction velocity of muscle fibers with surface EMG. *Electromyography and Clinical Neurophysiology* 25: 45-56.

Solomonow, M., R. Baratta, M. Bernardi, B. Zhou, Y. Lu, M. Zhu, and S. Acierno. 1994. Surface and wire EMG crosstalk in neighbouring muscles. *Journal of Electromyography and Kinesiology* 4: 131-142.

Solomonow, M., C. Baten, J. Smit, R. Baratta, H. Hermens, R. D'Ambrosia, and H. Shoji. 1990. Electromyogram power spectra frequencies associated with motor unit recruitment strategies. *Journal of Applied Physiology* 68: 1177-1185.

Stackhouse, C., P.A. Shewokis, S.R. Pierce, B. Smith, J. McCarthy, and C. Tucker. 2007. Gait initiation in children with cerebral palsy. *Gait & Posture* 26: 301-308.

Stålberg, E. 1966. Propagation velocity in human muscle fibers in situ. *Acta Physiologica Scandinavica Supplementum* 287: 1-112.

Staudenmann, D., I. Kingma, A. Daffertshofer, D.F. Stegeman, and J.H. van Dieen. 2006. Improving EMG-based muscle force estimation by using a high-density EMG grid and principal component analysis. *IEEE Transactions on Biomedical Engineering* 53: 712-719.

Stephens, J.A., and A. Taylor. 1972. Fatigue of maintained voluntary muscle contraction in man. *Journal of Physiology* 220: 1-18.

Strommen, J.A., and J.R. Daube. 2001. Determinants of pain in needle electromyography. *Clinical Neurophysiology* 112: 1414-1418.

Stulen, F.B., and C.J. De Luca. 1981. Frequency parameters of the myoelectric signal as a measure of muscle conduction velocity. *IEEE Transactions on Biomedical Engineering* 28: 515-523.

Sutherland, D.H. 2001. The evolution of clinical gait analysis part l: kinesiological EMG. *Gait & Posture* 14: 61-70.

Tam, H.W., and J.G. Webster. 1977. Minimize electrode motion artifact by skin abrasion. *IEEE Transactions on Biomedical Engineering* 24: 134-139.

Tang, A., and W.Z. Rymer. 1981. Abnormal force–EMG relations in paretic limbs of hemiparetic human subjects. *Journal of Neurology, Neurosurgery and Psychiatry* 44: 690-698.

Tanino, Y., S. Daikuya, T. Nishimori, K. Takasaki, and T. Suzuki. 2003. M wave and H-reflex of soleus muscle before and after electrical muscle stimulation in healthy subjects. *Electromyography and Clinical Neurophysiology* 43: 381-384.

Tanji, J., and M. Kato. 1973. Recruitment of motor units in voluntary contraction of a finger muscle in man. *Experimental Neurology* 40: 759-770.

Terao, Y., and Y. Ugawa. 2002. Basic mechanisms of TMS. *Journal of Clinical Neurophysiology* 19: 322-343.

Thorstensson, A., H. Carlson, M.R. Zomlefer, and J. Nilsson. 1982. Lumbar back muscle activity in relation to trunk movements during locomotion in man. *Acta Physiologica Scandinavica* 116: 13-20.

Thorstensson, A., A.J. Karlsson, J.H.T. Viitasalo, P. Luhtanen, and P.V. Komi. 1976. Effect of strength training on EMG of human skeletal muscle. *Acta Physiologica Scandinavica* 98: 232-236.

Trimble, M.H., and D.M. Koceja. 1994. Modulation of the triceps surae H-reflex with training. *International Journal of Neuroscience* 76: 293-303.

Troni, W., R. Cantello, and I. Rainero. 1983. Conduction velocity along human muscle fibers in situ. *Neurology* 33: 1453-1459.

Trontelj, J.V. 1993. Muscle fiber conduction velocity changes with length. *Muscle & Nerve* 16: 506-512.

Tsuruike, M., D.M. Koceja, K. Yabe, and N. Shima. 2003. Age comparison of H-reflex modulation with the Jendrassik maneuver and postural complexity. *Clinical Neurophysiology* 114: 945-953.

Tucker, K.J., and K.S. Türker. 2007. Triceps surae stretch and voluntary contraction alters maximal M-wave magnitude. *Journal of Electromyography and Kinesiology* 17: 203-211.

Upton, A.R., A.J. McComas, and R.E. Sica. 1971. Potentiation of "late" responses evoked in muscles during effort. *Journal of Neurology, Neurosurgery and Psychiatry* 34: 699-711.

Van Boxtel, A., P. Goudswaard, G.M. van der Molen, and W.J. van den Bosch. 1983. Changes in electromyogram power spectra of facial and jaw-elevator muscle during fatigue. *Journal of Applied Physiology* 54: 51-58.

Van Der Hoeven, J.H., and F. Lange. 1994. Supernormal muscle fiber conduction velocity during intermittent isometric exercise in human muscle. *Journal of Applied Physiology* 77: 802-806.

Van Der Hoeven, J.H., T.P. Links, M.J. Zwarts, and T.W. Van Weerden. 1994. Muscle fiber conduction velocity in the diagnosis of familial hypokalemic periodic paralysis—invasive versus surface determination. *Muscle & Nerve* 17: 898-905.

Van Der Hoeven, J.H., T.W. Van Weerden, and M.J. Zwarts. 1993. Long-lasting supernormal conduction velocity after sustained maximal isometric contraction in human muscle. *Muscle & Nerve* 16: 312-320.

van Eijden, T.M., and M.C. Raadsheer. 1992. Heterogeneity of fiber and sarcomere length in the human masseter muscle. *Anatomical Record* 232: 78-84.

van Vugt, J.P.P., and J.G. van Dijk. 2001. A convenient method to reduce crosstalk in surface EMG. *Clinical Neurophysiology* 112: 583-592.

Vaughan, V.G. 1989. Effects of upper limb immobilization on isometric muscle strength, movement time, and triphasic electromyographic characteristics. *Physical Therapy* 69: 119-129.

Verrier, M.C. 1985. Alterations in H reflex magnitude by variations in baseline EMG excitability. *Electroencephalography and Clinical Neurophysiology* 60: 492-499.

Vestergaard-Poulsen, P., C. Thomsen, T. Sinkjaer, and O. Henriksen. 1995. Simultaneous 31P-NMR spectroscopy and EMG in exercising and recovering human skeletal muscle: a correlation study. *Journal of Applied Physiology* 79: 1469-1478.

Vint, P.F., and R.N. Hinrichs. 1999. Longer integration intervals reduce variability and improve reliability of EMG derived from maximal isometric exertions. *Journal of Applied Biomechanics* 15: 210-220.

Vint, P.F., S.P. McLean, and M. Harron. 2001. Electromechanical delay in isometric actions initiated from nonresting levels. *Medicine and Science in Sports and Exercise* 33: 978-983.

Voss, E.J., J. Harlaar, and G.J. Van Ingen Schenau. 1991. Electromechanical delay during knee extensor contractions. *Medicine and Science in Sports and Exercise* 23: 1187-1193.

Walk, D., and M.A. Fisher. 1993. Effects of cutaneous stimulation on ipsilateral and contralateral motoneuron excitability: an analysis using H reflexes and F waves. *Electromyography and Clinical Neurophysiology* 33: 259-264.

Wallace, R.K., P.J. Mills, D.W. Orme-Johnson, M.C. Dillbeck, and E. Jacobe. 1983. Modification of the paired H reflex through the transcendental meditation and TM-Sidhi program. *Experimental Neurology* 79: 77-86.

Wallinga-De Jonge, W., F.L. Gielen, P. Wirtz, P. De Jong, and J. Broenink. 1985. The different intracellular action potentials of fast and slow muscle fibres. *Electroencephalography and Clinical Neurophysiology* 60: 539-547.

Walmsley, R.P. 1977. Electromyographic study of the phasic activity of peroneus longus and brevis. *Archives of Physical Medicine and Rehabilitation* 58: 65-69.

Walter, C.B. 1984. Temporal quantification of electromyography with reference to motor control research. *Human Movement Science* 3: 155-162.

Walthard, K.M., and M. Tchicaloff. 1971. Motor points. In *Electrodiagnosis and electromyography,* ed. Licht, S. (3rd ed., pp. 153-170). New Haven, CT: Elizabeth Licht.

Walton, C., J. Kalmar, and E. Cafarelli. 2003. Caffeine increases spinal excitability in humans. *Muscle & Nerve* 28: 359-364.

Wank, V., U. Frick, and D. Schmidtbleicher. 1998. Kinematics and electromyography of lower limb muscles in overground and treadmill running. *International Journal of Sports Medicine* 19: 455-461.

Wee, A.S. 2006. Correlation between the biceps brachii muscle bulk and the size of its evoked compound muscle action potential. *Electromyography and Clinical Neurophysiology* 46: 79-82.

Weresh, M.J., R.H. Gabel, R.A. Brand, and D.S. Tearse. 1994. Popliteus function in ACL-deficient patients. *Iowa Orthopaedic Journal* 14: 85-93.

Westad, C., R.H. Westgaard, and C.J. De Luca. 2003. Motor unit recruitment and derecruitment induced by brief increase in contraction amplitude of the human trapezius muscle. *Journal of Physiology* 552: 645-656.

Weytjens, J.L.F., and D. van Steenberghe. 1984. Spectral analysis of the surface electromyogram as a tool for studying rate modulation: a comparison between theory, simulation, and experiment. *Biological Cybernetics* 50: 95-103.

Williams, D.M., S. Sharma, and M. Bilodeau. 2002. Neuromuscular fatigue of elbow flexor muscles of dominant and non-dominant arms in healthy humans. *Journal of Electromyography and Kinesiology* 12: 287-294.

Windhorst, U., T.M. Hamm, and D.G. Stuart. 1989. On the function of muscle and reflex partitioning. *Behavioral and Brain Sciences* 12: 629-681.

Winkel, J., and K. Jørgensen. 1991. Significance of skin temperature changes in surface electromyography. *European Journal of Applied Physiology* 63: 345-348.

Winter, D.A. 1991. *The biomechanics and motor control of human gait: normal, elderly and pathological* (pp. 1-143). 2nd ed. Waterloo, ON: University of Waterloo Press.

Winter, D.A. 2005. *Biomechanics and motor control of human movment.* 3rd ed. Hoboken, NJ: Wiley.

Winter, D.A., A.J. Fuglevand, and S.E. Archer. 1994. Crosstalk in surface electromyography: theoretical and practical estimates. *Journal of Electromyography* 4: 15-26.

Winter, D.A., and A.Q. Quanbury. 1975. Multichannel biotelemetry systems for use in EMG studies, particularly in locomotion. *American Journal of Physical Medicine* 54: 142-147.

Winter, D.A., and H.J. Yack. 1987. EMG profiles during normal human walking: stride-to-stride and inter-subject variability. *Electroencephalography and Clinical Neurophysiology* 67: 402-411.

Wolf, S. 1983. *Guide to electronic measurements and laboratory practice.* Englewood Cliffs, NJ: Prentice-Hall.

Woods, J.J., and B. Bigland-Ritchie. 1983. Linear and nonlinear surface EMG/force relationships in human muscles. An anatomical/functional argument for the existence of both. *American Journal of Physical Medicine* 62: 287-299.

Wootten, M.E., M.P. Kadaba, and G.V. Cochran. 1990. Dynamic electromyography. II. Normal patterns during gait. *Journal of Orthopaedic Research* 8: 259-265.

Wu, G., W. Liu, J. Hitt, and D. Millon. 2004. Spatial, temporal and muscle action patterns of Tai Chi gait. *Journal of Electromyography and Kinesiology* 14: 343-354.

Yaar, I., and L. Niles. 1992. Muscle fiber conduction velocity and mean power spectrum frequency in neuromuscular disorders and in fatigue. *Muscle & Nerve* 15: 780-787.

Yamada, M., K. Kumagai, and A. Uchiyama. 1991. Muscle fiber conduction velocity studied by the multi-channel surface EMG. *Electromyography and Clinical Neurophysiology* 31: 251-256.

Yang, J.F., and P.J. Whelan. 1993. Neural mechanisms that contribute to cyclical modulation of the soleus H-reflex in walking in humans. *Experimental Brain Research* 95: 547-556.

Yang, J.F., and D.A. Winter. 1984. Electromyographic amplitude normalization methods: improving their sensitivity as diagnostic tools in gait analysis. *Archives of Physical Medicine and Rehabilitation* 65: 517-521.

Yates, S.K., and W.F. Brown. 1979. Characteristics of the F response: a single motor unit study. *Journal of Neurology, Neurosurgery and Psychiatry* 42: 161-170.

Young, C.C., S.E. Rose, E.N. Biden, M.P. Wyatt, and D.H. Sutherland. 1989. The effect of surface and internal electrodes on the gait of children with cerebral palsy, spastic diplegic type. *Journal of Orthopaedic Research* 7: 732-737.

Yu, B., D.A. Gabriel, L.A. Nobel, and K-N. An. 1999. Determination of the optimum cutoff frequency for a low-pass digital filter. *Journal of Applied Biomechanics* 15: 318-329.

Zecca, M., S. Micera, M.C. Carrozza, and P. Dario. 2002. Control of multifunctional prosthetic hands by processing the electromyographic signal. *Critical Reviews in Biomedical Engineering* 30: 459-485.

Zehr, E.P. 2002. Considerations for use of the Hoffmann reflex in exercise studies. *European Journal of Applied Physiology* 86: 455-468.

Zehr, E.P., and D.G. Sale. 1994. Ballistic movement: muscle activation and neuromuscular adaptation. *Canadian Journal of Applied Physiology* 19: 363-378.

Zigmond, M.J., F.E. Bloom, S.C. Landis, J.L. Roberts, and L.R. Squire. 1999. *Fundamental neuroscience.* New York: Academic Press.

Zipp, P. 1982. Recommendations for the standardization of lead positions in surface electromyography. *European Journal of Applied Physiology* 50: 41-54.

Zuniga, E.N., and D.G. Simons. 1969. Nonlinear relationship between averaged electromyogram potential and muscle tension in normal subjects. *Archives of Physical Medicine and Rehabilitation* 50: 613-620.

Zwarts, M.J., and L. Arendt-Nielsen. 1988. The influence of force and circulation on average muscle fibre conduction velocity during local muscle fatigue. *European Journal of Applied Physiology and Occupational Physiology* 58: 278-283.

Zwarts, M.J., and D.F. Stegeman. 2003. Multichannel surface EMG: basic aspects and clinical utility. *Muscle & Nerve* 28: 1-17.

Author Index

Subject Index

Note: The italicized *f* and *t* following page numbers refer to figures and tables, respectively.

About the Authors

Gary Kamen, PhD, is a professor in the department of kinesiology at the University of Massachusetts at Amherst. He has 30 years of experience in the field of kinesiology, including research in basic electromyography, neuromuscular physiology, motor control, exercise neuroscience, motor unit physiology, and numerous electromyographic applications. Through research studies, he has demonstrated the importance of motor unit firing rate for maximal force production in older adults, thus proving the importance of neural activation for muscular strength.

Kamen has published over 75 articles in the field of electromyography, motor unit recording techniques, motor control, and other concepts related to this book. He also published one of the first texts in exercise science. He is a fellow of both the American College of Sports Medicine and the American Association for Kinesiology and Physical Education, as well as a member of several organizations, including the Society for Neuroscience, the International Society for Electrophysiology and Kinesiology, and the International Society of Biomechanics.

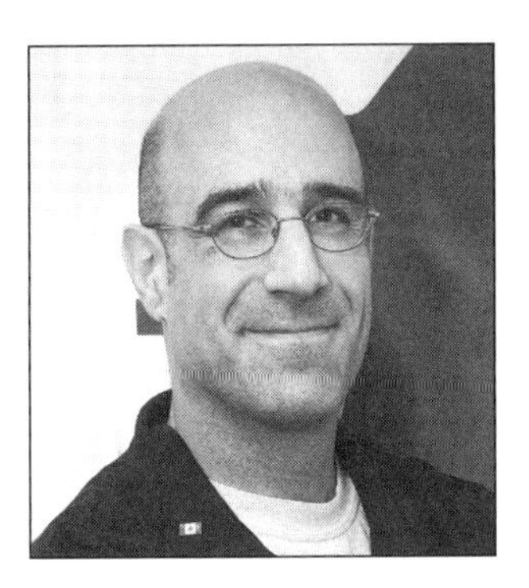

David A. Gabriel, PhD, is a professor in the department of physical education and kinesiology at Brock University in St. Catharines, Ontario. He has 20 years of experience conducting research related to kinesiology, rehabilitation, and clinical neurophysiology. This includes surface and indwelling electromyographic techniques as well as computer modeling and simulation of the EMG signal. From this research he has been able to solve difficult problems in EMG data collection, reduction, analysis, and interpretation.

Gabriel published a series of papers on a novel signal processing method for documenting subtle changes in the surface EMG signal and how those changes can be related to motor unit firing patterns. He is also widely published in other areas, including reliability of the surface EMG signal for both kinesiological and clinical studies and modeling and simulation of the surface of the EMG signal.

Gabriel is associate editor for the *Journal of NeuroEngineering and Rehabilitation*, an editorial board member for the *Journal of Electromyography and Kinesiology*, vice president and president-elect of the International Society for Electrophysiology and Kinesiology, a fellow of the American College of Sports Medicine, and a member of the Institute of Electrical and Electronics Engineers.